A Level Chemistry for OCR

Questions and Answers
with explanations and mind maps

Year 2

Pasan Witharana

Q & A

OCR Chemistry A
2021 onwards

ADVANCED LEVEL CHEMISTRY QUESTIONS AND ANSWERS

Copyright © 2022 Pasan Darshana Witharana. All rights reserved.
First paperback edition printed 2022 in the United Kingdom
A catalogue record for this book is available from the British Library.

ISBN 978-1-913606-80-0

No part of this book shall be reproduced or transmitted in any form or by any means, electronic or mechanical, including photocopying, recording, or by any information retrieval system without the written permission of the publisher.

Published by BLACK SPRING PRESS GROUP
Edited by Michael L Whybrow
Designed by Mu-Ching Tsai and Hayley Wood
Set by Mu-Ching Tsai
Proofread by Thivanka Witharana and Victoria Thornton

Acknowledgements

I am extremely grateful to the publishers, The Black Spring Press Group for their support with this project.

I am also grateful to Michael Whybrow, specialist examiner and editor, for his help with editing this book.

I am thankful to my parents and sister who helped me from the beginning to the end to make this book possible.

This task could not have been completed without their help.

Introduction

Module 5

Revision guide

Questions

Long answer questions

Multiple choice questions

Module 6

Revision guide

Practical procedures

Questions

Long answer questions

Multiple choice questions

Answers

Module 5

Module 6

Periodic table and Data Sheet

Index

CONTENTS

8

14

44

46

92

116

141

146

148

204

230

298

358

362

Introduction

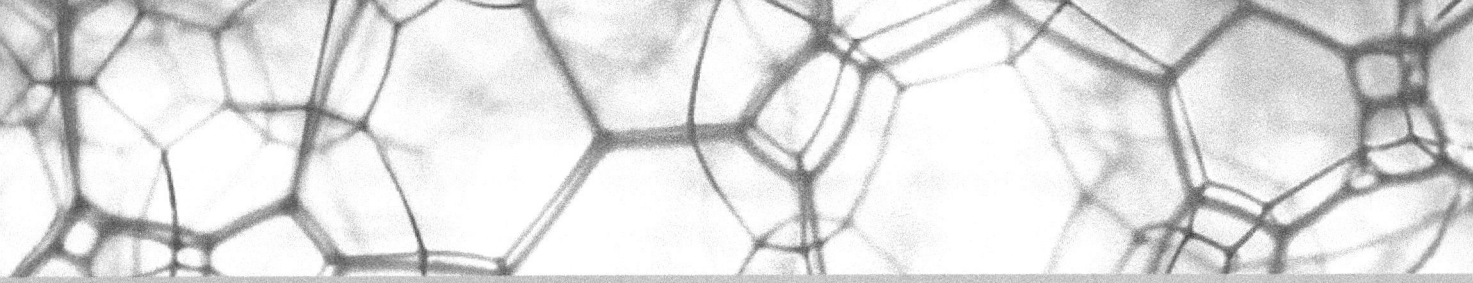

This book is aimed at Second Year Advanced Level Chemistry students.

Answering questions is one of the best ways to revise for any exam. It will test your understanding of different topics and will teach you how to phrase your knowledge as a correct answer. Often when we study, we absorb a great deal of information, but when it comes to answering questions, many students struggle to identify what points to include in their responses. The only way to know what to write in the exam is to get enough practice beforehand.

This book is written in such a way that you will learn the content you need to know for the second year of Advanced Level Chemistry, just by studying the questions and answers in the next few pages.

In your exam, every question will have a designated answer space. You should keep in mind that filling this space alone, will not score you marks. It is very important to know what to include in your answer so that you do not waste the opportunity to demonstrate your understanding of the subject. This guide will help you write all of the important points you need in your answer, to enable you to score maximum marks for each question.

The book has been divided into three sections:

- Revision guide consisting of mind maps
- Questions, including long answers and multiple choice
- Answers to questions.

When attempting the questions, it is always useful to simulate exam conditions. This will help you to work within time limits.

Explanations have been included where necessary. One of the key ways to learn information and to fully remember it, is to understand it properly. You are more likely to forget vital answers if you just try to memorise something that does not make much sense to you.

Three Exam Papers

H432-01 Periodic table elements and physical Chemistry

2 hour 15 minutes
100 marks
15 Multiple Choice Questions
6-7 Long Answer Questions (Number of questions variable)

Assesses content from modules 1, 2, 3 and 5

H432-02 Synthesis and analytical techniques

2 hour 15 minutes
100 marks
15 Multiple Choice Questions
6-7 Long Answer Questions (Number of questions variable)

Assesses content from modules 1, 2, 4 and 6

H432-03 Unified Chemistry

1 hour 30 minutes

70 marks
No Multiple Choice Questions
6-7 Long Answer Questions (Number of questions variable)

Assesses content from modules 1 to 6

A guide To Your Exams

Chemistry Q & A | 11

MODULE 5

Physical Chemistry And Transition Elements

- Reaction rates and equilibrium (quantitative)
- pH and buffers
- Enthalpy, entropy and free energy
- Redox and electrode potentials
- Transition elements

Revision Guide
Physical Chemistry and Transition Elements

Reaction rates

Different reactions progress at different rates. These reaction rates can be measured using various methods. Some reactions, for example, lead to the formation of a gas. The volume of gas formed at specific time intervals may be measured to help work out the rate of reaction.

Example: Magnesium reacts with hydrochloric acid to form magnesium chloride and hydrogen gas. The following apparatus can be used to collect the gas produced. $Mg + 2HCl \rightarrow MgCl_2 + H_2$

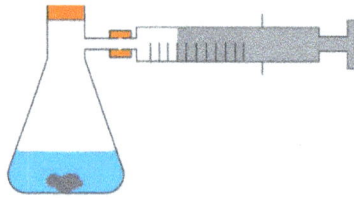

Figure 5.0 The experimental set up of magnesium metal in hydrochloric acid in a conical flask with attached gas syringe to collect the hydrogen gas

Rate-concentration graphs

During a reaction, two or more reactants react together to form the product(s). For example, consider the following reaction:
$$A + B \rightarrow C$$

Here, A and B react to form C. If C is a gas, by measuring how much of the gas is given off in a set period of time, the reaction rate can be measured. It is important to highlight that by changing the concentration of A and / or B, the reaction rate may be altered. However, doubling the concentration of A may have a different effect to doubling the concentration of B. The effect on the rate of changing the concentration of a reactant will depend on the order with respect to that reactant. For example, if A is a zero order reactant, changing its concentration will have no effect on the rate as shown by the first graph below.

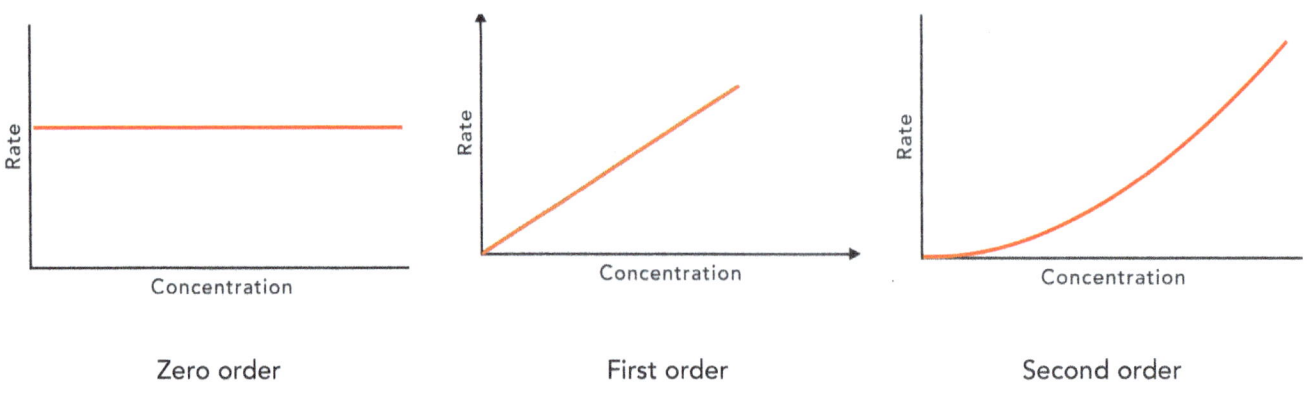

Figure 5.1 Rate-concentration graphs for zero, first and second order reactants

Orders of reactants

Changing the concentration of a reactant may affect the rate of a reaction. If the concentration of a first order reactant is doubled, the rate of reaction will double. Doubling the concentration of a 2nd order reactant will increase the reaction rate by four times. Changing the concentration of a zero order reactant will have no effect on the reaction rate.

Half-life

Half-life of a reactant is the time taken for the concentration of that reactant to decrease by half.

Consider a substance X with 16 particles reacting with another substance. It takes 3 minutes for the number of particles of X to go down to 8. The half-life of substance X is 3 minutes. As the reaction proceeds, for each half-life, the number of particles will reduce by half. Therefore, starting at 16 particles, it takes four half-lives for the number of particles to go down to one.

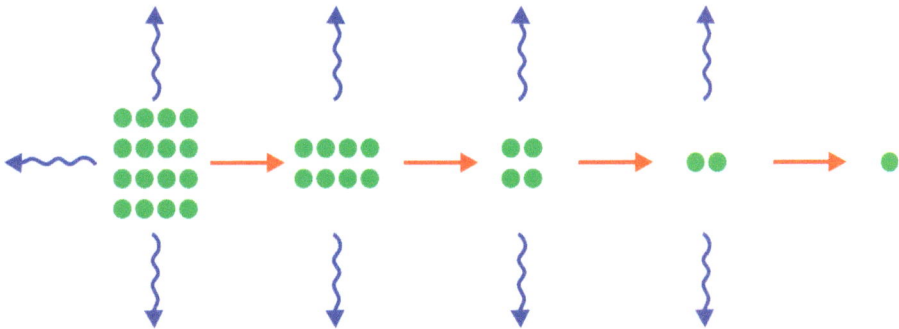

Figure 5.2 A diagram representing the decreasing number of particles during a reaction showing a constant half-life, where the amount of particles half during every time interval

Concentration-time graphs

As a reaction progresses, the amount remaining of a certain reactant progressively decreases over time. The way this decrease happens depend on the order with respect to that reactant.

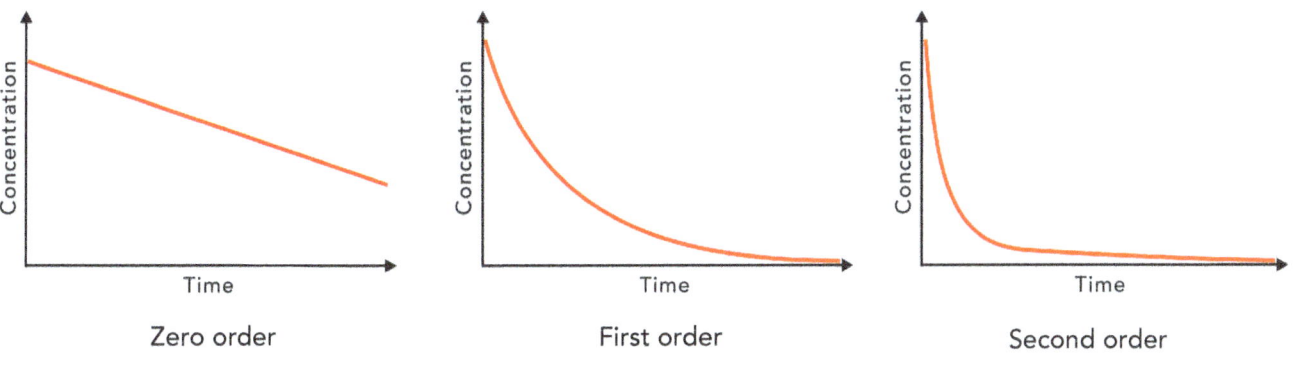

Figure 5.3 Concentration-time graphs for zero, first and second order reactants

As you can see from the third graph, the cocentration of a second order reactant goes down very fast initially. After that, the rate at which it goes down decreases. First order reactant: Concentration halves in equal time intervals. First order reactants have a constant half-life. Zero order reactant: Concentration decreases at a constant rate.

The rate equation

The rate equation can be used to work out the rate of reaction. Consider the following reaction where hydrogen and nitrogen monoxide are reacting to form water and nitrogen.

$2H_2(g) + 2NO(g) \rightarrow 2H_2O(g) + N_2(g)$

The rate equation for this reaction is as follows.

rate = $k[NO(g)]^2 [H_2(g)]$

It is evident that the order with respect to NO is 2 and the order with respect to H_2 is 1. This means that doubling the concentration of NO while keeping the concentration of H_2 the same will increase the rate of reaction by four times and doubling the concentration of H_2 while keeping the concentration of NO the same will increase the rate of reaction by two times. (You will not be expected to memorise the orders of different reactants for each reaction, as they are determined by experimental data).

Any reactant whose order is 0 will not be included in the rate equation (because it has no effect on the reaction rate).

Increasing the temperature will increase the value of the rate constant.

As the temperature goes up, the particles of the reactants will collide more frequently with increasing amounts of energy. A higher proportion of these collisions will exceed the activation energy leading to more successful collisions. Hence the reaction rate will increase.

The rate constant (k)

A constant is something which does not change. For a given reaction, the rate constant will always stay the same at a given temperature. The rate constant decides whether the reaction will be fast or slow. The bigger the rate constant, the faster the reaction.

Along with the rate constant, the concentrations of the reactants will also decide the rate of a reaction.

To work out the rate constant, you need experimental data. The rate constant cannot be determined just by using a chemical equation.

Arrhenius equation

$$k = Ae^{-\frac{E_a}{RT}}$$

- rate constant: k
- pre-exponential factor: A
- activation energy: E_a
- gas constant: R (given on the Data Sheet)
- temperature (in kelvin): T

Equation provided on the Data Sheet

The Arrhenius equation explains the effect of temperature (T) on the rate constant (k) and therefore the rate of reaction. A, pre-exponential factor is a constant which relates temperature to the rate constant (k).

Important notes:

- Temperature has to be measured in kelvin.
- E_a - activation energy - the minimum energy required for the reaction to take place. For the purpose of this equation, E_a has to be measured in joules per mole (not kilojoules per mole).
- e is a mathematical constant with the value 2.718. This can be typed on your calculator using the e^x button.

The same equation can be expressed in a different form as shown below.

$$\ln k = -\frac{E_a}{RT} + \ln A$$

Equation provided on the Data Sheet

In the exam, you may get questions based on graphs drawn using this equation. If $\ln k$ is plotted on the y-axis and $1/T$ on the x-axis, the gradient of such a graph is equal to $-E_a / R$. The y-intercept of the graph will be equal to $\ln A$.

Revision Guide
Physical Chemistry and Transition Elements

Reaction rates

- **Rate constant is affected by changes in temperature.**
- **The rate constant, k, decides the speed of a given reaction.**
- **Arrhenius equation can be used to work out the rate constant at different temperatures**
- **Half-life of a 2nd order reactant, increases with time**
- **Half-life of a 1st order reactant is constant**
- **Half-life of a reactant is the time it takes for its concentration to halve**
- **Half-life of a 0 order reactant decreases with time**
- **Units:** $mol\ dm^{-3}\ s^{-1}$
- **Rate of reaction:** Change in concentration of reactant or product per unit time
- **A rate-concentration graph can help deduce the order with respect to a reactant**
- **Concentration-time graphs can be plotted for a reaction**
  - Gradient is equal to the rate of reaction at that point
  - The gradient of the graph can be found at any point
  - Initial rate of reaction is the rate of reaction when the reaction has just started (time = 0 seconds)
  - Can be worked out by measuring the gradient of the graph using a tangent line touching point, t=0
- **The concentration of a reactant can affect the reaction rate in various amounts**
  - The order of a reactant is the power to which the concentration of the reactant is raised in the rate equation
  - When the concentration of a 1st order reactant is tripled, the reaction rate triples
  - When the concentration of a 2nd order reactant is tripled, the reaction increases by a factor of nine
  - Changes in concentration of a 0 order reactant does NOT affect the reaction rate

The equilibrium constant, K_c.

Equilibrium constant can either be K_c or K_p. K_c is used when working with concentrations (rather than partial pressures) and K_p is used when working with partial pressures (and therefore gases).

Consider the following equation:
$aA(g) + bB(g) \rightleftharpoons cC(g) + dD(g)$

For this equilibrium, K_p would be:

$$K_p = \frac{P_C^c \times P_D^d}{P_A^a \times P_B^b}$$

To work out the partial pressure, a balanced equation is needed. In the above formula, the numerator is obtained by multiplying the partial pressure of species C to the power of c (the coefficient of C in the balanced equation) by the partial pressure of species D to the power of d (the coefficient of D in the balanced equation).

Partial Pressures

Partial pressure is measured in pascals (Pa). If gas X has a partial pressure of 5.0×10^3 kPa, this means that gas X would exert a pressure of 5.0×10^3 kPa if it alone occupied the same volume at the same temperature.

Partial pressure increases with volume. Therefore, it can be thought of as the concentration equivalent of K_c.

Mole fractions

As the name suggests, mole fractions tells us the volume occupied by a given gas compared to the total volume. For example, if we have 1 mole of nitrogen and 3 moles of oxygen, the mole fraction of nitrogen would be 1/4.

The mole fraction is linked to the partial pressure of a gas. Partial pressure can be worked out by multiplying the mole fraction of a gas by the total pressure. This will tell you the pressure exerted by that gas alone.

Mole fractions may also be given as a percentage. For example, about 78% of the air is nitrogen. This can also be given as a fraction as 78/100 or as a decimal as 0.78. All of them mean the same thing.

Consider another example where nitrogen and oxygen are contained in a room. Oxygen occupies 25% of the volume and the total pressure exerted by the gases is 200 kPa. Based on this information, it is possible to work out the partial pressure of nitrogen. The mole fraction of nitrogen would be 75%. Partial pressure is calculated by multiplying the mole fraction by the total pressure: $0.75 \times 200 = 150$ kPa. Therefore, $P_N = 150$ kPa.

Revision Guide — Physical Chemistry and Transition Elements

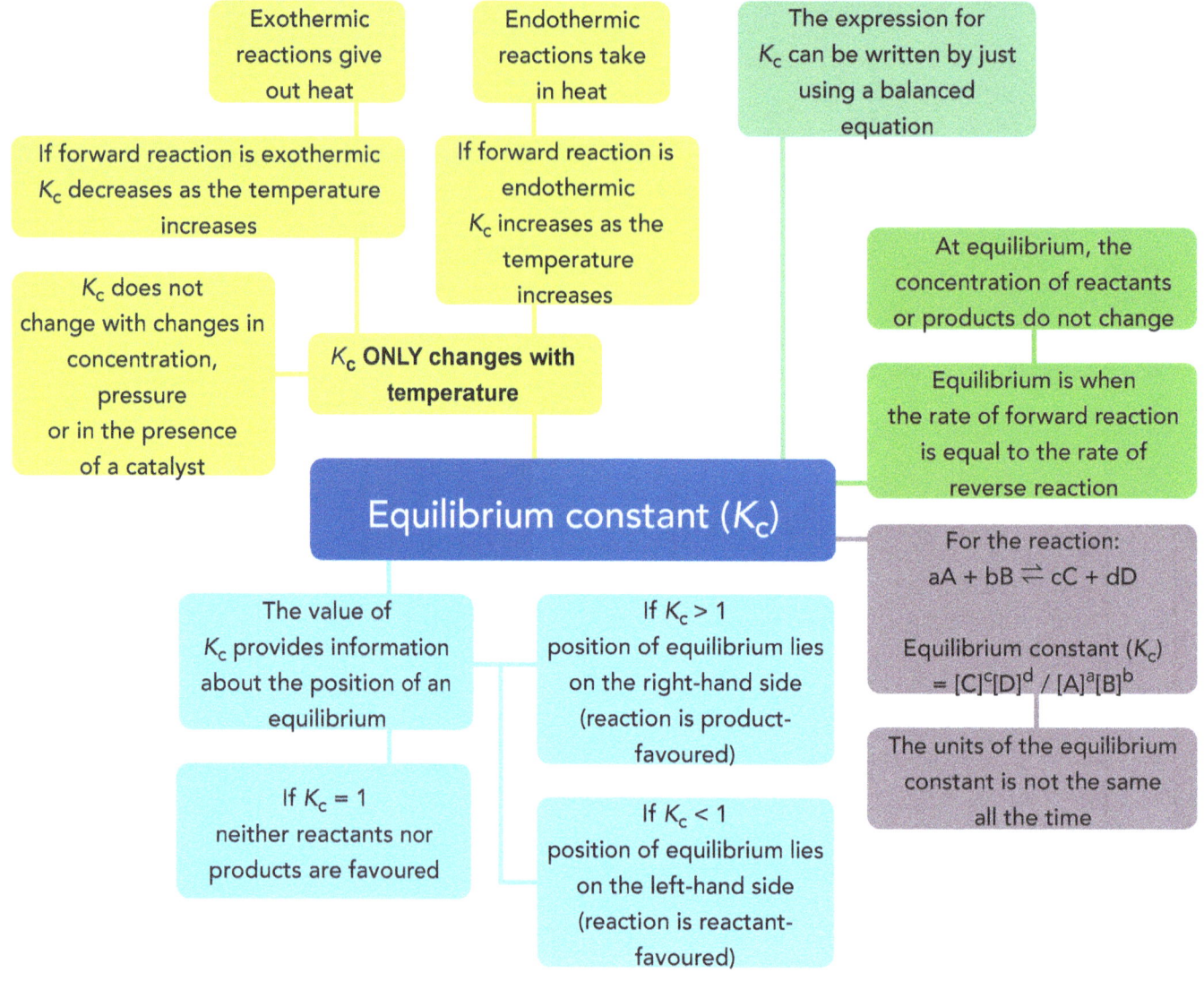

Homogeneous and Heterogeneous

A homogeneous equilibrium is one where everything is in the same phase (ie. solids, liquids or gases).
A heterogeneous equilibrium is one where everything is not in the same phase and therefore there could be a combination of solids, liquids or gases.

When calculating K_p of a heterogeneous equilibrium, all other phases except gases should be ignored, just like you ignore solids when working with K_c.

When writing expression for K_c and K_p, solids and liquids are not included.

Physical Chemistry and Transition Elements

Revision Guide 5

The difference between an alkali and a base

An alkali is a soluble base. All alkalis are bases but not all bases are alkalis.

When an alkali is dissolved in water, it forms hydroxide ions (OH⁻). These solutions will have a pH greater than 7.

Sodium hydroxide (NaOH), potassium hydroxide (KOH), calcium hydroxide (Ca(OH)$_2$) and magnesium hydroxide (Mg(OH)$_2$) are all examples of alkalis.

Strong acids and strong bases

Some of the common strong acids are:

Sulfuric acid	H_2SO_4
Nitric acid	HNO_3
Hydrochloric acid	HCl

Some common strong bases include:

Sodium hydroxide	NaOH
Potassium hydroxide	KOH
Calcium hydroxide	$Ca(OH)_2$

State symbols

Solids (s) : Examples include metals and precipitates from reactions.

Liquids (l): Include molten elements / compounds. Liquid is a state of matter.

Aqueous (aq): Solution where water is the solvent. Most acids and bases are used in the aqueous form in the laboratory.

Gas (g): Includes water vapour and elements such as nitrogen and hydrogen.

Dissociation of acids

When an acid is added to an aqueous solution, it dissociates. This means it breaks down into its ions. For example, when HCl (hydrochloric acid) is added to water, it breaks down into H⁺ and Cl⁻ ions as shown below:

HCl(aq) → H⁺(aq) + Cl⁻(aq)

HCl is a strong acid. Strong acids dissociate completely. Therefore, for a monobasic (contains one H^+ ion) strong acid (eg. HCl) the concentration of H^+ ions is taken to be equal to the concentration of the acid itself.

Weak acids dissociate partially as shown below:

$CH_3COOH(aq) \rightleftharpoons H^+(aq) + CH_3COO^-(aq)$

In a weak acid, only few of the molecules dissociate. K_a values indicate the strength of an acid. Weaker the acid, smaller the K_a value.

Note that the rate of dissociation in a weak acid is equal to the rate at which the ions recombine - this is shown by the equilibrium sign.

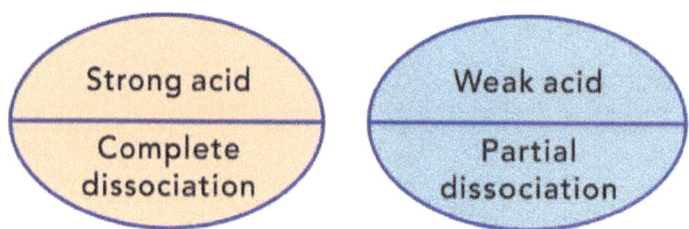

Figure 5.4 Dissociation of strong and weak acids

Buffer solutions

A buffer solution minimises pH changes on addition of small amounts of an acid or a base. Note that buffer systems cannot *prevent* pH changes.

A buffer solution may be formed from:

- a weak acid and a salt of the weak acid. Example: CH_3COOH and CH_3COONa
OR
- excess of a weak acid and a strong alkali. Example: CH_3COOH and NaOH (Note that in this case, NaOH reacts with some of the CH_3COOH to form CH_3COONa. This leaves some CH_3COOH that did not react along with the CH_3COONa that was formed)

The following equilibrium exists in a buffer solution consisting of CH_3COOH and CH_3COONa.

$$CH_3COOH \rightleftharpoons H^+ + CH_3COO^-$$

On addition of alkali
H^+ reacts with OH^- to form water, equilibrium shifts to the right as H^+ concentration decreases dissociating more CH_3COOH and forming more H^+ minimising pH changes.

On addition of acid
CH_3COO^- reacts with added acid / H^+
Equilibrium shifts to the left, forming CH_3COOH

pH titration curves

The pH titration curve for a given titration depends on the nature of the acid and base used.
Note that the change in pH of a weak acid-weak base titration is too gradual at the equivalence point (the point at which equal amounts of acid and base react resulting in neutralisation). Therefore, **no indicator is suitable for a weak acid-weak base titration**. Indicators used in titrations are considered weak acids.

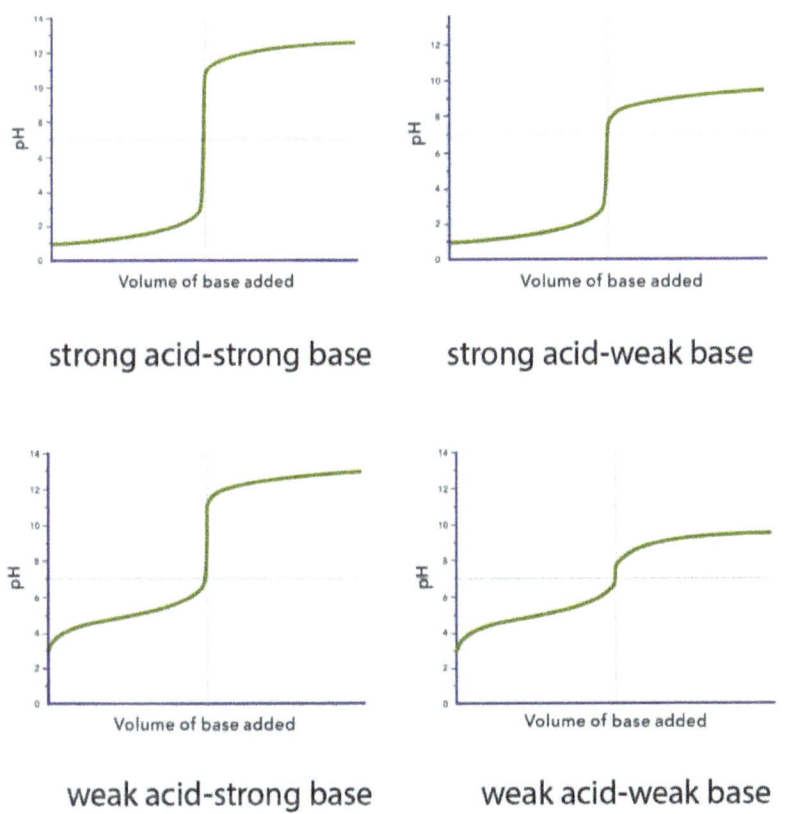

Figure 5.5
pH-titration curves for four titrations are given in the figure shown. Note that in all of the titrations given here, the base is being added from the burette (acid is placed in the conical flask). As the base is added, the pH rises.

When a strong acid is used in the titration, the initial pH is around 1 and for a weak acid, its around 3.

For a titration involving a strong base, the final pH reaches to about 13 and for a weak base it reaches to about 10.

Hydrogen ion = Proton

A hydrogen ion (H^+) is often referred to as a proton. Imagine a hydrogen atom. It has one electron, one proton and no neutrons. The only proton is inside the nucleus, as usual. When the hydrogen ion is formed, it loses its only electron. The only thing that remains now is the proton.

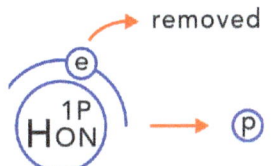

Figure 5.6 Hydrogen ion forming

Revision Guide — Physical Chemistry and Transition Elements

Conjugate acid-base pairs

The following figure shows two conjugate acid-base pairs. H_2O and NH_3 are the bases. H_3O^+ and NH_4^+ are the acids. The proton (H^+) of NH_4^+ combines with H_2O to form H_3O^+. In the process, NH_4^+ loses a proton.

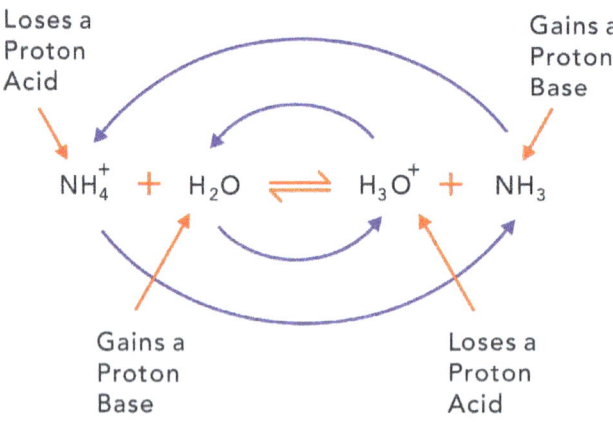

Figure 5.7 Conjugate acid-base pairs

Acids and bases

- A buffer solution can be made from a weak acid and a salt of the weak acid
- A buffer solution help minimise pH changes. It does NOT prevent pH changes
- K_a values could take the form of long numbers. It can be made more manageable using the pK_a
- $pK_a = -\log K_a$
- $K_a = 10^{-pK_a}$

- The concentration of H^+ ions is the same as the concentration of the acid (only for strong acids)
- For a strong monobasic acid HA, $[H^+(aq)] = [HA(aq)]$
- The acid dissociation constant (K_a) measure how well an acid dissociates
- For an acid HA, $$K_a = \frac{[H^+(aq)][A^-(aq)]}{[HA(aq)]}$$
- [HA(aq)] here means the concentration of the acid

- In an acid-alkali neutralisation reaction, H^+ from the acid and OH^- from the alkali react together forming H_2O
- An alkali releases OH^- ions. It is what makes it alkaline / basic
- Low K_a = weaker acid
- High K_a = stronger
- For example, for CH_3COOH:

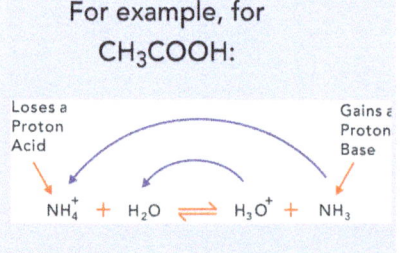

Chemistry Q & A

Physical Chemistry and Transition Elements — Revision Guide 5

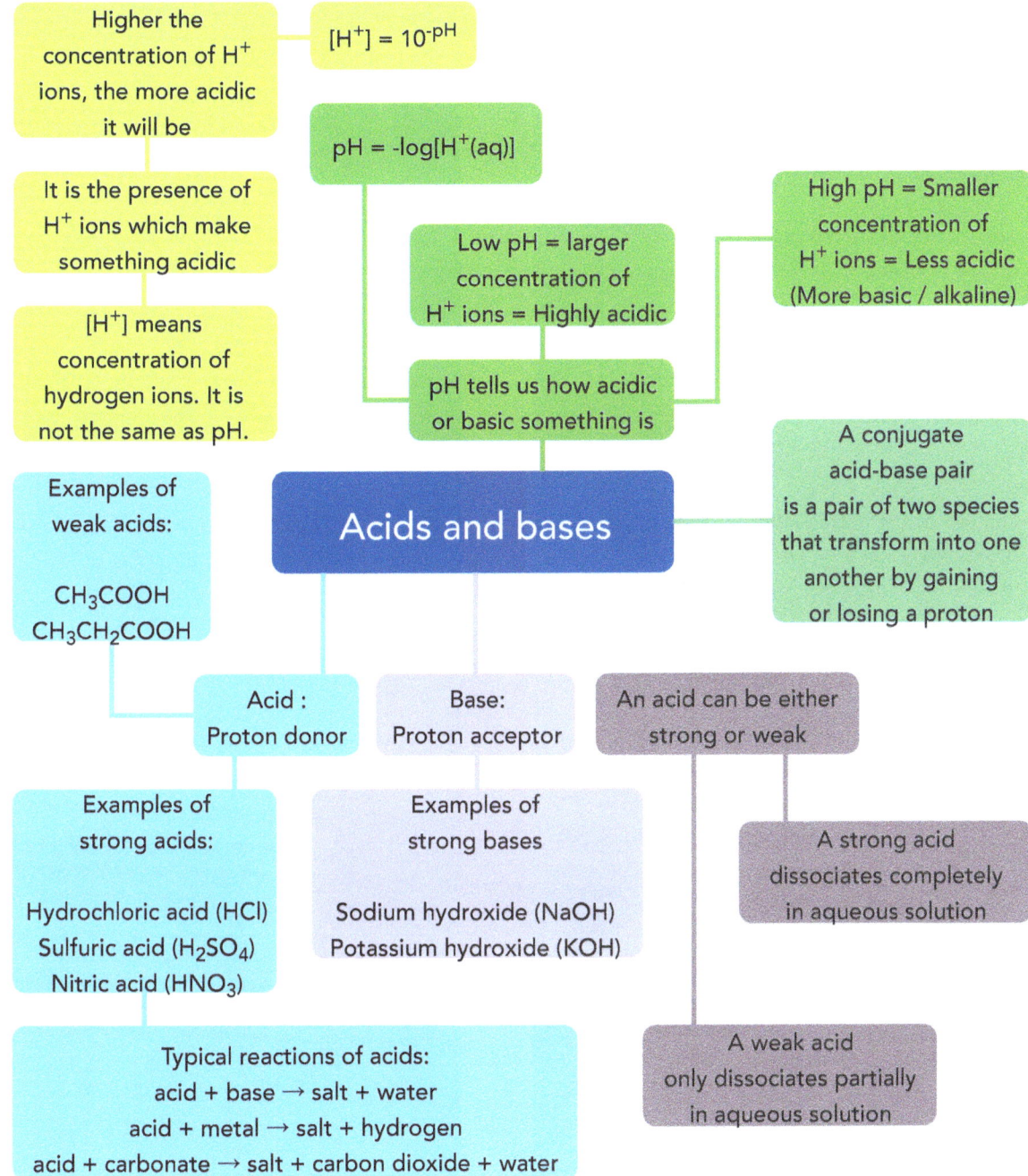

Standard state of an element

The standard state of an element is the state (solid, liquid or gas) in which an element is usually found under standard conditions. For example, hydrogen is found as a gas (g), carbon is found as a solid (s), bromine is liquid (l).

Standard conditions

Standard conditions include set values for temperature (298 K or 25 °C), pressure (100 kPa) and concentration of 1 mol dm^{-3}.

Revision Guide — Physical Chemistry and Transition Elements

When a reaction is taking place, differences in temperature, pressure and concentration can influence its rate, energy given out, etc. However, if two reactions are carried out under the same conditions, you can compare other factors influencing the reaction. Standard conditions allow such comparisons to be made.

Born-Haber cycles

The following figure shows the Born-Haber cycle of aluminium oxide. Note that in the formula of aluminium oxide, Al_2O_3, there are two aluminium ions and three oxygen ions. Therefore, all enthalpy changes relating to aluminium has been multiplied by two and that of oxygen has been multiplied by three.

In forming Al_2O_3, each aluminium atom loses three outer shell electrons becoming Al^{3+}. This involves three ionisation energies. On the other hand, each oxygen atom gains two electrons to become O^{2-}. This involves two electron affinities.

1. Enthalpy change of formation involves formation of Al_2O_3 from aluminium and oxygen in their standard states. Oxygen occurs as O_2. Therefore to get three oxygen atoms, it has to be written as ³⁄₂ O_2.

2. After both of the aluminium atoms go through the first, second and the third ionisation energies, six electrons have been removed - three from each of the two oxygen atoms. (2 x 3 = 6).

3. Each of the three oxygen atoms take in one electron each, as they go through the first electron affinity. This leaves three electrons unaccounted for.

4. Each of the three O⁻ ions take in one electron each, as they go through the second electron affinity, leaving no extra electrons.

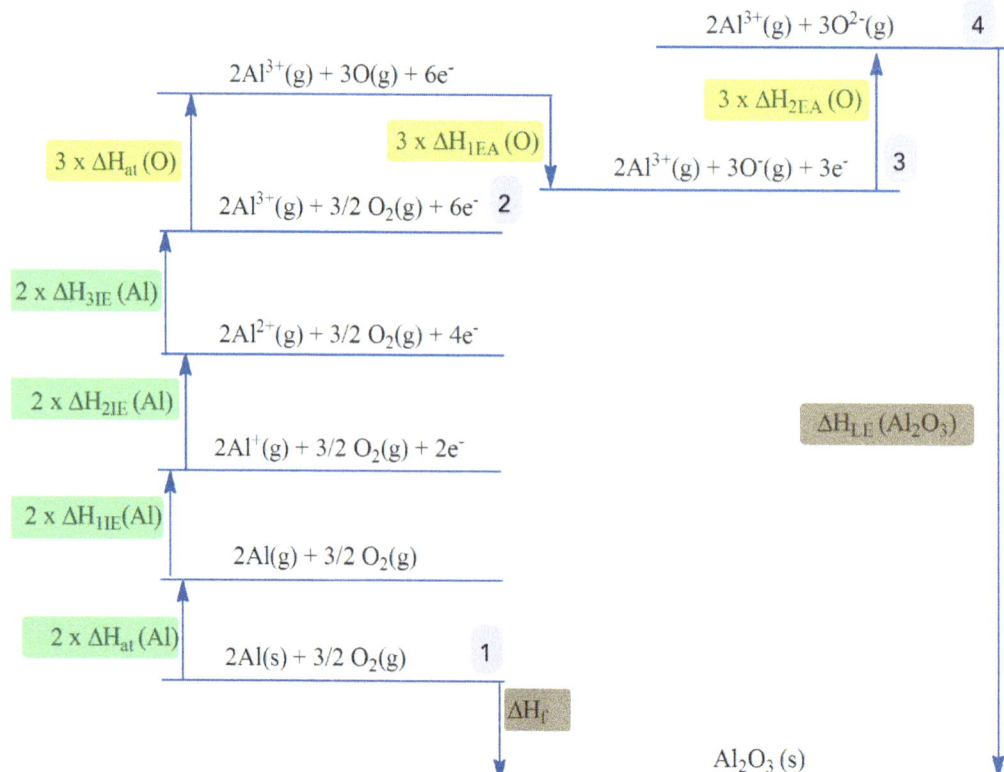

Figure 5.8 Aluminium oxide Born-Haber cycle

Enthalpy changes

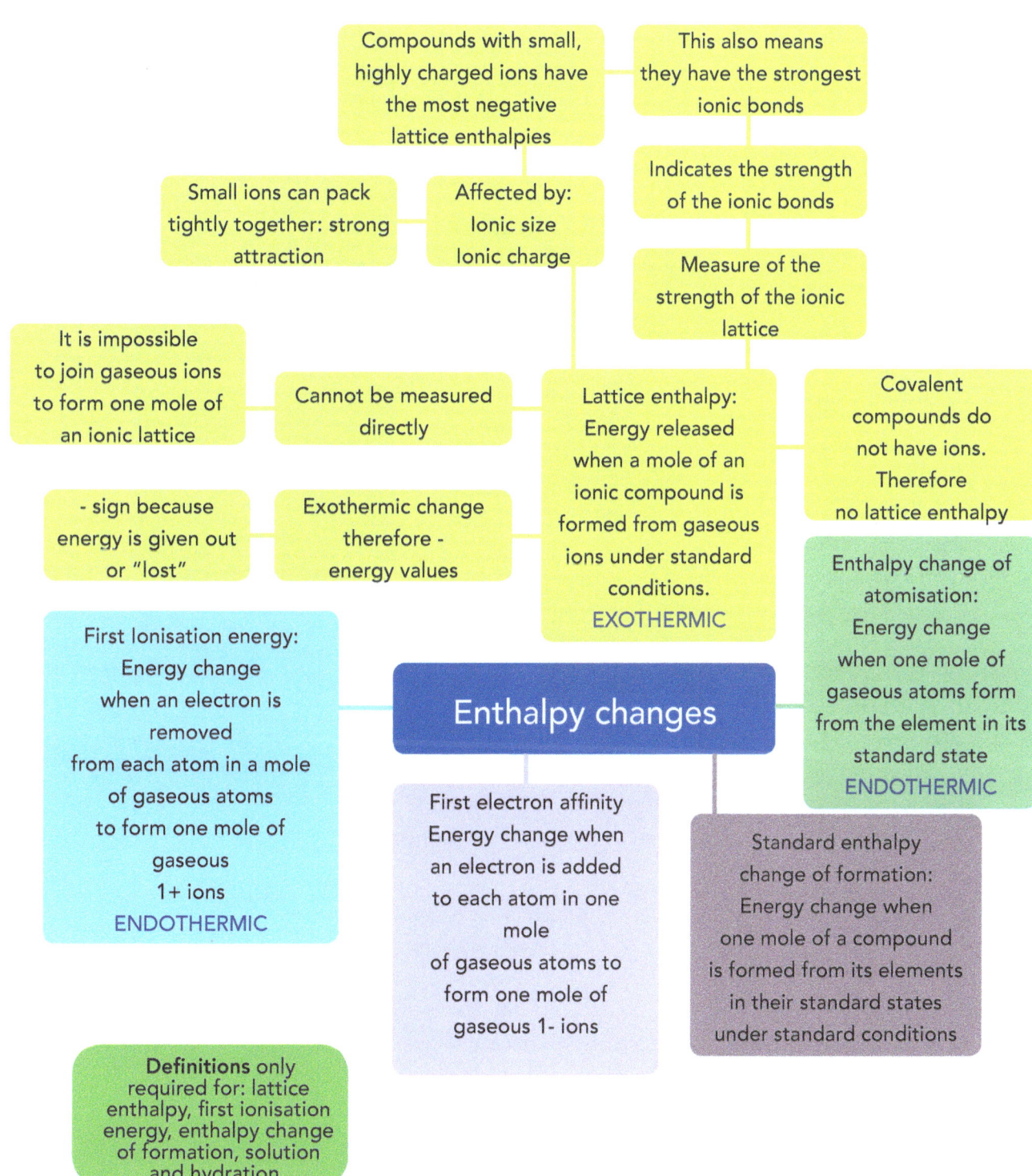

Revision Guide
Physical Chemistry and Transition Elements

Enthalpy changes

- **This attracts the positive ions in the compound**
  - The O atom in a water molecule is slightly negatively charged
  - Hydration involves ions bonding with water molecules

- **These attract the negative ions in the compound**
  - The H atoms in a water molecule are slightly positively charged

- **If a certain amount of energy is released when a compound is made from given species, the same amount of energy will be needed to break it into those species.**
  - Therefore, lattice enthalpy is also a measure of the energy needed to break down an ionic lattice

- **Depends on:**
  - » Energy needed to break down ionic lattice into gaseous ions
  - » Energy needed to hydrate the gaseous ions

- **Hess's law states that if a reaction can occur in more than one path, the total energy change in all those paths will be the same (if the initial and final conditions**
  - Born-Haber cycles: Arrows point up - ENDOTHERMIC; Arrows pointing down - EXOTHERMIC
  - Born-Haber cycles follow Hess' law

- **Standard enthalpy change of solution:** Energy change when one mole of a compound is completely dissolved in water under standard conditions

- State symbols are very important in enthalpy changes

- **Standard enthalpy change of hydration:** Energy change when one mole of isolated gaseous ions dissolve in water. **EXOTHERMIC**
  - Factors affecting its value: Ionic size, Ionic charge
  - Same as lattice enthalpy

- An exothermic reaction releases energy to its' surroundings, therefore there is an energy loss

- First electron affinity is mostly exothermic. Second electron affinity is endothermic.

- An endothermic reaction takes in energy from the surrounding. (reactants gains energy)

- Second electron affinity involves adding an electron (negative in charge) to an already negative ion. As electrons repel, it requires energy to overcome this.

- A test tube containing an endothermic reaction will feel cold because it takes in heat from your skin (your skin loses heat)

Chemistry Q & A

Physical Chemistry and Transition Elements — Revision Guide 5

Entropy

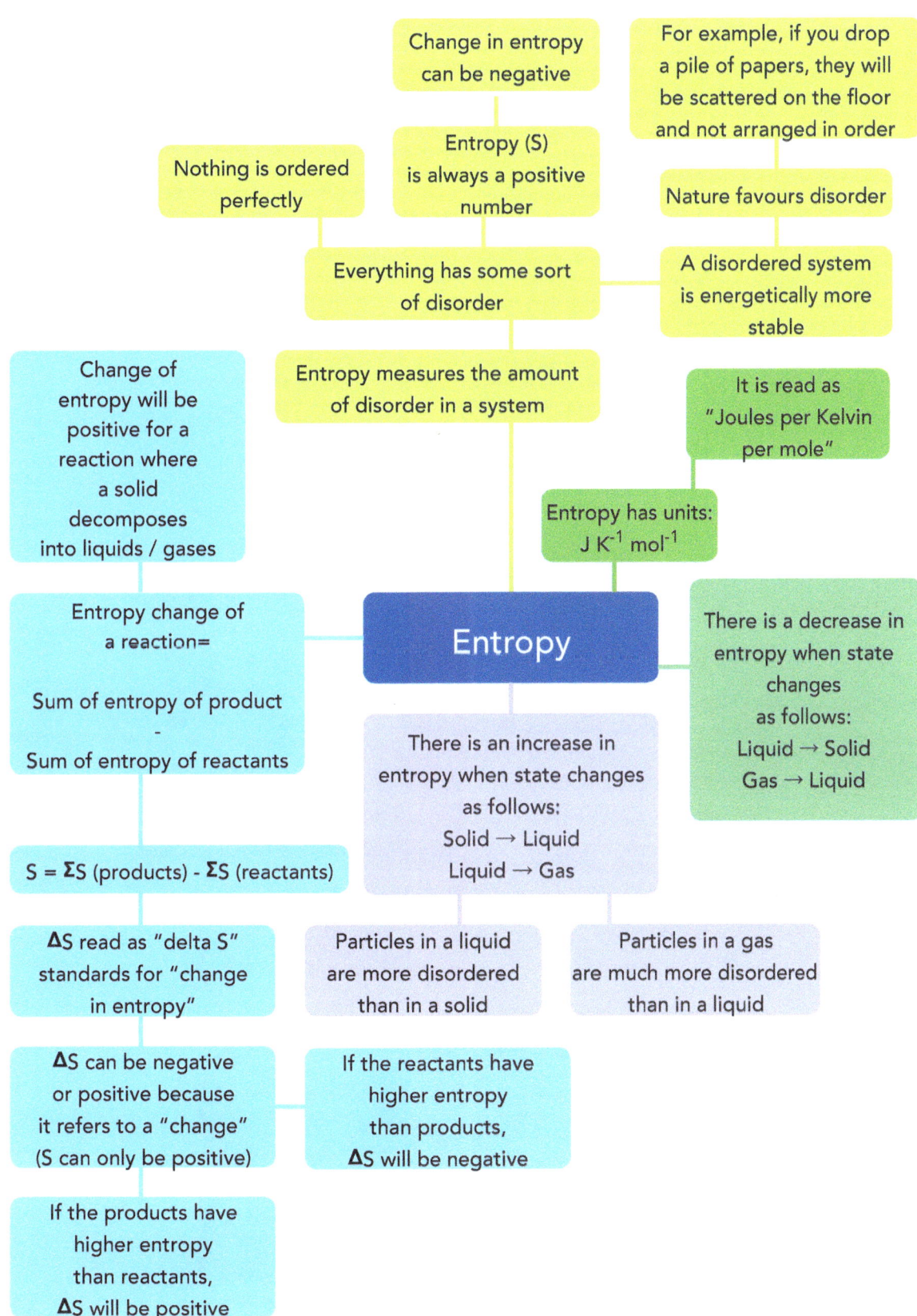

Revision Guide
Physical Chemistry and Transition Elements

Free energy

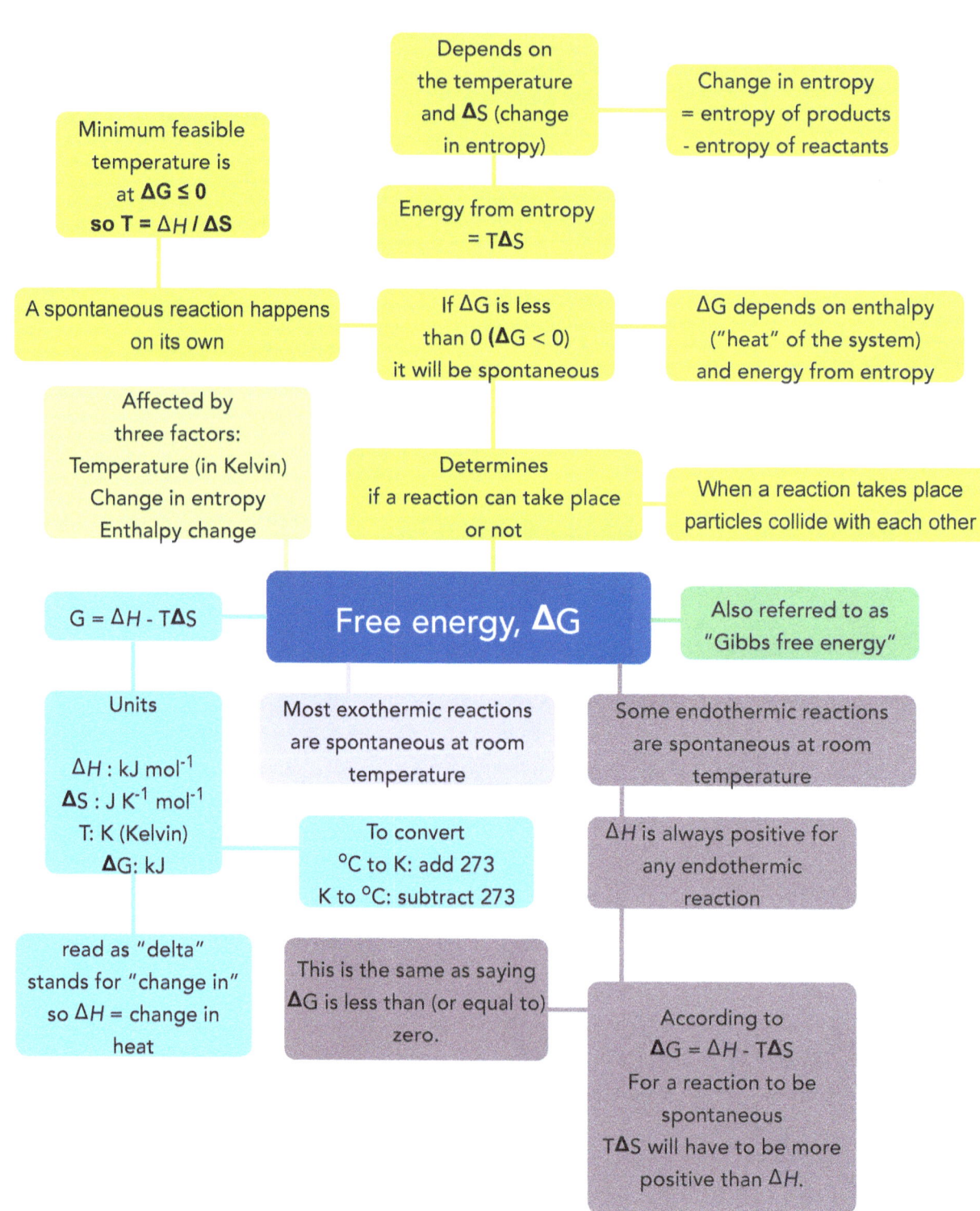

Chemistry Q & A

Physical Chemistry and Transition Elements — Revision Guide 5

Cells and half-cells

- Fe^{2+} is formed when Fe (iron) loses two electrons
- Fe^{3+} is formed when Fe (iron) loses three electrons

For example: Iron has two oxidation states Fe^{2+} and Fe^{3+}

For simple ions, its oxidation number is the same as its ionic charge

Each time an element loses an electron its oxidation number increases by one

The oxidation state of any element on its own is 0

The oxidation state of the element Na is 0. When Na combines with another ion, it becomes Na^+. Now its oxidation number is +1

Half-cell: consists of an element in two oxidation states

Between these two different oxidation states, there is an equilibrium

Element is changing from one oxidation state to the other by gaining or losing electrons

For example, the hydrogen half-cell has the equilibrium:
$2H^+ (aq) + 2e^- \rightleftharpoons H_2 (g)$

H oxidises to H^+ by losing an electron.
H^+ reduces to H by gaining an electron.
These two reactions are in equilibrium

A reactions goes on between the ions and the electrons in the two half-cells. This is the overall cell reaction

Hydrogen is a diatomic gas. It always exist as H_2. When writing hydrogen gas in equations it is always written as H_2

To balance H_2 on the right side, you need $2H^+$ ions on the left and also 2 electrons

In a metal/metal ion half-cell, a metal is dipped into a solution of its ions

In a non-metal/non-metal ion half-cell, a non-metal (example: hydrogen gas) is in contact with its ions (H^+)

A metal ion/metal ion half-cell consist of the same element in different oxidation states. Example: Fe^{2+}/Fe^{3+}. This half-cell requires a platinum electrode.

Standard electrode potential of a cell is the e.m.f of a half-cell compared to the standard hydrogen half-cell measured under standard conditions

An electrochemical cell is formed by joining together two half-cells with different electrode potentials

The half-cells are connected using a wire and a salt bridge

Wire carries the electrons, salt bridge carries ions

One half-cell gives out electrons while the other gains those electrons

The half-cell with the more negative / less positive electrode potential loses the electrons acting as the negative terminal while the other acts as the positive terminal

Revision Guide — Physical Chemistry and Transition Elements

Oxidation numbers

The oxidation number of a simple ion is the same as its charge. For example, in the compound $MgCl_2$, the oxidation number of Mg is +2 and that of Cl is -1. The oxidation number of Al in Al_2O_3 is +3 because the charge of the aluminum ion is 3+. The oxidation number of O is -2.

It is worth noting that compound ions do not have an oxidation number. However, the elements within the ion do. For example, the dichromate ion, $Cr_2O_7^{2-}$ does not have an oxidation number. However, in $Cr_2O_7^{2-}$ the oxidation number of Cr is +6 and that of O is -2.

Another example of a compound ion is nitrate, NO_3^-. Although this ion does not have an oxidation number, the oxidation number of N in nitrate is +5 and that of O is -2.

Electrochemical cells

An electrochemical cell is formed by combining two half-cells with different electrode potentials.

Each half-cell has its own equilibrium consisting of electrons and ions. The wire carries these electrons between the two half-cells and the salt bridge carries the ions.

The half-cell with the more negative electrode potential becomes the negative terminal. The other becomes the positive terminal. The negative terminal supplies the electrons and the positive one receives them.

For example in an electrochemical cell made from combining the following:

$Ag^+(aq) + e^- \rightleftharpoons Ag(s)$ $E^\ominus = +0.80\ V$
$Cu^{2+}(aq) + 2e^- \rightleftharpoons Cu(s)$ $E^\ominus = +0.34\ V$

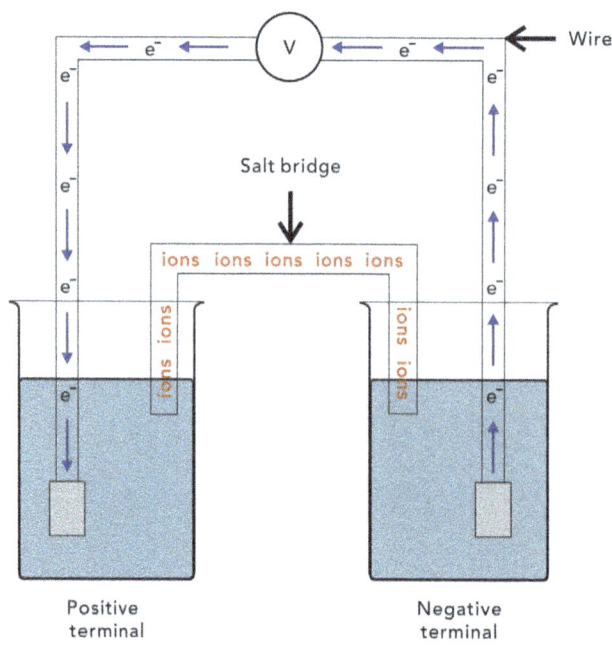

Figure 5.9 Flow of electrons in an electrochemical cell

The half-cell with the greater value will undergo reduction as it accepts electrons. In this case, the positive terminal is the silver half-cell as it accepts electrons. The copper half-cell will make up the negative terminal as it loses electrons (oxidation).

Looking at the copper equilibrium, we can see that the electrons (e⁻) are found on the left hand side. As a result, the equilibrium will move from right to left resulting in the release of electrons. On the other hand, Ag^+ will accept electrons and the equilibrium will move from left to right.

Combining the two half-equations gives us the overall equation. Since we know that the copper half-cell provides the electrons, we can write it with the electrons on the products side as follows:

$Ag^+(aq) + e^- \rightleftharpoons Ag(s)$
$Cu(s) \rightleftharpoons Cu^{2+}(aq) + 2e^-$

The next step is to balance the electrons on both sides. The copper half-cell has two electrons. Therefore, we will double the number of species in the silver half-cell.

$2Ag^+(aq) + 2e^- \rightleftharpoons 2Ag(s)$
$Cu(s) \rightleftharpoons Cu^{2+}(aq) + 2e^-$

Finally, we will cancel the 2e⁻ as it appears on the left-hand side of the silver half-equation and the right-hand side of the copper half-equation.

$2Ag^+(aq) \rightleftharpoons 2Ag(s)$
$Cu(s) \rightleftharpoons Cu^{2+}(aq)$

Now, combine the two sides of both half-equations. This gives us the following overall equation.

$2Ag^+(aq) + Cu(s) \rightleftharpoons 2Ag(s) + Cu^{2+}(aq)$

Oxidation and Reduction

The following half-equation shows reduction as the species gains an electron. The oxidation number goes down by one.

$Fe^{3+} + e^- \rightarrow Fe^{2+}$
(Brown)   (Pale green)

The following is a redox reaction where both oxidation and reduction take place.

$Cr_2O_7^{2-} + 14H^+ + 6Fe^{2+} \rightarrow 2Cr^{3+} + 7H_2O + 6Fe^{3+}$
(Orange)   (Green)
+6 -2   +1   +2   +3   +1 -2   +3

Chromium has been reduced from +6 to +3. Iron has been oxidised from +2 to +3.

Revision Guide
Physical Chemistry and Transition Elements

The following is an example of a **disproportionation reaction**. A disproportionation reaction is one where the same species gets both oxidised and reduced. The same equation is shown below along with the oxidation numbers.

$2Cu^+ \rightarrow Cu + Cu^{2+}$
(Colourless)   (Brown precipitate of copper) (Blue solution of copper(II))

$2Cu^+ \rightarrow Cu + Cu^{2+}$
   +1   0   +2

Cu^+ ion has been reduced to Cu (oxidation number goes from +1 to 0) and oxidised to Cu^{2+} (oxidation number goes from +1 to +2) at the same time.

Fuel cells

- In this equilibrium, reaction to one side will be oxidation and the other reduction
- Each half-cell is a equilibrium
- Half-cell with the most negative / least positive electrode potential act as the negative terminal. The other as the positive terminal.
- Involves two half-cells with different electrode potentials
- Standard cell potential (E cell) = e.m.f of cell measured in volts
- Standard cell potential ($E^{\ominus}_{cell}$) = $E^{\ominus}$ (positive terminal) - $E^{\ominus}$ (negative terminal)
- Electrochemical cells - also known as fuel cells
- Used as a source of electrical energy
- Uses energy from the reaction of a fuel with oxygen to create a voltage

$2H_2O(l) + 2e^- \rightleftharpoons H_2(g) + 2OH^-(aq)$ $E^{\ominus} = -0.83$ V
$1/2\ O_2(g) + H_2O(l) + 2e^- \rightleftharpoons 2OH^-(aq)$ $E^{\ominus} = +0.40$ V

- You are not required to know the half-cell equations given here.
- Hydrogen system is more negative (-0.83 V). It gives out electrons.
- Oxygen system accepts those electrons
- The equilibrium moves to the right as it accepts the electrons. While using oxygen, it forms OH^- ions.
- These OH^- ions are used by the hydrogen system as its equilibrium moves to the left.
- The hydrogen equilibrium moves to the left in order to release the electrons (electrons are on the left hand side of the equilibrium)
- $H_2(g)$ is on the right-hand side of the equilibrium. As the reaction proceeds left, it is used up. Therefore it has to be supplied for this fuel cell to function.

Physical Chemistry and Transition Elements — Revision Guide 5

Transition elements

Definition: A d-block element that forms an ion with a partially filled d sub-shell. Zn and Sc are NOT transition elements despite being in the d-block.

Electron configuration (filling and losing)
- On the other hand, when electrons are being filled, 4s electrons are filled first. So for transition elements: 4s filled first, 4s lost first (before 3d)
- In transition metals, when ions are formed, the 4s electrons are always removed first before the 3d electrons

Catalysts
Transition elements and their compounds are used as catalysts.

Examples:
- Cobalt
- Titanium
- Vanadium
- Iron

Chromium and Copper
- Chromium and Copper are transition elements
- Cr and Cu are exceptions when it comes to their electronic configuration
- Their 4s sub-shells hold only one electron (not fully filled)
- Cr: $1s^2\ 2s^2\ 2p^6\ 3s^2\ 3p^6\ 4s^1\ 3d^5$
- Cu: $1s^2\ 2s^2\ 2p^6\ 3s^2\ 3p^6\ 4s^1\ 3d^{10}$

Coloured compounds
They form coloured compounds.

Oxidation states
- Transition elements can have more than one oxidation state in their compounds
- For example, iron (Fe) can exist as Fe^{2+} or Fe^{3+}
- On the other hand, other metals such as sodium (Na) will have only one oxidation state (+1)

Reactions with NaOH(aq)
Aqueous solutions of transition metals react with NaOH(aq) to form coloured precipitates.

$Mn^{2+}(aq) + 2OH^-(aq) \rightarrow Mn(OH)_2(s)$
Pink solution → white/cream precipitate (turns brown when exposed to air)

$Fe^{2+}(aq) + 2OH^-(aq) \rightarrow Fe(OH)_2(s)$
Pale green solution → Green precipitate (turns rusty brown at the surface when exposed to air)

$Fe^{3+}(aq) + 3OH^-(aq) \rightarrow Fe(OH)_3(s)$
Pale yellow solution → Rusty brown precipitate

$Cu^{2+}(aq) + 2OH^-(aq) \rightarrow Cu(OH)_2(s)$
Pale blue solution → blue precipitate

$Cr^{3+}(aq) + 3OH^-(aq) \rightarrow Cr(OH)_3(s)$
Dark blue solution → green precipitate dissolves in excess NaOH to form green solution

Revision Guide — Physical Chemistry and Transition Elements

Precipitation reactions

This section outlines some more precipitation reactions of transition metal ions.

Reactions with aqueous ammonia

$[Cu(H_2O)_6]^{2+}(aq) + 2NH_3(aq) \rightarrow [Cu(H_2O)_4(OH)_2](s) + 2NH_4^+(aq)$ Pale blue solution to pale blue precipitate

The precipitate dissolves in excess ammonia.

$[Cu(H_2O)_6]^{2+}(aq) + 4NH_3(aq) \rightarrow [Cu(NH_3)_4(H_2O)_2]^{2+}(aq) + 4H_2O(l)$ Pale blue solution to deep blue solution

Note that in the exam, it is acceptable to write $[Cu(H_2O)_6]^{2+}$ as $Cu^{2+}(aq)$ and $[Cu(H_2O)_4(OH)_2]$ as $Cu(OH)_2(s)$. Same applies for other complex formulae.

$[Fe(H_2O)_6]^{2+}(aq) + 2NH_3(aq) \rightarrow [Fe(H_2O)_4(OH)_2](s) + 2NH_4^+(aq)$ Pale green solution to green precipitate

$[Fe(H_2O)_6]^{3+}(aq) + 3NH_3(aq) \rightarrow [Fe(H_2O)_3(OH)_3](s) + 3NH_4^+(aq)$ Pale yellow solution to rusty brown precipitate

$[Mn(H_2O)_6]^{2+}(aq) + 2NH_3(aq) \rightarrow [Mn(H_2O)_4(OH)_2](s) + 2NH_4^+(aq)$ Pink solution to white/cream precipitate

$[Cr(H_2O)_6]^{3+}(aq) + 3NH_3(aq) \rightarrow [Cr(H_2O)_3(OH)_3](s) + 3NH_4^+(aq)$ Dark blue solution to green precipitate

With excess ammonia, the precipitate dissolves.

$[Cr(H_2O)_6]^{3+}(aq) + 6NH_3(aq) \rightarrow [Cr(NH_3)_6]^{3+}(aq) + 6H_2O(l)$ Dark blue solution to purple solution

Reactions with aqueous sodium hydroxide

The transition elements mind map outlines the reactions of aqueous transition metal complexes with aqueous sodium hydroxide. This section describes these reactions using their complex formula.

$[Cu(H_2O)_6]^{2+}(aq) + 2OH^-(aq) \rightarrow [Cu(H_2O)_4(OH)_2]^{2+}(s) + 2H_2O(l)$ Pale blue solution to blue precipitate

$[Fe(H_2O)_6]^{2+}(aq) + 2OH^-(aq) \rightarrow [Fe(H_2O)_4(OH)_2](s) + 2H_2O(l)$ Pale green solution to green precipitate

$[Fe(H_2O)_6]^{3+}(aq) + 3OH^-(aq) \rightarrow [Fe(H_2O)_3(OH)_3](s) + 3H_2O(l)$ Pale yellow solution to rusty brown precipitate

$[Mn(H_2O)_6]^{2+}(aq) + 2OH^-(aq) \rightarrow [Mn(H_2O)_4(OH)_2](s) + 2H_2O(l)$ Pink solution to white/cream precipitate

Physical Chemistry and Transition Elements — Revision Guide 5

$[Cr(H_2O)_6]^{3+}(aq) + 3OH^-(aq) \rightarrow [Cr(H_2O)_3(OH)_3](s) + 3H_2O(l)$ Dark blue solution to green precipitate

The precipitate dissolves in excess sodium hydroxide.

$[Cr(H_2O)_3(OH)_3](s) + 3OH^-(aq) \rightarrow [Cr(OH)_6]^{3-}(aq) + 3H_2O(l)$ Green precipitate to green solution

Ligand substitution

$[Cu(H_2O)_6]^{2+}(aq)$ reacts with $NH_3(aq)$ forming a deep blue solution. It can also react with $Cl^-(aq)$ to form a yellow solution.

$[Cr(H_2O)_6]^{3+}(aq)$ reacts with an excess of ammonia to form a purple solution of $[Cr(NH_3)_6]^{3+}$

These are ligand substitution reactions where one ligand gets replaced by another.

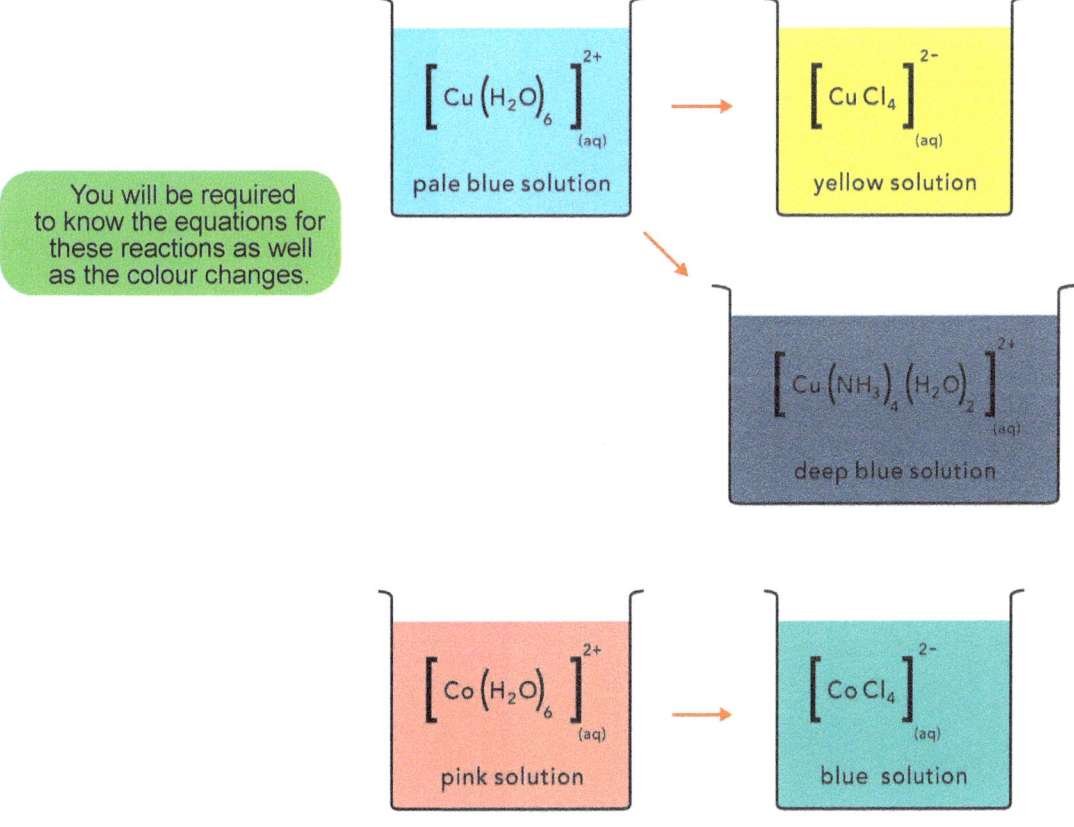

You will be required to know the equations for these reactions as well as the colour changes.

Figure 5.10 Ligand substitution reactions of copper and cobalt

Revision Guide — Physical Chemistry and Transition Elements

- In a healthy human, carbon monoxide will not be present in their blood
- However, the body needs oxygen, carbon monoxide can kill a person easily
- Carbon monoxide can replace oxygen from the Fe^{2+} binding site in haem. This is ligand substitution
- Oxygen is carried in human blood bonded to Fe^{2+}. The Fe^{2+} is found within a "haem group".

- Different ligands have different sizes. The number of ligands that can fit around a transition metal ion is partly decided by the size of the ligands
- Following the ligand substitution, the shape of the complex ion may change
- A reaction in which one ligand is replaced by another ligand

Ligand substitution

Aqueous copper(II) ions react with ammonia

$[Cu(H_2O)_6]^{2+}(aq) + 4NH_3(aq) \rightleftharpoons [Cu(NH_3)_4(H_2O)_2]^{2+}(aq) + 4H_2O(l)$
Pale blue solution   Deep blue solution

Initially, on addition of small amount of ammonia:
Pale blue precipitate (of $Cu(OH)_2$ forms)

When more is added:
Pale blue precipitate dissolved forming deep blue solution

Aqueous copper(II) ions react with concentrated hydrochloric acid

$[Cu(H_2O)_6]^{2+}(aq) + 4Cl^-(aq) \rightleftharpoons [CuCl_4]^{2-}(aq) + 6H_2O(l)$
Pale blue solution   Yellow solution

Aqueous chromium(III) ions react with an excess of ammonia

$[Cr(H_2O)_6]^{3+}(aq) + 6NH_3(aq) \rightleftharpoons [Cr(NH_3)_6]^{3+}(aq) + 6H_2O(l)$
Dark blue solution   Purple solution

Complex ions

- Stereoisomers have the same structural formula but have a different arrangement of the atoms in space

- Stereoisomerism involving transition metals are:
  cis-trans
  optical

- Some complex ions show stereoisomerism

- All the things making up a complex ion is shown inside square brackets

- Charge of the complex ion is shown outside the square brackets

- 6 coordinated complex ions
  - octahedral shape

- 4 coordinated complex ions
  - square planar shape
  or
  - tetrahedral shape

- A transition metal ion bonded to one or more ligands by coordinate bonds

Complex ion

- Cis-platin is a drug used in cancer treatment

- $[PtCl_2(NH_3)_2]$

- Binds to the DNA of cancerous cells and stops them from reproducing

- Coordination number: No. on coordinate bonds to the metal ion

- In a coordinate bond, one of the bonded atoms provides both electrons for the shared pair

- Ligand: *A molecule or ion that donates a pair of electrons to the transition metal ion forming a coordinate bond*

- A monodentate ligand: donates only one pair of electrons to the metal ion

- A bidentate ligand: donates two pairs of electrons to the metal ion

- A hexadentate ligand: donates six pairs of electrons to the metal ion

- ethane-1,2-diamine $(NH_2CH_2CH_2NH_2)$ "en" is a common bidentate ligand

- Each "en" forms two coordinate bonds with the metal ion

Complex ions and shapes

Octahedral (Bond angle: 90°)

$[Cu(H_2O)_6]^{2+}$, $[Fe(H_2O)_6]^{3+}$

Tetrahedral (Bond angle: 109.5°)

$[CuCl_4]^{2-}$ and $[CoCl_4]^{2-}$

Square planar (Bond angle: 90°)

$[PtCl_2(NH_3)_2]$

Oxidising and reducing agents

An oxidising agent oxidises another substance. In the process, it is reduced. A reducing agent reduces another substance. In the process it is oxidised.

Oxidation is the loss of electrons and reduction is the gain of electrons (OILRIG).

Physical Chemistry and Transition Elements — Revision Guide 5

Constants you need to know

Throughout this module, you will come across some constants. Most of these constants do not have set units.

The table below outlines these constants. Please note, square brackets means "concentration". Therefore, [Cl⁻] means the concentration of chloride ions.

Constant	Formula	Units
Acid dissociation constant (K_a)	For an acid HA $HA(aq) \rightleftharpoons H^+(aq) + A^-(aq)$ $K_a = [H^+(aq)][A^-(aq)]/[HA(aq)]$	$mol\ dm^{-3}$
Equilibrium constant (K_c)	For equilibrium $aA + bB \rightleftharpoons cC + dD$ $K_c = [C]^c[D]^d / [A]^a[B]^b$	No set unit (Depends on the expression)
Rate constant (k)	$k = rate / [reactant1]^{order}[reactant2]^{order}$ (equation varies from one reaction to another)	No set unit (Depends on the expression)
Ionic product of water (K_w)	$K_w = [H^+(aq)][OH^-(aq)]$ At 25°C, $K_w = 1.00 \times 10^{-14}\ mol^2 dm^{-6}$	$mol^2 dm^{-6}$

Significant Figures

A measurement given to higher number of significant figures is more precise than a measurement given to a lower number of significant figures.

For example, 4.345g is more precise than 4.3g. However, if the measuring equipment only measures to a fewer number of significant figures, any calculations based on that measurement should also be given to the same number of significant figures or less. Consider the following example.

Q. A sample of magnesium was found to have a mass of 5g. Calculate the amount, in mol, present in the sample.

A. Moles = mass(g) / molar mass = 5 / 24.3 = 0.2 mol.

In the above example, if you enter 5 / 24.3 into the calculator, it would give the answer as 0.20576... However, you would write the answer to this question as 0.2, as the mass is given to one significant figure. Therefore, you cannot be more precise than that.

In the exam, answers are usually given to three significant figures. Before deciding on the correct number of figures to use, always read the question.

(Tip: Both 4.32 and 0.00432 are given to 3 sf. The zeros before the first non-zero number does not count as significant figures).

Decimal Places

Care should be taken with decimal places when taking measurements using measuring equipments. For example, when you weigh a sample, if the weighing scale gives the reading to two decimal places, you should report the measurement to two decimal places. That is, 2.45g should never be written as 2.450g. The first number is given to two decimal places and the latter to 3 decimal places.

The same applies to other measuring equipments. Note: 10-mL measuring cylinders are always read to 2 decimal places (e.g. 3.45 mL) and 100-mL measuring cylinders are always read to 1 decimal place (e.g. 3.5 mL).

Physical Chemistry and Transition Elements

Revision Guide 5

Module 5 Questions

Module 5 involves a lot of calculations. When doing calculations, it is good practice to show the steps in your working. This will help you avoid any mistakes and also ensure you score the marks you deserve.

Some of these questions require you to know formulae. If you suspect a formula may be needed to answer any given question, write it down before you proceed any further. This will help your thinking.

pH calculations require you to use the log function in your calculator. Make sure you know how to use it properly. When typing numbers on your calculator, brackets may be used to separate numbers and symbols. Brackets can be very helpful, especially when working with fractions.

Module 5

QUESTIONS
Physical Chemistry and Transition Elements

1. The rate at which a reaction takes place is linked to the half-lives of the reactants involved. A student reacted the **compounds P** and **Q** together to form **R** and concluded that the initial concentration of **compound P** was 0.500 mol dm^{-3} and its half-life was 23 seconds. The half-life of **compound P** stayed constant for the duration of the reaction. However, the half-life of **compound Q** decreased as the reaction progressed.

    a. Find the concentration of **compound P**, 3 minutes and 27 seconds after the reaction has started. [2]

    b. State the orders with respect to **compounds P** and **Q**. [2]

    c. Write the rate equation for this reaction. [2]

2. State the effect of changing the concentration of a zero order reactant on the reaction rate. [1]

3. As a reaction progresses, bonds of the reactants are broken and are rearranged to form the products. Some reactions form gases or coloured products. These can be used as clues to work out the rate of the reaction.

    a. Describe how you would investigate the reaction rates of various reactions when different products are formed [4]

    b. Describe how the reaction rate can be determined from a concentration-time graph at any given time. [1]

    c. Describe how the initial rate of reaction can be calculated from a concentration-time graph. [1]

    A 200 cm^3 solution of butanol was found to have a concentration of 4.00 x 10^{-2} mol dm^{-3}.

    d. Calculate the concentration of this solution in g dm^{-3}. [3]

    e. Explain why butanol is soluble in water. [1]

4. Transition elements may form complex ions with other atoms or molecules. Copper(II) ions form a complex ion with chloride ions.

    a. Write the formula of the complex ion formed by copper(II) ions and chloride ions. [1]

    b. State the shape of this complex ion and the bond angle between two adjacent ligands. [2]

5. $[Co(H_2O)_6]^{2+}$ is a complex ion

    a. State the coordination number of this complex ion. [1]
    b. State the shape of this complex ion. [1]
    c. Write the electronic configuration, in terms of sub-shells, for the Co^{2+} ion. [1]
    d. Predict the H-O-H bond angle in this complex ion. Explain your answer. [3]

6. Some reactions progress faster than others. Reaction rates can be determined by taking into account the concentrations of the reactants or products.

    a. Describe the term "rate of reaction". [1]

    b. Describe the term "rate constant". [1]

    c. Describe and explain the effect of temperature on the rate constant and the rate of a reaction. [2]

7. Describe the effect of changing the concentration of reactants of different orders on the reaction rate. [2]

8. Describe how you could determine the rate of a reaction at any point during the course of the reaction. State its units. [2]

9. The rate of a reaction is 0.5 mol dm^{-3} s^{-1}. Calculate the change in the rate of reaction, if the concentration was doubled for a first order reactant. [1]

10. A student carried out a reaction and plotted the following graph for the thermal decomposition of **X**:

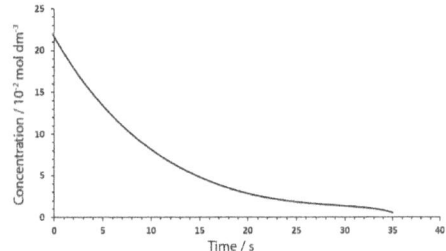

   a. i. Describe the term half-life. [1]

   ii. Determine the half-life of **X**. Show three half-lives on the graph. [2]

   iii. State the order with respect to **X**. [1]

   iv. Describe how the half-life of **X** may change as you double its concentration. [1]

   b. Calculate the rate of the reaction after 8 seconds. State its units. [5]

11. An aqueous solution of copper (II) ions containing $[Cu(H_2O)_6]^{2+}$ reacts with an excess of aqueous ammonia:

   a. Write the electronic configuration of Cu. [1]
   b. Write an equation for this reaction. [2]
   c. State the type of the reaction. [1]
   d. Describe any colour changes which may be observed during this reaction. [2]
   e. State the shape of the major product of the reaction. [1]

12. Describe the changes in concentration of zero order, first order and second order reactants with time, after a reaction has started. [3]

13. A student reacted **P**, **Q** and **R** together to form the products **A** and **C**. The rate equation for the reaction is shown below.

   rate = $k[P]^2[R]$

   a. Draw a rate-concentration graph for **P**. [1]

   b. Draw a concentration-time graph for **R**. [1]

   c. Draw a rate-concentration graph for **Q**. [1]

14. Describe the changes in half-life of zero, first and second order reactants with time, as a reaction progresses. [3]

QUESTIONS
Physical Chemistry and Transition Elements

15. Hydrazoic acid is a weak acid with the formula HN_3.

    A chemist prepared a buffer solution by mixing together 100 cm³ of 6.05×10^{-2} mol dm⁻³ HN_3 ($K_a = 1.90 \times 10^{-5}$ mol dm⁻³) and 250 cm³ of 2.00×10^{-2} mol dm⁻³ KOH. The solution was made up to 2.00 dm³.

    a. Write an equation for the dissociation of hydrazoic acid in water. [2]

    b. Hydrazoic acid is an extremely explosive liquid at room temperature and pressure. Suggest why the chemist can safely prepare the buffer solution mentioned above. [1]

    c. Explain why a buffer is formed when the two solutions of HN_3 and KOH are mixed together. [2]

    d. Calculate the pH of this buffer solution. Give your answer to a suitable number of decimal places. [6]

16. The effect of the concentration of a reactant on the reaction rate will depend on the order with respect to that reactant.

    a. Describe the "order with respect to a reactant". [1]

    b. State why you may exclude the concentration of a zero order reactant from a rate equation. [1]

17. E(s), F(aq) and G(aq) reacts to form S(aq) and T(aq) in an exothermic reaction. The rate-determining step for the reaction was found to be: $F + 2G \rightarrow S + Q$.

    a. Determine the rate equation for the reaction. [1]

    b. Draw a rate-concentration graph for F. [1]

    c. Draw a concentration-time graph for G. [1]

    d. Draw a concentration-time graph for E. [1]

18. Nitrogen monoxide reacts with oxygen to form nitrogen dioxide as shown in the following equation:

    $2NO(g) + O_2(g) \rightarrow 2NO_2(g)$

    The reaction was carried out three times using different concentrations of NO and O_2. The initial rate of reaction was calculated. The following table shows the results of this experiment.

	[NO] / mol dm⁻³	[O_2] / mol dm⁻³	Rate / mol dm⁻³ s⁻¹
Experiment 1	0.36	0.15	0.36
Experiment 2	0.36	0.30	0.72
Experiment 3	0.72	0.15	1.44

    a. Deduce the orders with respect to NO and O_2. Give your reasoning. [4]

    b. Write the rate equation for the reaction. [1]

    c. Determine the units of the rate constant. [1]

    d. Calculate the rate constant. [1]

    e. Calculate the rate of reaction, when NO has a concentration of 2.02×10^{-3} mol dm⁻³ and O_2 has a concentration of 1.05×10^{-3} mol dm⁻³. [1]

19. Describe the effect of changing the concentration of a second order reactant on the reaction rate. [1]

Physical Chemistry and Transition Elements — QUESTIONS 5

20. The equilibrium constant, K_c, indicates the position of equilibrium of an equilibrium reaction.

    a. Describe the effect of changing the temperature on the equilibrium constant, K_c. [2]

    b. A student says, "to work out an expression for K_c, experimental data are required, just like for the rate constant, K.". Comment on the statement. [1]

    c. State the effect of changing the concentration, pressure or having a catalyst on K_c. [1]

21. Most reactions take part in a series of steps. One of the steps in the mechanism may be referred to as the rate-determining step.

    a. Describe the term, "rate-determining step". [1]

    b. A student states, "if a reactant is included in the rate equation, it will form part of the rate-determining step". Comment on his statement. [1]

22. **Compound A** and **compound B** react to form **compound C**, as shown in the following equation.

    $$A + B \rightarrow C$$

    This reaction is carried out three times using different concentrations of **A** and **B**. The initial rate of reaction was calculated. The following table shows this information.

	[A] / mol dm^{-3}	[B] / mol dm^{-3}	rate / mol dm^{-3} s^{-1}
Experiment 1	0.0210	0.0150	0.0320
Experiment 2	0.0420	0.0150	0.0320
Experiment 3	0.0210	0.0600	0.512

    a. Deduce the order with respect to **A** and **B**. Give your reasoning. [4]

    b. Write the rate equation for the reaction. [1]

    c. Calculate the rate constant and determine its units. [2]

23. Describe a Brønsted-Lowry base. [1]

24. Write balanced equations along with their ionic equations for the reactions between the following species. Include state symbols.

    a. Aqueous hydrochloric acid and aqueous sodium hydroxide. [2]

    b. Solid sodium and aqueous nitric acid. [2]

    c. Solid calcium oxide and aqueous sulfuric acid. [2]

    d. Solid copper carbonate and aqueous sulfuric acid. [2]

25. Describe the term "dynamic equilibrium". [2]

26. When a reaction has reached equilibrium, the position of equilibrium does not change. The equilibrium constant, K_c, differs from one equilibrium reaction to the next.

    a. For the equilibrium, $aA + bB \rightleftharpoons cC + dD$, write down an expression for K_c. [1]

    b. A student states "the units of K_c are always mol dm^{-3}." Comment on the student's statement. [1]

    c. What information does the value of K_c, give about the position of equilibrium? [3]

QUESTIONS

Physical Chemistry and Transition Elements

27. A student dissolved 2.56 g of NaOH in 23 cm³ of water. Calculate its concentration in mol dm⁻³. Give your answer to a suitable number of significant figures. [3]

28. This question is about reaction mechanisms.

    a. Describe the term "reaction mechanism". [1]

    b. Describe the term "intermediate". [2]

    c. Describe how you could work out the rate-determining step using the rate equation. [2]

29. Born-Haber cycles can be used to calculate unknown enthalpy changes of ionic compounds.

    a. The following Born-Haber cycle can be used to work out the lattice enthalpy of caesium chloride.

    i. Complete the Born-Haber cycle by writing down the species present at each of the stages. Include state symbols in your answer. [2]

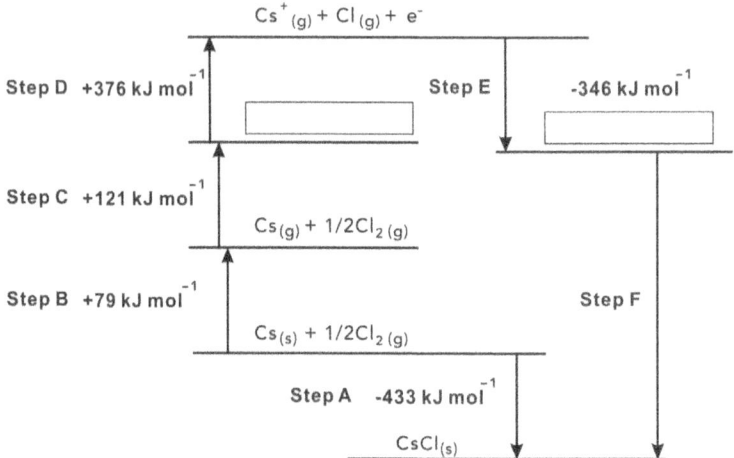

    ii. Name the enthalpy changes involved in the following stages of the Born-Haber cycle

    Step A

    Step D

    Step E. [2]

    b. Calculate the lattice enthalpy of caesium chloride. [2]

    c. Describe and explain why the lattice enthalpy of magnesium oxide (MgO) is more exothermic than that of caesium chloride (CsCl). [3]

30. The following equation shows the reaction between chlorine dioxide and hydroxide ions.

    $2ClO_2 + 2OH^- \rightarrow ClO_3^- + ClO_2^- + H_2O$

    The rate equation for this reaction is given below.

    rate = $k[ClO_2]^2[OH^-]$

Physical Chemistry and Transition Elements

QUESTIONS 5

a. When the rate of reaction is 1.75 x 10⁻³ mol dm⁻³ s⁻¹, ClO₂ had a concentration of 3.20 x 10⁻² mol dm⁻³ and OH⁻ had a concentration of 2.90 x 10⁻³ mol dm⁻³.

   Calculate the rate constant, k. Determine its units. [4]

b. State whether the expression for the rate constant could be determined just by using a balanced equation without any other data. [1]

31. Strong acids react differently to weak acids. Hydrochloric acid is an example of a strong acid.

    a. Describe a Brønsted-Lowry acid. [1]

    b. i. What happens when an acid is added to water? [1]

       ii. Write an equation to show the dissociation of hydrogen chloride gas in water. [2]

    c. Write a balanced equation to show the reaction between hydrochloric acid and magnesium. [2]

    d. State an example of a weak acid. [1]

32. In the Haber process, ammonia is made using nitrogen and hydrogen. Ammonia is a very valuable chemical, used in the production of pharmaceuticals, fertilisers and cleaning products.

    a. Write a balanced equation for the formation of ammonia from nitrogen and hydrogen. [2]

    This is an equilibrium reaction where the forward reaction has a ΔH value of -92 kJ mol⁻¹.

    b. A compromised temperature is used for the Haber process. Explain with reference to the rate constant and the equilibrium constant. [3]

    c. State one factor that does not affect the value of the rate constant. [1]

    d. State the formula of the ammonium ion and ammonium carbonate. [2]

    e. A student heated the equilibrium mixture used in the Haber process. Describe and explain the effect on the position of equilibrium. [2]

    f. State the ΔH value for the reverse reaction. [1]

33. State the effect of changing the concentration or pressure on the equilibrium constant, K_c. [1]

34. Chemical reactions are often represented using a single equation. However, in reality, these reactions take place in multiple stages. The slowest step in a reaction is known as the rate-determining step. It is so called because the rate of the entire reaction will depend on the rate at which the rate-determining step occurs.

    Carbon monoxide reacts with nitrogen dioxide to form carbon dioxide and nitrogen monoxide.

    The rate equation for this reaction is as follows.

    rate = $k[NO_2]^2$

    a. Write a balanced equation for the reaction between carbon monoxide and nitrogen dioxide. [2]

    b. State the order with respect to carbon monoxide in this reaction. [1]

c. Suggest a two-step mechanism for the reaction. [2]

d. A student found the initial rate of reaction to be 7.54×10^{-9} mol dm^{-3} s^{-1} when 4.00×10^{-3} mol dm^{-3} of NO$_2$ was used. Calculate the rate constant and and determine its units. [4]

35. The H$^+$ concentration of a solution is related to its pH.

    a. State the equation for calculating pH. [1]

    b. i. At a pH of 6.63, the ionic product of water, K_w, is 5.48×10^{-14}. Determine the concentration of OH$^-$ ions. [2]

    ii. At 25°C the pH of water is 7. State the concentration of H$^+$ and OH$^-$ ions of water at this pH. [1]

    c. Hydrogen ions and hydroxide ions form from water as shown in the following equilibrium.

    $$H_2O\ (l) \rightleftharpoons H^+(aq) + OH^-(aq)$$

    Increasing the temperature of water, increases the ionic product of water. Determine if the forward reaction of the above equilibrium is endothermic or exothermic. Explain your reasoning. [3]

36. Strong alkalis dissociate fully in water. NaOH is an example of a strong alkali.

    a. Describe how you would calculate the pH of a strong base if its concentration is known. [3]

    b. A solution of NaOH has a concentration of 0.035 mol dm^{-3}. Calculate its pH. [2]

37. Nitric acid is a highly corrosive mineral acid. It is used in rockets and also to artifically age pine.

    a. State the oxidation number of nitrogen in nitric acid. [1]

    b. A sample of nitric acid has a concentration of 2.42×10^{-3} mol dm^{-3}. Calculate its pH. [2]

38. Buffer solutions help minimize changes in pH when an acid or base is added.

    a. Describe how you would make a buffer solution. [2]

    b. Describe the difference between a strong and a weak acid. Give an example for each. [2]

    c. Describe how you would calculate the pH of a buffer solution using the K_a value. [3]

39. A student was given an unknown sample of a hydrated cobalt(II) salt and asked to determine its molecular formula.

    The student was also given a sample of sodium phosphate, Na$_3$PO$_4$. 3.00 g of the cobalt salt was reacted with an excess of sodium phosphate. The equation for the reaction is given below.

    $$3Co^{2+}(aq) + 2PO_4^{3-}\ (aq) \rightarrow Co_3(PO_4)_2(s)$$

    The precipitate was isolated and its weight was found to be 1.54 g. Assume that all the Co^{2+} ions have been converted into Co$_3$(PO$_4$)$_2$(s).

    The student took a second 3.00 g sample of the cobalt salt and heated it in a crucible to remove the water of crystallisation. 1.64 g of the anhydrous salt remains.

Physical Chemistry and Transition Elements

QUESTIONS 5

a. State the oxidation number of phosphorus in $Co_3(PO_4)_2$. [1]

b. Determine a possible formula of the hydrated cobalt(II) salt. Your answer must show **all** your working. [8]

40. A student carried out two titrations using two different acids and bases. Sketch a titration pH curve, with pH on the y-axis and volume of base added on the x-axis for each titration. Explain the shape of both of the graphs.

    You may use the following table to help answer this question.

Strong acid	Weak acid	Strong base	Weak base
HCl	HCOOH	NaOH	NH_3
HBr	HF	KOH	CH_3NH_2
$HClO_4$	HNO_2	$Ba(OH)_2$	NH_4OH

    a. HF and NH_4OH. [3]

    b. H_2SO_4 and NH_4OH. [3]

    One of his colleagues carried out two more titrations. Sketch a titration pH curve for the following titrations.

    c. HNO_3 and $Ba(OH)_2$. [1]

    d. HNO_2 and KOH. [1]

41. Describe the significance of the value of the acid dissociation, K_a, constant. [1]

42. Boric acid, H_3BO_3 may be used as an antiseptic for minor burns. Its acidity makes it suitable for this purpose. An acid such as sulfuric acid, H_2SO_4 is not a good choice. (For this question, both acids need to be treated as monobasic acids).

    a. 1.7×10^{-3} mol dm^{-3} solutions of boric acid and sulfuric acid have different pH values

    Explain why their pH values are different and calculate the pH of 1.7×10^{-3} mol dm^{-3} solutions of boric acid and sulfuric acid. Give your answer correct to three significant figures.

    (pK_a of boric acid is 9.23). [5]

    b. A buffer solution may be made by dissolving sodium citrate in acetic acid (CH_3COOH).
       i. Describe the function of a buffer solution. [1]
       ii. Describe and explain how a buffer system based on acetic acid allows it to control the pH. [2]

43. The K_a of acetic acid is 1.75×10^{-5} mol dm^{-3}.

    a. What does the acid dissociation constant, K_a measure? [1]

    b. State an expression for the acid dissociation constant K_a of an acid HA. State the units of K_a [2]

    c. i. State the equations for calculating pK_a from K_a and K_a from pK_a. [2]

       ii. Calculate the pK_a of acetic acid. [1]

QUESTIONS
Physical Chemistry and Transition Elements

44. State an expression for the ionic product of water, K_w. [1]

45. Acetic acid is an example of a weak acid. It reacts with water as shown in the following equation:

    $CH_3COOH\ (aq) + H_2O\ (l) \rightleftharpoons H_3O^+(aq) + CH_3COO^-(aq)$

    a. Write an expression linking $[H^+]$ to pH. [1]

    b. i. Describe a conjugate acid-base pair. [1]

       ii. Identify the conjugate acid-base pairs in the equilibrium. [2]

    c. A student says, "K_c of the equilibrium will be around 1". Comment on the student's statement. [3]

46. Find the pH of the solutions with the following H^+ concentrations. Give your answers to two decimal places.

    a. 0.00000453 mol dm^{-3} c. 3.56 x 10^{-10} mol dm^{-3}

    b. 0.00270 mol dm^{-3} d. 5.98 x 10^{-5} mol dm^{-3} [4]

47. Explain how a buffer solution works in terms of conjugate acid-base pairs. You may use HA to represent the acid in your explanation. [3]

48. There are many buffer systems in the healthy human blood that helps to control its pH.

    a. Name a buffer system that helps to control the pH of human blood. [1]

    b. Explain how this buffer system works. [4]

49. As a reaction progresses, the concentration of the reactants decrease. The concentrations of some reactants decrease faster over time compared to others.

    The Cannizzaro reaction is used to disproportionate an aldehye. It involves reacting potassium hydroxide with an aldehyde. A student carried out a reaction using 2.5 mol dm^{-3} potassium hydroxide. In this reaction, potassium hydroxide follows first order kinetics. The student found the half-life of potassium hydroxide to be 120 s.

    a. Some time after the reaction has started, the student calculated the concentration of potassium hydroxide to be 0.156 mol dm^{-3}. Calculate how long it has been since the reaction began. [2]

    The aldehyde used in this reaction was C_6H_5CHO. The aldehyde followed second order kinetics.

    b. Write the rate equation for this reaction. [2]

    c. Another student suggested that increasing the temperature of the reaction would increase the value of the rate constant. Is the student right? Explain. [1]

    They then reacted hydrochloric acid and potassium hydroxide. In this reaction 25.0 cm^3 of hydrochloric acid of concentration 0.84 mol dm^{-3} and 35.0 cm^3 of potassium hydroxide of concentration 0.600 mol dm^{-3} were used. The temperature went up by 5.2 °C.

    Specific heat capacity of solution = 4.18 J g^{-1} K^{-1}

    Density of solution = 1.00 g cm^{-3}

d. Calculate the enthalpy change of neutralisation for the above reaction. [5]

50. Describe how you would calculate [H$^+$] from the pH for a strong base. [1]

51. A sample of barium iodide contains 31.6 g of barium and 58.4 g of iodine.

    a. Find the empirical formula of this compound. (Ba = 137.3 ; I = 126.9). [2]

    b. Calculate the percentage by mass of barium present in the sample. [1]

52. A student wanted to produce some calcium oxide for a project. 20.0 g of calcium carbonate was heated inside a test tube and the gas produced was collected. 8.20 g of calcium oxide was produced.

    a. Write a balanced equation for the above reaction. [1]

    b. Calculate the atom economy of the reaction. [2]

    c. Calculate the percentage yield of the reaction. [4]

    d. Suggest a test that could be carried out to confirm the identity of the gas produced. Identify any compounds responsible for any colour changes. [3]

53. The lattice enthalpies of NaI and Na$_2$O are -682 kJ mol^{-1} and -2481 kJ mol^{-1} respectively.

    a. Write an equation to illustrate the lattice enthalpy of NaI. [1]

    b. What does the size of the lattice enthalpy of a compound indicate? [1]

    Some enthalpy changes associated with CsF are given in the following table.

Enthalpy change	Energy / kJ mol^{-1}
Atomisation of caesium	76.5
Atomisation of fluorine	79.4
Lattice enthalpy of CsF	-756.9
First electron affinity of fluorine	X
First ionisation energy of caesium	375.7
Enthalpy change of formation of CsF	-553.5

    c. i. Draw a Born-Haber cycle for the formation of CsF. Show the enthalpy changes involved. [4]

       ii. Calculate the missing enthalpy change, X. [3]

54. Define standard enthalpy change of solution. [2]

55. Standard enthalpy changes of hydration of ions depend on certain factors. Some ions tend to behave differently to others.

    a. Define standard enthalpy change of hydration. [2]

    b. State two factors that affect the standard enthalpy change of hydration. Explain. [4]

QUESTIONS — Physical Chemistry and Transition Elements

    c. Write an equation to show the enthalpy change of hydration of magnesium. [2]

    d. Arrange the following ions in order of the size of their enthaly change of hydration from the most exothermic to the least exothermic.

        Mg^{2+} I^- Sr^{2+} Al^{3+} [2]

    e. "Standard enthalpy change of hydration is endothermic for gaseous ions with a larger radii". Comment on this statement. [1]

56. Nitrogen monoxide reacts with hydrogen to form nitrogen and water, as shown below:

$$\_\_NO(g) + \_\_H_2(g) \rightleftharpoons N_2(g) + \_\_H_2O(g)$$

A scientist mixes 0.32 mol of NO with 0.25 mol of H_2. The mixture is left for some time till it reaches equilibrium at a constant temperature.

Once the equilibrium has been reached, 0.12 mol of H_2 was found to remain.

    a. Balance the equation above by writing numbers in the three spaces given. [1]

    b. i. Write an expression for, K_c. [1]

       ii. Deduce the units of K_c. [1]

       iii. Determine the value of K_c for this equilibrium. Show all of your working. [4]

57. Describe the terms, "equivalence point" and "end point". [2]

58. Define enthalpy change of neutralisation. [2]

59. Lithium fluoride has a lattice enthalpy of -1005 kJ mol^{-1}.

    a. Define lattice enthalpy. [2]

    b. Describe the effect of ionic charge and ionic radii on lattice enthalpy. [2]

    c. Explain why lattice enthalpy cannot be measured directly. [1]

    d. Explain why covalent compounds do not have an associated lattice enthalpy. [1]

60. Adding or removing electrons from an atom or ion involves energy changes. Some processes give out energy and others take in energy.

    a. Write equations to illustrate second ionisation energy of aluminium and strontium. [2]

    b. State the type of enthalpy change involved in second ionisation energy. [1]

61. Draw a Born-Haber cycle for sodium chloride. Label the enthalpy changes involved on your diagram. [4]

62. Ionic compounds consist of positive and negative ions attracted to each other by electrostatic forces.

Metals tend to lose electrons and non-metals tend to gain electrons.

    a. State the type of enthalpy that indicates the strength of the electrostatic forces of attraction between ions in an ionic compound. [1]

b.

   i. Write an equation for the first electron affinity of sulfur. [2]

   ii. Explain why the first electron affinity of an element may be exothermic. [1]

63. Describe the term entropy. [1]

64. Diatomic elements consist of discrete molecules formed from two covalently bonded atoms. There are several diatomic elements. N_2, O_2, F_2, Cl_2, Br_2 and I_2 are six of these.

   a. State the formula of another diatomic molecule consisting of a single element. [1]

   b. Write equations to illustrate the enthalpy change of atomisation of oxygen, fluorine and lithium. [2]

   c. State the enthalpy change involved in enthalpy change of atomisation. Explain. [2]

65. Transition elements and their compounds are widely used as catalysts. Catalysts help make processes more economical. Manganese is an example of a transition element. It forms Mn^{2+} ions.

   a. Describe transition elements using manganese as an example. [2]

   b. Describe one ligand substitution reaction of chromium in the +3 oxidation state. Your answer should include any observations and balanced equations. [3]

   c. State the shape and coordination number of the complex ion show below: [1]

$$\left[\begin{array}{c} NH_3 \\ H_3N\cdots Co\cdots NH_3 \\ Cl \quad NH_3 \\ NH_3 \end{array}\right]^{2+}$$

66. Sodium is a very reactive metal and is the sixth most abundant element in the Earth's crust. The ionisation energies of sodium are given in the following table.

Ionisation Energy Number	Enthalpy / kJ mol^{-1}
1st	495.80
2nd	4562.00
3rd	6910.30

a.

   i. Define first ionisation energy. [2]

   ii. Write an equation to show the enthalpy change involved in the first ionisation energy for sodium. [2]

   iii. Describe and explain the differences among the first, second and third ionisation energies of sodium. [4]

QUESTIONS
Physical Chemistry and Transition Elements

    b. ^{22}Na is one of the radioactive isotopes of sodium. It has a half-life of 2.6 years. 26g of this isotope was present in a sample.

       i. Calculate the number of sodium atoms present in this sample. [2]

       ii. Calculate the number of atoms remaining after 9.1 years in this sample. [2]

67. Formation of an ionic compound involves atoms gaining and losing electrons. In a chemical reaction, the bonds in the reactants are broken and are rearranged to form the products.

    Three ionic compounds are listed here: NaCl, Al_2O_3, $CaCl_2$

    a. Formation of Al_2O_3 involves the third ionisation energy of aluminium. Write an equation for the third ionisation energy of aluminium. [1]

    b. Select the compound(s) from the above list whose formation involves second electron affinity. Explain. [2]

    c. State the types of enthalpy changes (ie. exothermic or endothermic) involved in the making and breaking of bonds. [2]

68. State the standard conditions used in chemistry. [1]

69. State a factor that should be taken into account when choosing an indicator for an acid-base titration. [1]

70. A new food production company wanted to determine the amount of iodate (IO_3^-) in a sample of salt. Their lab technician was given this task who reacted 25.0 cm^3 of iodate with iodide (I^-) ions under acidic conditions. The iodine which was formed was then reacted with thiosulfate ions. The equations for these reactions are as follows.

    Equation number 1: $IO_3^- + 5I^- + 6H^+ \rightarrow 3I_2 + 3H_2O$

    Equation number 2: $I_2 + 2S_2O_3^{2-} \rightarrow 2I^- + S_4O_6^{2-}$

    34.25 cm^3 of thiosulfate of concentration 0.150 mol dm^{-3} was needed to completely react with all the iodine present.

    a. Using the information provided, calculate the concentration of iodate ions in the salt. [4]

    b. Using oxidation numbers, describe what happens to the iodine in the first and the second equation. [4]

71. State Hess's law. [1]

72. Potassium fluoride was dissolved in water. The associated enthalpy changes are shown in the table below.

Process	Enthalpy change / kJ mol^{-1}
Lattice enthalpy of KF	-813
Enthalpy change of hydration of K^+	-320
Enthalpy change of hydration of F^-	-524
Enthalpy change of solution of KF	X

a. Draw a Born-Haber cycle showing the enthalpy changes. The enthalpy change of solution of KF is exothermic. [3]

b. Calculate the enthalpy change of solution of KF, X. [2]

c. State why the ionic radius of potassium is greater than that of fluoride. [1]

73. Define standard enthalpy change of formation. [2]

74. Describe a redox reaction. [1]

75. State the factors that determine the feasibility of a reaction. [2]

76. Even though some reactions may be endothermic, they can take place spontaneously. Explain. [2]

77. Atoms of a gas scattered over a room have a greater entropy than atoms of a liquid contained inside a jug.

   a. Describe the difference in entropy of a solid and a liquid. [1]

   b. Write an expression for change of entropy. [1]

   c. Entropy is always positive. However, *change in entropy* could take negative values. Explain. [2]

78. This question is about the Arrhenius equation, which can be used to work out the rate constant, k.

   The Arrhenius equation may be written in the form:

   $\ln k = \ln A - E_a / RT$

   A graph may be plotted with $\ln k$ on the y-axis and $1/T$ on the x-axis.

   a. State the relationships between the gradient and the y-intercept of the graph with the Arrhenius equation. [1]

   b. A graph was plotted for a reaction and the gradient was found to be -7490 and the y-intercept was 3.20. Calculate E_a and A. (R = 8.314 J K^{-1} mol^{-1}) [3]

79. Describe the change in entropy when a solid lattice dissolves. [1]

80. Describe the term, "free energy change". [1]

81. Calcium carbonate decomposes as shown in the following equation.

   $$CaCO_3(s) \rightarrow CaO(s) + CO_2(g) \qquad \Delta H = 178.2 \text{ kJ mol}^{-1}$$

   The table below shows the entropy values for each of the species.

	CaCO$_3$(s)	CaO(s)	CO$_2$(g)
S$^\ominus$ / J K^{-1} mol^{-1}	92.9	39.8	213.7

a. Calculate the change in entropy for the above reaction. [2]

b. Write a formula connecting free energy and entropy. State the units of each term. [2]

c. Calculate the Gibbs free energy for this reaction at 15 °C. Using your answer, state and explain whether the reaction would be feasible at this temperature. [3]

d. Calculate the minimum temperature in °C at which this reaction is feasible. [3]

82. Explain the change in entropy in a gaseous reaction where the reactants have three more gas particles than its products. [1]

83. Convert the following temperatures that are in °C to K and those that are in K to °C.

   i. 5 °C   ii. 18 K   iii. 17 K   iv. 17 °C   v. -273 °C   vi. 1 K   vii. 25 °C   [7]

84. Copper reacts with nitric acid, as shown in the following equation.

$$Cu(s) + HNO_3(aq) \rightarrow Cu(NO_3)_2(aq) + NO(g) + H_2O(l)$$

a. How would oxidation numbers show reduction of one species to another? [1]

b. How would oxidation numbers show oxidation of one species to another? [1]

c. State one species which has been oxidised and one that has been reduced in the above equation. Explain. [4]

85. In terms of entropy, explain the difference between a solid and a liquid. [1]

86. Describe a reducing agent. [1]

87. Enthalpy changes related to the formation of CuO is shown below along with a Born-Haber cycle.

Letter	Enthalpy change	Energy / kJ mol^{-1}
A	Enthalpy change of formation of CuO	-155
B	First ionisation energy of copper	x
C	Second electron affinity of oxygen	+790
D	Second ionisation energy of copper	+1960
E	Atomisation of oxygen	+249
F	Atomisation of copper	+339
G	First electron affinity of oxygen	-141
H	Lattice enthalpy of CuO	-4097

a. Write the correct letter in each of the boxes [4]

b. Using the Born-Haber cycle and the values given in the table above, calculate the first ionisation energy of copper [2]

88. Describe a half-cell. [1]

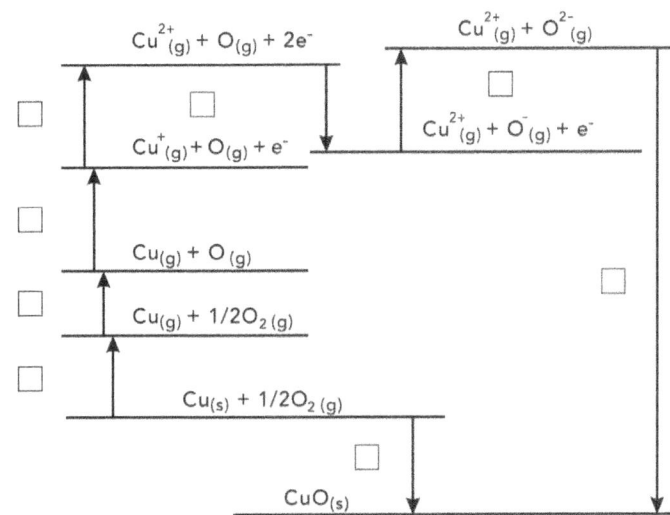

89. Describe the change in entropy in a gaseous reaction where the products have two more gas particles than its reactants. [1]

90. What does the value of free energy, ΔG, indicate about the feasibility of a reaction? [1]

91. Oxidation numbers may be used to work out whether a species has been oxidised or reduced in an equation. State the most common oxidation numbers of the following species:

    a. Oxygen in a compound

    b. A simple ion

    c. Hydrogen in a compound

    d. Fluorine in a compound

    e. An uncombined atom
    [5]

92. Describe an oxidising agent. [1]

93. What does the standard electrode potential indicate about the tendency of a half-cell to lose electrons? [1]

94. Determine if solid magnesium will react with dilute sulfuric acid. You are provided with the following data:

    $Mg^{2+}(aq) + 2e^- \rightleftharpoons Mg(s)$ $E^\ominus = -2.37$ V $2H^+(aq) + 2e^- \rightleftharpoons H_2(g)$ $E^\ominus = 0$ V

    Explain your answer. [2]

95. Draw a half-cell of Fe^{2+} and Fe^{3+} and write the equation for this half-cell. [3]

96. Write down the electron configuration of titanium. [1]

97. An electrochemical cell may be made by combining two half-cells with different electrode potentials.

    One of the half-cells provides the electrons driving an electric current around the cell.

QUESTIONS
Physical Chemistry and Transition Elements

Two redox equilibria are given below:

$Cu^{2+}(aq) + 2e^- \rightleftharpoons Cu(s)$ $E^\ominus = +0.34$ V
$Ag^+(aq) + e^- \rightleftharpoons Ag(s)$ $E^\ominus = +0.80$ V

   a. Draw a labelled diagram to show how this cell may be set up in a laboratory. [3]

   b. Calculate the cell potential of this electrochemical cell. [1]

   c. Write down the overall cell reaction. [1]

   d. State how electrons and ions are transferred in this system. [1]

98. An electrochemical cell can be made by combining half-cells of different electrode potentials.

    a. Describe the flow of electrons in an electrochemical cell. [1]

    b. Describe how you could measure the standard electrode potential of a half-cell. [1]

    c. Describe how you would calculate the standard cell potential, if the standard electrode potentials of each half-cell are known. [1]

99. Half-cells can be used to build an electrochemical cell. Half-cells may consist of metals and / or non-metals.

    a. Draw a half-cell consisting of a metal in contact with its ions in aqueous solution. Write an ionic equation for the half-cell you have drawn. [2]

    b. Draw a half-cell consisting of a non-metal and its ions in aqueous solution. Write an ionic equation for the half-cell you have drawn. [3]

    c. Describe how half-cells may be used to make an electrochemical cell. [1]

100. Explain why a prediction regarding the feasibility of a reaction made using standard electrode potentials may not be valid. [2]

101. Transition elements are found in the d block of the periodic table. These elements have special properties which set them aside from other metals.

     Vanadium(V) oxide is used as a catalyst in the contact process.

     a. Explain why vanadium can be described as a transition metal. [1]

     b. Write down the electron configuration of zinc, chromium and copper. [2]

     c. State the chemical formula of vanadium(V) oxide. [1]

     d. Sulfur dioxide reacts with oxygen to form sulfur trioxide. This is an equilibrium reaction.

        i. Write down a balanced equation for this equilibrium. [2]

        The forward reaction in the above equilibrium has an enthalpy value of -196 kJ mol^{-1}.

        ii. Describe and explain the effect of increasing the temperature on this equilibrium. State how the equilibrium constant, K_p, would be affected by an increase in temperature. [3]

        iii. Describe and explain how you could alter the pressure of this system to achieve a higher yield of sulfur trioxide. [2]

102. Transition elements and their compounds are often used as catalysts in industrial processes. Catalyts work by lowering the energy required for reactions to occur.

    a. State a characteristic of compounds formed by transition metals. [1]

    The Haber process is used to make ammonia via the following equilibrium.

    $N_2(g) + 3H_2(g) \rightleftharpoons 2NH_3(g)$

    b. Explain the effect of decreasing the pressure on the above equilibrium. [2]

    c. How does the presence of a catalyst affect this equilibrium? Explain in terms of K_p. [2]

    d. For this question, assume that the above reaction goes to completion in the forward direction. If 49.0 g of nitrogen (N_2) was used in the process, calculate the mass of ammonia (NH_3) formed in grams. [3]

    e. During a reaction involving hydrolysis of urea, 400 dm^3 of ammonia and 400 dm^3 of carbon dioxide was produced. Calculate the number of moles of each gas formed. [2]

103. Complex ions are formed when ligands bond to a central metal ion. The coordination number of a complex ion can vary.

    a. Describe the term coordination number. [1]

    b. State the name of the bond formed between a ligand and the central transition metal ion in a complex ion. [1]

104. Transition metals can form complex ions with ligands such as water. These ligands may be monodentate or multidentate. Transition metals and their compounds are often used as catalysts. Describe the following terms:

    a. Monodentate ligand [1]

    b. Catalyst [2]

    c. Ligand [1]

    d. Complex ion [1]

    e. Multidentate ligand [2]

    f. Hexadentate ligand [2]

105. Using an ionic equation, describe what you would observe when you react aqueous iron(II) ions with aqueous sodium hydroxide. [2]

106. Some complex ions with an octahedral shape can exist as *cis-trans* isomers. What sort of ligands and how many of each type of ligand should it have for it to have *cis-trans* isomers? [1]

107. Two isomers of the same compound have the same molecular formula. There are different kinds of isomerism, one of which is stereoisomerism.

    a. Describe stereoisomers. [2]

    b. State the two types of stereoisomers in transition element chemistry. [1]

108. Using an ionic equation, describe what you would observe when you react aqueous copper(II) ions with aqueous sodium hydroxide. [2]

QUESTIONS
Physical Chemistry and Transition Elements

109. Reacting sodium hydroxide with iron (III) ions in an aqueous solution results in the formation of a metal hydroxide.

    a. Write an ionic equation for this reaction including state symbols. [2]

    b. Describe what you would observe during this reaction. [2]

    c. A solution of NaOH has a concentration of 0.035 mol dm^{-3}. Calculate its pH. [3]

110. Transition metals and their compounds are often used as catalysts in industry. State an advantage of this use. [1]

111. A student was given a sample containing iron(II) ions. The student worked out that there were 2.55 g of iron (II) ions in the sample. The sample was then dissolved and a solution of volume 250 cm^3 was made. A 10.0 cm^3 portion was then taken and titrated with acidified potassium manganate(VII) to completely oxidise iron(II) ions.

    a. Calculate the number of moles of iron(II) ions present in the sample. [1]

    b. Write a balanced ionic equation for the reaction between iron(II) ions and acidified potassium manganate (VII). [2]

    c. 18.3 cm^3 of acidified potassium manganate (VII) was needed for the titration. Calculate the concentration of the acidified potassium manganate (VII) solution. [3]

    d. State the oxidising agent in this reaction and explain how it works. [2]

    e. Describe how the oxidation states of transition metal ions compare to other metal ions. [1]

112. Transition metal complexes have various shapes which often depend on the number and type of ligands surrounding the central metal ion.

    a. State an example of a complex ion with an octahedral shape. Draw this complex ion. [2]

    b. i. State the number of coordinate bonds in an octahedral complex. [1]

       ii. State the bond angles in a complex ion with an octahedral shape. [1]

113. Names of transition metal compounds often have a roman numeral within brackets next to the metal name. What does this tell us? [1]

114. Describe a precipitation reaction. [1]

115. What sort of ligands should a four-coordinate complex have for it to be able to form *cis-trans* isomers? [1]

116. This question is about bidentate ligands.

    a. Describe a bidentate ligand. [1]

    b. Name and draw a bidentate ligand. [2]

    c. Draw the 3D structures of optical isomers of a complex ion formed between nickel (II) ions and bidentate ligands. [2]

117. Transition metal complexes may show optical isomerism. Some of these complex ions may be used as drugs to help treat disease or reduce the symptoms of a disease.

Physical Chemistry and Transition Elements — QUESTIONS 5

a. State another name for optical isomers. [1]

b. State the name of an anti-cancer drug which is a transition metal complex. [1]

c. Describe optical isomers. [1]

118. Entropy measures the degree of disorder in a system. A gas has a higher entropy than a solid as its particles are more disordered.

   a. For each of the following reactions, predict whether the change in entropy will be positive or negative:

   i. $H_2O(l) \rightarrow 2H_2(g) + O_2(g)$ [1]

   ii. $2NH_3(g) + CO_2(g) \rightarrow H_2O(l) + NH_2CONH_2(l)$ [1]

   b. i. Titanium (IV) oxide reacts with carbon and chlorine, as shown below:

   $TiO_2(s) + C(s) + 2Cl_2(g) \rightarrow TiCl_4(l) + CO_2(g)$

   Using the data given in the following table, calculate the entropy change of reaction. [2]

	$TiO_2(s)$	$C(s)$	$Cl_2(g)$	$TiCl_4(l)$	$CO_2(g)$
$S^\ominus$ / J K^{-1}mol^{-1}	49.9	5.7	223	252	214

   ii. State the relationship between entropy and free energy. [1]

119. State the chemical name and chemical formula of the bidenetate ligand "en". [2]

120. The manganate ion has a tetrahedral shape and is used widely as an oxidising agent.

   a. i. Write a half-equation for the reduction of MnO_4^- ions in acidic solution. [2]

   ii. State any observations during the above reaction. [1]

   b. i. A student used 30.0 cm^3 of nitric acid to completely reduce 0.100 g of potassium manganate. Calculate the pH of the nitric acid used. [4]

   ii. Suggest why the answer to b(i) would be different if sulfuric acid was used. [1]

121. State the names of the shapes formed by four-coordinate complex ions. [1]

122. Write a balanced ionic equation for the redox reaction between aqueous Cu^{2+} ions and aqueous iodide ions. [2]

123. Describe the effect of carbon monoxide on human blood. State the type of reaction involved. [2]

124. The ligands in a complex ion may be substituted by other ligands. These reactions are often accompanied by characteristic colour changes.

   Write a balanced equation for a ligand substitution reaction involving a copper complex ion and chloride ions. State any colour changes observed. [3]

125. Aqueous solution of $[Cr(H_2O)_6]^{3+}$ reacts with an excess of ammonia. Describe the changes observed during this reaction with an equation. [2]

5 QUESTIONS — Physical Chemistry and Transition Elements

126. Haemoglobin is an important protein found in human blood. It gives blood its characteristic colour.

    a. State the function of haemoglobin. [1]

    b. Describe the role of iron in haemoglobin. [1]

    The carbonic acid buffer system is used to control the pH of blood. It has to be maintained between 7.35 and 7.45. The equilibrium for this buffer system is shown below.

    $H_2CO_3(aq) \rightleftharpoons H^+(aq) + HCO_3^-(aq)$

    c. A student says, "The carbonic acid buffer system prevents pH changes in the blood". Comment on the student's statement. [1]

    d. Describe and explain the effect of adding an acid to this buffer solution. [2]

    e. Describe the effect of adding an alkali to the position of equilibrium. [1]

    f. A sample of carbonic acid was found to have a concentration of 0.035 mol dm^{-3}. Its pH was measured as 3.27. (For this question, treat the acid as a monobasic acid).

        i. Calculate K_a. [2]

        ii. Calculate pK_a. [1]

127. This question is about the reactions of aqueous copper (II) ions.

    a. State the complex formula of aqueous copper (II) ions. [1]

    b. i. Write an equation for the reaction between aqueous copper (II) ions and excess aqueous ammonia. [2]

        ii. State all observations during the reaction. [2]

        iii. State both types of reaction happening here. [2]

    c. i. Write an equation for the reaction between aqueous copper (II) ions and concentrated hydrochloric acid. [2]

        ii. State any observations during the reaction. [1]

        iii. State the type of reaction. [1]

        iv. State the shape of the major product of the reaction and draw its 3D structure. Suggest why it has this shape. [3]

128. Write an ionic equation for the reaction between iron(II) ions and acidified manganate(VII) ions. [2]

129. State a use for the transition metal complex cis-platin and describe how it works. [2]

130. Oxidising agents oxidise other species by gaining electrons, and reducing agents reduce other species by losing electrons. What sort of an agent is MnO_4^-? [1]

131. Iron is a very important component of the human diet. Lack of iron can cause anaemia which may result in the affected person feeling tired faster than normal.

Physical Chemistry and Transition Elements
QUESTIONS 5

Iron is an example of a transition metal.

a. Using the electronic configuration of Fe^{2+}, explain why iron is classified as a transition metal. [2]

b. State a protein that is found in humans containing iron. [1]

132. Copper is found in many alloys. Using iodide ions and thiosulfate ions, the proportion of copper in an alloy can be determined.

0.870 g of brass was reacted forming a 250 cm³ solution containing Cu^{2+} ions. 25.0 cm³ of this solution was then reacted with iodide ions forming iodine. This iodine was titrated with 0.040 mol dm⁻³ sodium thiosulfate. 15.5 cm³ of thiosulfate was required to reach the end point. Starch was used in this titration.

a. Balance the equations given below: [2]

$Cu(s) \rightarrow Cu^{2+}(aq) + \_\_e^-$

$\_\_Cu^{2+}(aq) + \_\_I^-(aq) \rightarrow 2CuI(s) + I_2(aq)$

$\_\_S_2O_3^{2-}(aq) + I_2(aq) \rightarrow 2I^-(aq) + S_4O_6^{2-}(aq)$

b. Determine the proportion of copper in brass as a percentage. [5]

c. Explain the use of starch. [1]

133. Describe a ligand substitution reaction. [1]

134. The following data may be used to construct a Born-Haber cycle for magnesium chloride

Enthalpy change	Energy / kJ mol⁻¹
Enthalpy change of atomisation of chlorine	122
First ionisation energy of magnesium	736
Second ionisation energy of magnesium	1450
Enthalpy change of atomisation of magnesium	148
Enthalpy change of formation of $MgCl_2$	-641
First electron affinity of chlorine	y
Lattice enthalpy of $MgCl_2$	-2521

a. i. Draw a Born-Haber cycle for the formation of $MgCl_2$. Show the enthalpy changes involved. [4]

ii. Using the above table, calculate the first electron affinity of chlorine. [3]

b. i. Define the term lattice enthalpy [2]

ii. State two factors that affect the value of lattice enthalpy. [1]

135. Methanol is the simplest alcohol and is also known as "wood alcohol", because it used to be produced as a byproduct of the destructive distillation of wood.

Methanol decomposes as shown in the equation.

QUESTIONS
Physical Chemistry and Transition Elements

$CH_3OH(g) \rightarrow CO(g) + 2H_2(g)$ $\Delta H = 90.7$ kJ mol^{-1}

	$CH_3OH(g)$	$CO(g)$	$H_2(g)$
$S^\ominus$ / J K^{-1} mol^{-1}	238	198	131

    a. Calculate the change in entropy for the reaction. [1]

    b. Determine the minimum temperature in °C at which this reaction will be feasible. [3]

136. This question is about the reaction between iodine and thiosulfate ions.

    a. i. Write an equation for the reaction. [2]

       ii State the type of the reaction. [1]

    b. Starch can be used to identify the end point in the reaction between iodine and thiosulfate ions. State and explain any observations. [2]

137. Oxidation and reduction can be defined in various ways. One definition states that oxidation is the loss of electrons and reduction is the gain of electrons.

    a. Describe oxidation and reduction using oxidation numbers. [1]

    b. i. Write a half-equation for the oxidation of Fe^{2+} ions to Fe^{3+} ions. [1]

       ii. State any observations during the oxidation. [1]

    c. Aqueous iron(II) sulfate was reacted with aqueous bromine. During the reaction, iron gets oxidised and bromine gets reduced.

       Write a balanced ionic equation for the reaction. State any colour changes observed. [6]

138. The number of moles can be calculated provided the molar mass and mass in grams of a species are known. Write a formula for this. [1]

139. A student was asked to find out the amount of iron present in a sample. The sample was reacted with acidified potassium dichromate. All the iron in this sample was in the Fe^{2+} form. The net ionic equation for this reaction is given below.

$6Fe^{2+} + Cr_2O_7^{2-} + 14H^+ \rightarrow 6Fe^{3+} + 2Cr^{3+} + 7H_2O$

In the titration the student used potassium dichromate of concentration 0.0324 mol dm^{-3}. To reach the end point, 42.30 cm^3 of potassium dichromate was used. The sample weighed 2.357 g.

    a. State the formula of potassium dichromate. [1]

    b. Using oxidation numbers, state which species have been oxidised and reduced. [4]

    c. Determine the mass, in grams, of iron present in the sample. [4]

    d. Determine the percentage by mass of iron present in the sample. [2]

140. A student is asked to design an apparatus to determine ΔH for the combustion of hexane.

Hexane is a liquid at room temperature.

a. Write a balance equation for the complete combustion of hexane. [2]

During this experiment, the student heated 100 cm³ of water from the heat of combustion of 0.4502 g of hexane. The increase in temperature of the water was measured to be 51.3 °C.

The specific heat capacity of water is 4.18 J g⁻¹ K⁻¹.

The standard entropies of some species are given below:

	$C_6H_{14}(l)$	$O_2(g)$	$H_2O(g)$	$CO_2(g)$	$CO(g)$
$S^\ominus$ / J K⁻¹ mol⁻¹	295	205	189	214	198

b. Draw a laballed diagram of an apparatus which the student could use to determine the ΔH for the combustion of hexane. [4]

c. Calculate the minimum temperature, in °C, at which the reaction is feasible. [7]

141. An uknown subtance Y reacted with hydrochloric acid to produce magnesium chloride, water and carbon dioxide.

    a. Suggest the identity of Y. [1]

    b. Write a balanced equation for the above reaction. [2]

    A student was asked to calculate the rate of this reaction.

    The following table shows the concentration of HCl present at different time intervals.

Time / s	Cocentration of HCl / mol dm⁻³
0	0.200
5	0.120
10	0.075
15	0.050
20	0.0250

    c.
        i. Draw a concentration-time graph using the data given above. [3]

        ii. Calculate the mass, in g, of HCl present after 12 seconds if the volume of HCl was 20.0 cm³. [2]

        iii. Using your graph, determine the rate of reaction after 12 seconds. State its units. Show any working on the graph. [4]

        iv. State the order with respect to HCl. [1]

    d. For this reaction, an excess of substance Y was used. Explain why this was necessary. [1]

e. The rate of this reaction could also be worked out by measuring the volume of carbon dioxide given off. Draw a labelled diagram of an apparatus which would enable you to collect the CO_2 given off. [3]

142. This question is about buffer solutions.

A buffer solution may be made using an acid and a salt of the weak acid. A student wanted to prepare a buffer solution with a pH of 4.16 using ethanoic acid.

K_a for ethanoic acid = 1.74×10^{-5} mol dm^{-3}.

The student was provided with two solutions of ethanoic acid and sodium ethanoate, both with concentrations 2.50 mol dm^{-3}.

5.25 cm^3 of sodium ethanoate was used during the preparation. Calculate the volume of ethanoic acid required.

Give your answer to a suitable number of significant figures. [5]

143. N_2O_5 decomposes according to the following equation.

N_2O_5 (g) → $4NO_2$ (g) + O_2 (g)

R = 8.31 J mol^{-1} K^{-1}.

The following data shows the relationship between ln k and T.

ln k	T / K	1 / T
-6.19	333	
-7.41	250	
-8.65	200	
-9.89	167	

a. Complete the above table. [2]

b. i. Plot a graph with ln k on the y-axis and 1 / T on the x-axis. [3]

ii. Determine the activation energy of this reaction. State its units. [5]

144. A student was asked to determine the water of crystallisation of zinc sulphate using a sample.

The sample of zinc sulphate has a mass 18.5 g. This sample was heated until no change in mass was detected. 12.7 g of the solid remained after heating.

The equation for the reaction that took place is given below:

$ZnSO_4.xH_2O$ → $ZnSO_4$ + xH_2O

a. Draw the labelled diagram of a suitable apparatus which the student could use to heat the sample of zinc sulphate. [3]

b. Determine the water of crystallisation of zinc sulphate. [4]

QUESTIONS 5

Physical Chemistry and Transition Elements

145. The Boltzmann distribution, named after Ludwig Boltzmann, shows the proportion of molecules exceeding the activation energy in a given reaction.

    A reaction will only take place if the molecules exceed a given activation energy.

    The diagram below shows the Boltzmann distribution for a certain reaction.

    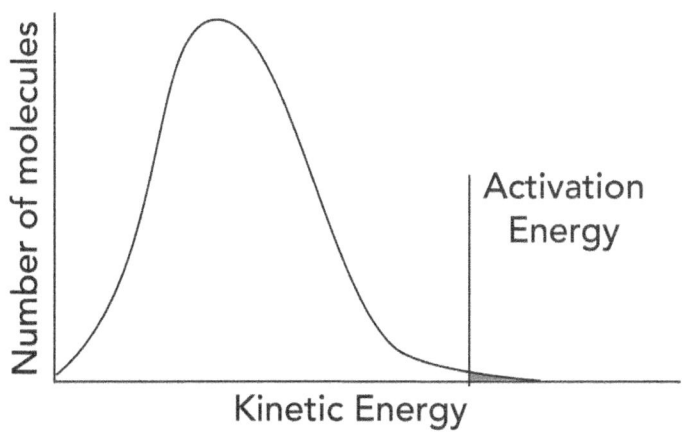

    a. State one way in which the above graph may be different for a different reaction at the same temperature. [1]

    b. i. On the above graph, draw another curve to show the effect of a higher temperature on the same reaction. [2]

    ii. Using your graph, describe and explain why the reaction rate is higher at an increased temperature. [3]

    c. Draw a Boltzmann distribution curve to illustrate the effect of a catalyst on a reaction. [1]

146. Enthalpy profile diagrams may be drawn for chemical reactions involving energy changes.

    A student wanted to carry out the following reaction and work out the enthalpy change of reaction.

    HCl (aq) + NaOH (aq) → NaCl (aq) + H$_2$O

    a. With use of a diagram, describe how the procedure may be carried out. [8]

    0.0500 dm^3 of NaOH and 0.0500 dm^3 of HCl of concentration 1.00 mol dm^{-3} was reacted. The temperature increased from 23.2 °C to 30.7 °C in 15 seconds.

    Specific heat capacity of water = 4.18 J g^{-1} K^{-1}.

    b. i. Calculate the enthalpy change of this reaction. Give your answer to a suitable number of significant figures. [5]

    ii. Determine the enthalpy change of neutralisation of this reaction. [3]

    iii. State three assumptions you have made in your calculation in part b(i). [3]

QUESTIONS
Physical Chemistry and Transition Elements

147. Polyethylene can be formed using hexene. Hight density polyethylene (HDPE) has a wide range of uses including its use in the manufacture of plastic containers.

    Hexene may be produced using the following reaction:

    $C_6H_6 (g) + 3H_2 (g) \rightleftharpoons C_6H_{12} (g)$

    A lab technician mixes together 0.340 mol C_6H_6 with 0.640 mol H_2 in a sealed container.

    The mixture is allowed to reach equilibrium.

    At equilibrium, the mixture contains 0.200 mol C_6H_6. The total pressure was found to be 400 kPa.

    Calculate Kp, showing all your working. Include units in your answer. [5]

148. A student wanted to form oxygen using hydrogen peroxide. The equation for this reaction is shown below.

    $2H_2O_2 (aq) \rightarrow 2H_2O (l) + O_2 (g)$

    15.0 cm³ of hydrogen peroxide solution of concentration 1.00 mol dm⁻³ was used and 70.0 cm³ of oxygen was produced.

    Calculate the percentage yield of the reaction. Give your answer to a suitable number of significant figures. [5]

149. An industrial chemist carried out a reaction at different temperatures and determined the respective rate constants. Using this information, the following graph was plotted.

    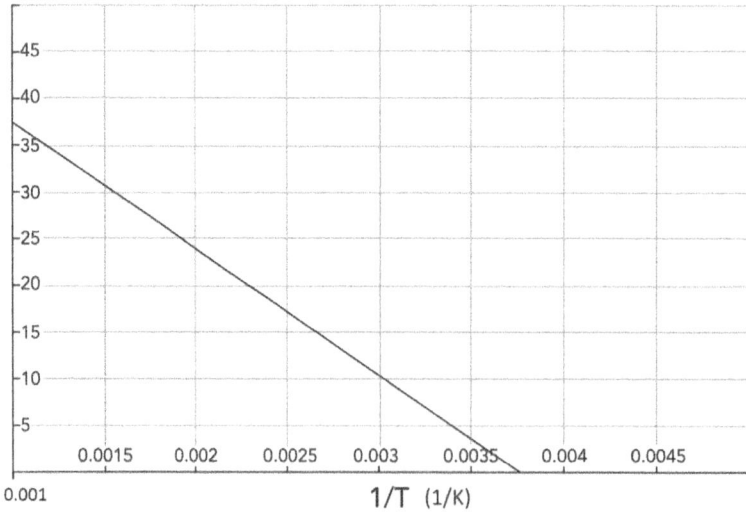

    (Please take the y-intercept of the above graph to be 50 for the purpose of any calculation and not 38 as shown)

    Gas constant = 8.314 J K⁻¹ mol⁻¹.

    a. Label the y-axis of the above graph. [1]

b. Using the information provided, determine the rate constant at 40 °C. You do not have to work out the units of the rate constant. Give your answer to a suitable number of significant figures. [8]

150. A student was provided with 20.0 mg of an unknown metal and a sample of nitric acid of known concentration. The student was asked to identify this metal as part of a practical assessment.

It was found that 20.0 cm³ of a gas was given out during the reaction between the metal and the nitric acid. The metal forms a 2+ ion.

a. Using a labelled diagram, explain how the student could measure the volume of gas given off during the reaction. [5]

b. Write an equation for this reaction and identify the metal using X to represent the metal. [5]

151. An industrial chemist was investigating the effect of temperature on a reaction. The reaction was carried out at 30 °C and then again at 45 °C. The activation energy of this reaction was found to be 55 kJ mol⁻¹.

Assuming A and the concentration of all the species are the same, determine the ratio of the reaction rates at 45 °C to 30 °C. [5]

$R = 8.314$ J K⁻¹ mol⁻¹.

152. A buffer solution was made using 70.0 cm³ of 0.200 mol dm⁻³ propanoic acid and 200 cm³ of 0.400 mol dm⁻³ sodium propanoate.

Calculate the pH of this buffer solution. [6]

K_a of propanoic acid = 1.32 × 10⁻⁵ mol dm⁻³.

153. A student was investigating the reactions of a transition metal **complex ion M**. 3.57g of the complex ion was reacted with aqueous sodium hydroxide forming 2.30g of a coloured **precipitate Y**. Complex ion **M** consists of a transition metal and one type of ligand. One mole of **complex ion M** reacts with aqueous sodium hydroxide to form a mole of the **precipitate Y**. Molar mass of **precipitate Y** = 103 g mol⁻¹.

M contains 32.5%, by mass, of the transition metal.

Determine the formula of **M**. State the colour of the **precipitate Y**.

Write a balanced ionic equation for the reaction that takes place between the complex ion M and aqueous sodium hydroxide. Show all your working. [10]

154. Give chemical explanations for the following statements.

a. The concentration of a sample of HCl with pH 2 is 1.00 × 10⁻² mol dm⁻³. [1]

b. For a weak acid, the K_a expression is simplified as, $Ka = \dfrac{[H^+(aq)]^2}{[HA(aq)]}$ [1]

c. A reaction where ΔH = -90 kJ mol⁻¹ and ΔS = -0.234 kJ K⁻¹ can take place spontaneously at 27 °C. [1]

d. When half-cell X with $E^\ominus$ = +1.36 V and half-cell Y with $E^\ominus$ = + 1.07 V were connected together, half-cell Y was oxidised. [1]

e. The purity of a sample can be determined by analysing the temperature at which it undergoes certain changes. [1]

f. Second electron affinity is endothermic. [1]

155. The following table gives the standard electrode potentials of six redox systems. You are required to use this information to answer the questions below.

Redox system	Equation	$E^\ominus$ / V
1	$F_2(g) + 2e^- \rightleftharpoons 2F^-(aq)$	2.87
2	$O_3(g) + 2H^+(aq) + 2e^- \rightleftharpoons O_2(g) + H_2O(l)$	2.07
3	$S_2O_8^{2-}(aq) + 2e^- \rightleftharpoons 2SO_4^{2-}(aq)$	2.01
4	$Cl_2(g) + 2e^- \rightleftharpoons 2Cl^-(aq)$	1.36
5	$Pb^{2+}(aq) + 2e^- \rightleftharpoons Pb(s)$	x
6	$Zn^{2+}(aq) + 2e^- \rightleftharpoons Zn(s)$	-0.76

a. Outline an experimental setup that could be used in the laboratory to measure the standard electrode potential of the 5th redox system.

In your answer you should include the details of the apparatus, solutions and the standard conditions required to measure standard electrode potential. [4]

b. Use the information given in the table to calculate the standard cell potential of an electrochemical cell made based on redox systems **1** and **6**. [1]

c. From the above table, predict the oxidising agent(s) that will oxidise $O_2(g)$ to $O_3(g)$. [1]

d. State the effect on the position of equilibrium of the 4th redox system, if the concentration of $Cl_2(g)$ is decreased. [1]

e. From **1** to **6**, state the redox system which has the greatest tendency to release electrons, shifting the position of equilibrium to the left. [1]

f. i. Using the information given in the table, predict whether $Zn^{2+}(aq)$ may react with $Cl^-(aq)$. [1]

ii. Such predictions based on standard electrode potentials may not be accurate. State two reasons why. [2]

156. Isocyanic acid, HOCN, is the simplest chemical compound containing the elements, carbon, hydrogen, nitrogen and oxygen.

HOCN dissolves in water very easily.

a. Suggest a structure for isocyanic acid. Draw the displayed formula of HOCN. [1]

b. i. When HOCN is dissolved in water, an equilibrium is set up.

Suggest an equation for the equilibrium reaction and identify the conjugate acid-base pairs.

HOCN(aq) + H_2O(l) ⇌ [2]

ii. Suggest a mechanism for the forward reaction in this equilibrium.

Your mechanism should use displayed formulae and curly arrows, and show all species present at equilibrium. [2]

c. 1.00 mole of isocyanic acid was mixed 5.00 x 10^{-2} mole of water and allowed to reach equilibrium.

The equilibrium mixture was found to contain 3.00 x 10^{-3} mol of the conjugate base of isocyanic acid.

Calculate the equilibrium constant, K_c. Show all your working. Include units in your answer. [4]

157. Iron (III) oxide reacts with carbon monoxide as shown in the following equation.

Fe_2O_3 (s) + 3CO (g) → 2Fe(s) + $3CO_2$(g) ΔS = 15.2 J K^{-1} mol^{-1}

You are provided with the following information.

	Fe_2O_3 (s)	CO (g)	Fe(s)	CO_2(g)	C (s)
$\Delta H_f^\ominus$ / kJ mol^{-1}	-824	-111	0.00	-394	0.00
$S^\ominus$ / J K^{-1} mol^{-1}	87.4	198	X	214	5.70

a. Calculate the standard entropy of Fe(s). [2]

b. Calculate the standard enthalpy change of the reaction. [2]

c. Determine whether this reaction is feasible at 3 °C. Show all your working. [4]

158. Trichloroethene, C_2HCl_3, is a sweet smelling, non-flammable liquid. Many years ago, it was used as an anaesthetic. However, it was later found to have its own disadvantages such as the promotion of heart attacks.

1,2-dichloroethane reacts with chlorine as shown in one of the equations below. (Only one of the following equations is correct).

Equation 1 $ClCH_2CH_2Cl + 2Cl_2 \rightarrow C_2HCl_3 + 3HCl$

Equation 2 $2ClCH_2CH_2Cl + 2Cl_2 \rightarrow C_2HCl_3 + C_2H_2Cl_2 + 3HCl + H_2$

Equation 3 $3ClCH_2CH_2Cl + 2Cl_2 \rightarrow C_2HCl_3 + 2C_2H_2Cl_2 + 3HCl + 2H_2$

MnO_4^- ions react with C_2HCl_3 as in the equation below. MnO_2 has a brownish black colour.

$2MnO_4^- + C_2HCl_3 \rightarrow H^+ + 2MnO_2 + 2CO_2 + 3Cl^-$

An industrial chemist carried out a titration reacting 25.0 cm^3 of 3.45 x 10^{-2} mol dm^{-3} of 1,2-dichloroethane with chlorine. 34.5 cm^3 of 5.00 x 10^{-2} mol dm^{-3} NaMnO$_4$(aq) was required for this investigation.

a. Draw the skeletal formula of trichloroethene. [1]

b. Suggest an experimental method that would allow you to determine which of the three equations is correct following the formation of the product, trichloroethene. State any observations. [3]

c. Using the information given above, determine which of the three equations is correct. Show **all** your working. [4]

159. A student carried out several acid-base titrations. The following graphs of volume of base added against pH was plotted for two of the titrations.

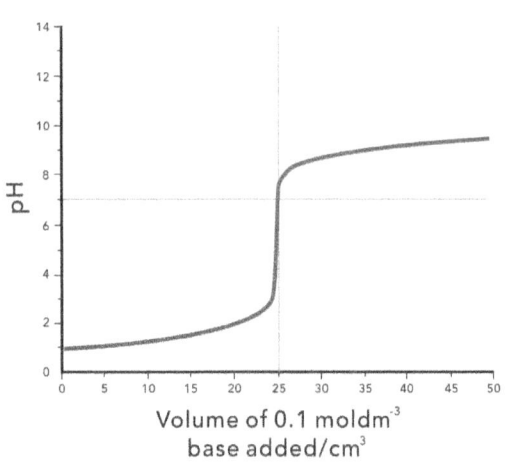

Titration 1

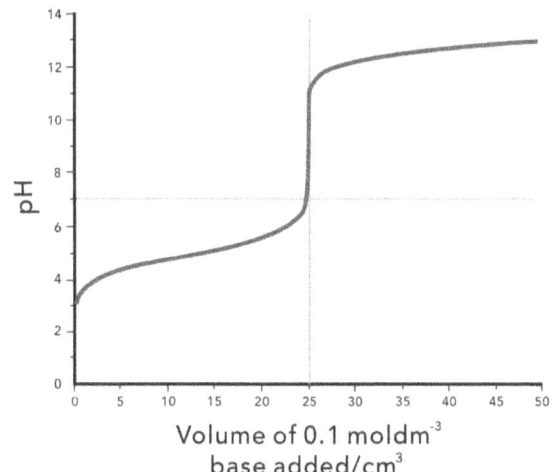

Titration 2

You are provided with the following information about different indicators.

Indicator	Colour		pH range
	Acid	Base	
Methyl Orange	Red	Yellow	3.2 - 4.4
Bromocresol Green	Yellow	Blue	3.8 - 5.4
Phenol Red	Yellow	Red	6.8 - 8.4
Phenolphthalein	Colourless	Pink	8.2 - 10.0

a. State the types of acids and bases used in titrations 1 and 2. [2]

b. From the information given above, select suitable indicators for titrations 1 and 2. Explain your reasoning. [4]

c. Outline an experimental setup that could be used in the laboratory to carry out the titrations. Describe how the student could continuously monitor the pH as the reaction progresses. In your answer, you should include details of the apparatus and a labelled diagram. [5]

d. At one point in one of the two titrations, the concentration of H^+ ions in the solution was 0.01 mol dm^{-3}. Calculate the amount, in mol, of OH^- ions present at this point. State which titration this was. [4]

160. This question is about rate equations.

    a. A chemist reacted the compounds, P and Q at a very high temperature of 927 °C to form M and N as shown in the following equation.

    P + Q → M + N

    The equation for the rate-determining step of the above reaction is given below:

    2P + S → T + N

    The chemist reacted S of concentration 0.500 mol dm^{-3} and found the rate of reaction to be 0.600 mol dm^{-3} s^{-1}.

    P and Q was made to react at different temperatures to obtain the following values.

ln k	T / K
1.570	1140
1.770	1180
1.980	1220
2.180	1270

    A = 1810 s^{-1}. R = 8.314 J mol^{-1} K^{-1}.

    Calculate the rate equation from the rate determining step and thus determine the concentration of P, when the rate of reaction is 0.600 mol dm^{-3} s^{-1}. Your answer should show all the relevant working and a labelled graph. [9]

    b. The same reaction was carried out with a catalyst. The activation energy with the catalyst present was 72% of the activation energy without the catalyst. Determine the rate constant for the above reaction with the catalyst present. Determine its units. [7]

161. A student reacted an unknown compounds A with HCl to produce a gas C, water and another compound D, as shown in the following equation.

    A(s) + 2HCl(aq) → C(g) + D(aq) + H_2O

    62.0 cm^3 of gas C was collected, measured at 25 °C and a pressure of 98 kPa. It was found to weigh 0.108g.

    The student was told that the gas could either be sulfur dioxide, chlorine, carbon dioxide or ammonia.

    R = 8.314 J mol^{-1} K^{-1}

5 QUESTIONS — Physical Chemistry and Transition Elements

a. Determine which gas was formed during the reaction. [4]

b. Compound **A** was suspected to be a carbonate. The student was given another sample of a solution of compound **A** and sodium sulfate. The samples were unlabelled. Barium nitrate was added to one of the samples and a white precipitate was observed. The student concluded the sample to be sodium sulfate. Explain why the student might be wrong and how the samples could be correctly identified. [3]

c. HCl used in the above reaction had a pH of 0.9. 50.0 cm³ of the acid was used. Compound **A** was present in excess. Calculate the percentage yield of the reaction. [6]

162. You are provided with the following data relating to sodium chloride.

Enthalpy change	Energy / kJ mol^{-1}
Lattice enthalpy of NaCl	-776
Enthalpy change of hydration of sodium	-406
Enthalpy change of solution of NaCl	5.00

a. Use the data in the table to calculate a value for the enthalpy change of hydration of chlorine. Draw a Born-Haber cycle in your answer. [5]

b. A student dissolved 7.00 g of sodium chloride in 30.0 g of water. The initial temperature of the water is 296 K. Determine the final temperature in K, of the solution. Assume that only the 30.0 g of water changed in temperature.

Specific heat capacity of water is 4.18 J K^{-1} g^{-1}. [5]

163. Give chemical explanations for the following statements.

a. Fe(OH)$_2$(s) turns rusty brown at the surface when exposed to air. [1]

b. [Cr(H$_2$O)$_3$(OH)$_3$] is a neutral complex. [1]

c. Scandium cannot be described as a transition metal. [1]

d. 1-chloro-3-nitrobenzene is the likely product in the chlorination of nitrobenzene. [1]

e. A person exposed to air containing CO may die within hours. [1]

164. A student reacted aqueous cobalt(II) ions with concentrated hydrochloric acid to form a complex ion with a tetrahedral shape in a ligand substitution reaction. The pink solution turns into a blue solution. This is an equilibrium reaction.

a. Write a balanced equation for the reaction. [2]

b. The student adds water to the equilibrium mixture. Describe and explain the changes that are taking place including any observations. [3]

c. Draw the 3D shape of the major product of the reaction in part (a). [2]

d. i. Write an expression for the equilibrium constant, K_c, of the reaction. [2]

   ii. Deduce the units for K_c. [3]

Physical Chemistry and Transition Elements — QUESTIONS 5

165. This question is about transition metal complexes / complex ions.

    a. Aqueous chromium has the formula $[Cr(H_2O)_6]^{3+}$.

        i. State the observations when aqueous ammonia is slowly added until excess is present. [1]

        ii. Explain why the ion has a 3+ charge. [1]

        iii. State the shape of this complex ion. [1]

    b. $[Fe(H_2O)_4(OH)_2]$

        i. State and explain the oxidation number of iron. [2]

        ii. State the colour and the state of this complex. [1]

    c. $Fe^{3+}(aq)$

        i. State the complexed formula of the ion. [1]

        ii. State the name of two reagents that may turn this solution into a rusty brown precipitate. [1]

166. Nitryl fluoride, NO_2F, is a colourless gas and a strong oxidizing agent. Nitryl fluoride can be prepared by reacting nitrogen dioxide with fluorine.

    The rate equation for this reaction is: rate = $k[NO_2][F_2]$ $k = 38$ mol^{-1} dm^3 s^{-1}

    a. Suggest a two step reaction mechanism for this reaction and label the rate-determining step in your mechanism. [4]

    b. State the overall order of the reaction. [1]

    c. A student investigated the effect of changing the concentration of NO_2 on the reaction rate and plotted the following graph. The concentration of F_2 was kept fixed at X mol dm^{-3}.

    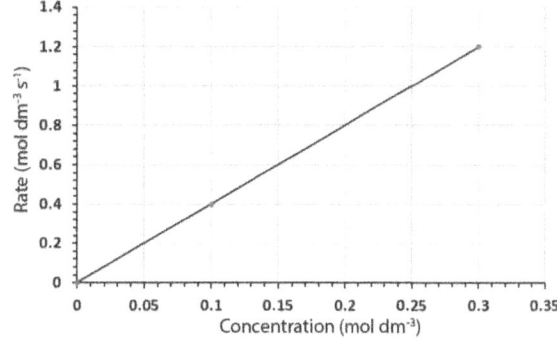

    Assuming the concentration of F_2 is X mol dm^{-3}, determine the mass in grams of F_2 that would be required for the rate of reaction to be 0.5 mol dm^{-3} s^{-1} if F_2 occupied a volume of 300 cm^3. [5]

QUESTIONS — Physical Chemistry and Transition Elements

167. HNO_3 was reacted with **compound X** as shown in the following equation.

    HNO_3 (aq) + X → P + Q + R(g)

    A chemist investigated the change in reaction rate with change in concentration of HNO_3. His results are shown in the following table. Throughout the investigation, the concentration of X was kept fixed.

Concentration / mol dm^{-3}	Rate of reaction / mol dm^{-3} s^{-1}
0.10	0.10
0.15	0.25
0.20	0.30
0.25	0.50
0.30	0.80
0.35	1.44

    a. The order with respect to HNO_3 is either 0, 1 or 2. Determine its order in the above reaction. Show all your reasoning. [4]

    b. Describe how you could determine the concentration of H^+ ions in the solution as the reaction is in progress. [2]

    c. A student said "Using different concentrations of X for the above investigation would still give the same set of results and would allow determination of the order with respect to nitric acid". Comment on his statement. [2]

    d. When R was bubbled through lime water, a white precipitate was observed. A student added warm NaOH (aq) to X and a gas M was produced. M turned damp red litmus paper blue. Suggest an identity for **M** and **X**. [2]

168. A reaction has a rate constant of 1.02×10^{-10} s^{-1} at 290 K and an activation energy of 112 kJ mol^{-1}.

    a. Calculate its rate constant at 315 K. [4]

    b. The rate equation of the above reaction is one of the following.

    rate = $k[NO_2]$ rate = $k[NO_2]^2$ rate = $k[NO_2]^2[NO]$

    State which of the above is a possible rate equation. Explain. [3]

169. Hydrogen peroxide reacts with iodide ions and hydrogen ions as shown in the following equation.

    $H_2O_2(aq) + 3I^-(aq) + 2H^+(aq) \rightarrow 2H_2O(l) + I_3^-(aq)$

    A student worked out the rate equation to be rate = $k[H_2O_2][I^-][H^+]$

    The following rate mechanism was suggested for the reaction.

    Rate-determining step: ____ + ____ + ____ → ____ + HOI (aq)

_____ + I⁻(aq) + _____ → _____ + I₂(aq)

_____ + _____ → _____

a. Complete the reaction mechanism and identify the rate-determining step. [6]

b. Determine the units of the rate constant, k. [2]

170. Gaseous iodine monochloride reacts with hydrogen to produce two gases as shown in the following equation:

$$H_2(g) + 2ICl(g) \rightarrow I_2(g) + 2HCl(g)$$

The following two-step mechanism was proposed for the above reaction.

slow step: _____ (g) + _____ (g) → HI(g) + HCl(g)

fast step: _____(g) + ICl(g) → _____ (g) + _____ (g)

a. Complete the reaction mechanism. [4]

b. i. Determine the rate equation of the reaction. [2]

ii. When the volume of each of the reactants in the rate-determining step is 480 cm³, the rate of reaction was found to be 0.250 mol dm⁻³ s⁻¹. Calculate the rate constant. [3]

iii. Determine the units of the rate constant. [2]

c. State the overall order of the reaction. [1]

171. Barium chloride is a water-soluble salt. Like other salts of barium, it is toxic and results in a yellow-green colour during the flame test.

Enthalpy change	Energy / kJ mol⁻¹
Enthalpy change of atomisation of barium	180
First ionisation energy of barium	503
Second ionisation energy of barium	965
Enthalpy change of atomisation of chlorine	122
Enthalpy change of formation of barium chloride	-859
First electron affinity of chlorine	-349
Lattice enthalpy	x

a. i. Draw a labelled Born-Haber cycle for the formation of barium chloride from its constituent elements. [3]

ii. Using your Born-Haber cycle determine the lattice enthalpy of barium chloride. [2]

QUESTIONS
Physical Chemistry and Transition Elements

    b. Describe a use of barium chloride in the laboratory. [2]

    c. The lattice enthalpy of barium oxide is -3054 kJ mol^{-1}. Explain. [2]

172. Sodium thiosulfate reacts with hydrochloric acid to form aqueous sodium chloride, solid sulfur, gaseous sulfur dioxide and water. A student was asked to investigate the effect of using sodium thiosulfate of different concentrations on the reaction rate.

    a. i. Write a balanced equation for the reaction that is taking place. [3]

       ii. Write its ionic equation. [2]

    b. The student was provided with sodium thiosulfate of different concentrations. Describe a simple procedure that may be carried out to determine the reaction rates when different concentrations of sodium thiosulfate are used. [5]

    c. The student was provided with the following set of data by the teacher.

Concentration of thiosulfate / mol dm^{-3}	0.100	0.080	0.060	0.040	0.020
Rate of reaction / mol dm^{-3} s^{-1}	0.0255	0.0200	0.0149	0.0097	0.0045

       i. Deduce the order with respect to sodium thiosulfate. Your answer should include a graph. [4]

       ii. State another factor other than concentration that can affect the rate of reaction. [1]

173. You are provided with the following standard electrode potential data.

Equation	$E^{\ominus}$ / V
$Fe_2O_3(s) + 3H_2O(l) + 2e^- \rightleftharpoons 2Fe(OH)_2(s) + 2OH^-(aq)$	-0.86
$HSO_4^-(aq) + 3H^+ + 2e^- \rightleftharpoons SO_2(aq) + 2H_2O(l)$	+0.16
$NO_3^-(aq) + 2H^+(aq) + e^- \rightleftharpoons NO_2(g) + H_2O(l)$	+0.80
$ClO_2(g) + H^+ + e^- \rightleftharpoons HClO_2(aq)$	+1.19
$CoO_2(s) + 4H^+(aq) + e^- \rightleftharpoons Co^{3+} + 2H_2O$	+1.42

    a. Identify the strongest reducing agent from the species in the table above. Explain your reasoning. [3]

    b. i. A cell was made by using iron(III) oxide and chlorine dioxide. Using the data provided, calculate its standard cell potential. [2]

       ii. Write an equation for the overall cell reaction. [3]

       iii. Identify which half-cell would be the positive electrode. Give a reason for your answer. [2]

c. CoO_2 may act as an oxidising agent. Identify which element in CoO_2 is reduced during the process. Explain your reasoning. [2]

174. The following equation shows the reaction between iron(II) ions and dichromate(VI) ions.

$$6Fe^{2+} + Cr_2O_7^{2-} + 14H^+ \rightarrow 6Fe^{3+} + 2Cr^{3+} + 7H_2O$$

a. Identify one species that is oxidised and one that is reduced. Explain your reasoning. [4]

A student was supplied with an impure sample of $FeSO_4.7H_2O$. The sample was dissolved in water, making the solution up to 250 cm³. An excess of acid was added to 25.0 cm³ of the solution. It required 10.0 cm³ of a 0.0350 mol dm⁻³ solution of $K_2Cr_2O_7$ to react completely.

b. A student wanted to work out the rate of the above reaction at different time intervals. Describe a method that could be used. [4]

c. Calculate the mass of $FeSO_4.7H_2O$ in the original sample. [5]

175. P reacts with Q as shown in the following equation:

$$P + Q \rightarrow W + X$$

The rate-determining step for the reaction is : ___ P + Q → W + H

You are provided with the following graph showing the changes in concentration of one of the reactants with time.

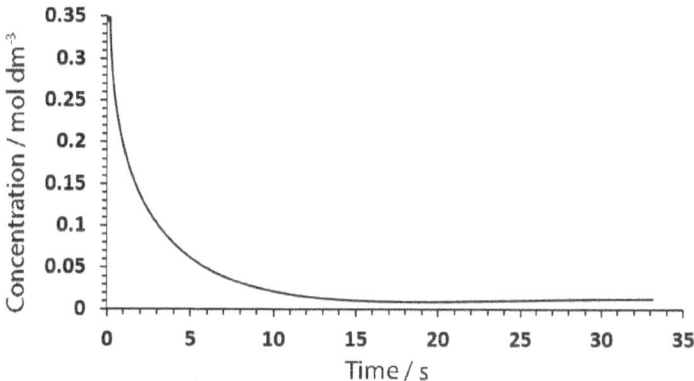

a. Complete the rate-determining step. Explain your reasoning. [3]

b. What term is given to the product H of the rate-determining step? [1]

c. Determine the initial rate of the above reaction. State its units. [4]

The concentration of Q at the start of the reaction was 0.35 mol dm⁻³.

d. Determine the rate constant of the reaction and its units. [5]

176. A first order reaction was carried out and the change in concentration of a reactant with time was investigated. The results are given in the following table.

QUESTIONS
Physical Chemistry and Transition Elements

Concentration / mol dm^{-3}	Time / s
0.060	0.0
0.050	0.8
0.035	2.3
0.024	4.0
0.014	6.4
0.005	11

   a. Draw a concentration-time graph using the data given in the table. [3]

   b. Determine the rate of reaction after 2.4 seconds of the reaction starting. [5]

   c. Using your graph, show that the half-life of the reaction is constant. Determine the half-life. [3]

   d. Determine the rate constant of the reaction using the half-life. State its units. [3]

   e. Show that the rate of reaction is the same as (or closer to) your answer for part(b), using your rate constant from part(d). [3]

177. Transition elements and their compounds are often used as catalysts in reactions.

   a. Give two examples of transition metals / compounds acting as catalysts. Write an equation for each reaction they catalyse. [4]

   b. i. Name a transition metal complex that is used in the treatment of disease and draw its displayed formula. [2]

   ii. Draw and name an isomer of the transition metal complex in b(i). Name the type of isomerism. [3]

178. $[Cr(C_2O_4)_2(H_2O)_2]^-$ is a complex ion. $C_2O_4^{2-}$ is a ligand in this complex ion. Two of the oxygen atoms in this ligand has a - charge each.

   a. i. Draw the 3D structures for the two cis-trans isomers of the complex ion and name the type of each isomer. [2]

   ii. Name the shape of the complex ion. [1]

   b. Determine the oxidation state of chromium in $[Cr(C_2O_4)_2(H_2O)_2]^-$. Explain your reasoning. [3]

   c. Identify and name the types of ligands in this ion. [2]

179. Iron forms $[FeCl_4]^-$ complex ions with chlorine ligands. A student was told that the Cl-Fe-Cl bond angle is not 90°.

   a. i. Draw the complex ion. [1]

ii. State and explain the Cl-Fe-Cl bond angle. [3]

iii. State the name of the shape of [FeCl₄]⁻. [1]

iv. State the name of the shape given to a four-coordinate complex whose bond angles are 90°. [1]

b. State and explain the oxidation number of iron in [FeCl₄]⁻. [2]

180. The following diagram shows a Born-Haber cycle for the formation of aluminium oxide

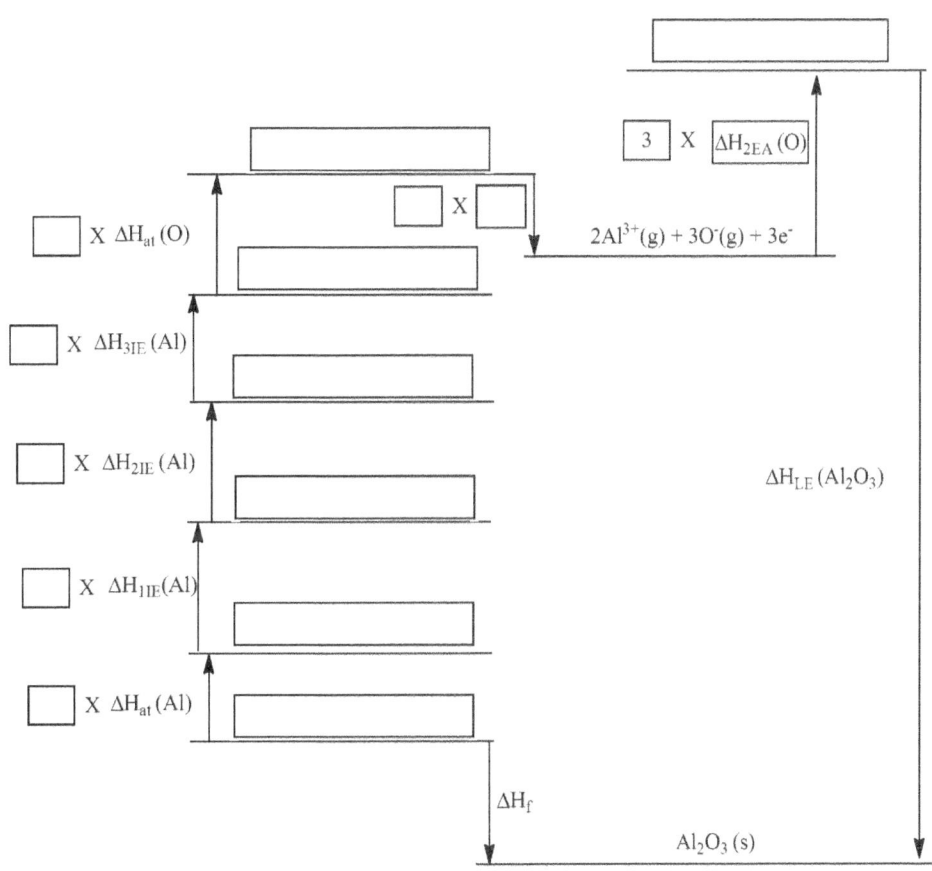

Use the data in the table to answer the question.

Enthalpy change	Energy / kJ mol⁻¹

QUESTIONS
Physical Chemistry and Transition Elements

Enthalpy change of formation of aluminium oxide	-1676
Enthalpy change of atomisation of aluminium	326
Enthalpy change of atomisation of oxygen	249
First ionisation energy of aluminium	578
Third ionisation energy of aluminium	2745
First electron affinity of oxygen	-142
Second electron affinity of oxygen	844
Lattice enthalpy of aluminium oxide	-15461

    a. Complete the Born-Haber cycle. [5]

    b. Determine the second ionisation energy of aluminium. [4]

181. Describe test-tube tests to identify each of the following species. Write an ionic equation with state symbols to show the reaction taking place. State any observations.

    a. SO_4^{2-} [4]

    b. Cu^{2+} [3]

    c. Br^- [5]

    d. Mn^{2+} [3]

    e. NH_4^+ [3]

    f. I^- [5]

182. A student wanted to determine the concentration of a sample of potassium permanganate. 0.340 g of sodium oxalate was dissolved in a flask. 27.30 cm³ of potassium permanganate was required to fully react with the dissolved sodium oxalate. The equation for the reaction is as follows:

$$5Na_2C_2O_4(aq) + 2KMnO_4(aq) + 8H_2SO_4(aq) \rightarrow 2MnSO_4(aq) + K_2SO_4(aq) + 5Na_2SO_4(aq) + 10CO_2(g) + 8H_2O(l)$$

    a. i. State the colour changes observed during the reaction. [2]

        ii. Determine the concentration of potassium permanganate. [4]

        iii. Write an ionic equation for the above reaction. [2]

    b. Draw a labelled diagram to show the set-up of an apparatus which the student could use to carry out the investigation. Identify $KMnO_4(aq)$ on your diagram. [4]

183. A student titrated 25.0 cm³ of 0.203 mol dm⁻³ CH_3COOH with 0.250 mol dm⁻³ NaOH. (K_a of CH_3COOH = 1.77 × 10⁻⁵)

    a. Calculate the initial pH. [3]

    b. Determine the volume of NaOH required to fully react with CH_3COOH, in cm³. [3]

c. i. Sketch a graph of pH against volume of base added. [1]

   ii. State how you would continuously monitor the changes in pH as the base is added. [1]

d. Suggest why no indicator is suitable for a weak acid-weak base titration. [1]

184. Sucrose is a naturally occurring carbohydrate that can be extracted from cane and beet sugar. A student hydrolysed a sample of sucrose in the laboratory.

   a. Determine the rate constant of the reaction at 27 °C for the hydrolysis of sucrose provided that the rate constant is equal to 1.03×10^{-3} mol dm^{-3} s^{-1} at 40 °C. The activation energy of the reaction is 110 kJ mol^{-1}. [4]

   b. A catalyst was introduced for the same reaction which increased the rate constant by 1.5 times when the reaction was carried out at 27 °C. Draw a Boltzmann distribution curve to show the effect of the catalyst. You should label the diagram to show the change in activation energy using numbers specific to this reaction. [5]

185. A student wanted to find out if solid lead will react with chromium (III) ions.

   The following half-equations are provided.

   Cr^{3+} (aq) + e$^-$ ⇌ Cr^{2+} (aq) $E^\ominus$ = -0.41 V Pb^{2+} (aq) + 2e$^-$ ⇌ Pb (s) $E^\ominus$ = -0.13 V

   a. i. Write a balanced ionic equation for the overall reaction. [2]

   ii. Determine if the reaction will be feasible. [2]

   b. The student then constructed an electrochemical cell based on the same two equilibria.

   i. Determine the overall cell reaction. [2]

   ii. Calculate the standard cell potential of the reaction. [1]

186. A buffer solution was made by mixing 1.0 dm^3 of 0.050 mol dm^{-3} ethanoic acid and 0.30 dm^3 of 0.020 mol dm^{-3} calcium hydroxide and making the mixture up to 1.5 dm^3. Determine the pH of the buffer solution. Your answer should include a balanced equation. K_a of ethanoic acid = 1.74×10^{-5} mol dm^{-3}. [7]

187. This question is about a buffer solution made by mixing ethanoic acid and sodium ethanoate.

   a. State the chemical formula of sodium ethanoate. [1]

   b. i. Determine the ratio of ethanoic acid to sodium ethanoate required to make a buffer solution of pH 5.5. pK_a for ethanoic acid is 4.76. [5]

   ii. Calculate the amount, in moles, of ethanoic acid required to make a buffer solution of pH 5.5 using 20.0 cm^3 of 0.0600 mol dm^{-3} sodium ethanoate. [2]

188. You are given the following equilibrium between the compounds **P**, **Q** and **R**.

   $$2P(g) + Q(g) \rightleftharpoons 2R(g) \quad \Delta H = -70 \text{ kJ mol}^{-1}$$

   a. Write an expression for the equilibrium constant K_c for this equilibrium and state its units. [2]

b. i. The mixture was left to reach equilibrium in a container of volume 0.75 dm³ at a temperature of X° C. At equilibrium, the mixture contained 0.25 mol of P, 0.10 mol of Q and 0.27 mol of R. Calculate a value of the equilibrium constant. [2]

ii. The same mixture was then allowed to reach equilibrium in a container of volume 0.50 dm³. State and explain whether the amount of R in the mixture will increase, decrease or stay the same. [3]

c. The same mixture was allowed to reach equilibrium at a temperature of Y° C, in the container of volume 0.75 dm³. A student found out that the value of K_c has decreased. State and explain which of the temperatures, out of X and Y is higher. [3]

189. This question is about ammonium bromide.

   a. State its chemical formula. [1]

   b. The standard enthalpy change of solution of ammonium bromide is +16.78 kJ mol⁻¹. Explain why it dissolves spontaneously in water even though the enthalpy change of solution is endothermic. [2]

   c. i. A student dissolved 5.00 g of ammonium bromide in 70.0 g of water. The initial temperature was 25.0 °C. Calculate an estimate for the final temperature of the solution. Specific heat capacity of water is 4.20 J K⁻¹ g⁻¹. [5]

   ii. Suggest one reason why your answer to c(i) is only an estimate. [1]

190. 7.45 g of solid potassium chloride, KCl, were dissolved in 90.0 cm³ of water. The temperature of the solution increased from 23.0 °C to 27.5 °C. The specific heat capacity of the solution is 4.18 J g⁻¹ K⁻¹.

   a. Calculate the enthalpy change of solution of KCl. [3]

   b. State one reason as to why this is only an approximation. [1]

191. The concentration of a sample of phenol, C_6H_5OH, is 0.072 mol dm⁻³. pK_a of phenol is 9.80.

   a. Calculate its pH. [4]

   b. State an assumption you have made in your calculation. [1]

192. A solution of KOH has a concentration of 0.082 mol dm⁻³. Calculate its pH at 25 °C. [3]

193. The following equilibrium is established when aqueous ammonia is added to water.

$$NH_3 \,(aq) + H_2O \,(l) \rightleftharpoons NH_4^+ \,(aq) + OH^-(aq)$$

   a. Write an expression for the equilibrium constant, K_c. [2]

   b. Calculate K_c if the equilibrium concentrations are as follows: NH_3 = 0.03 mol dm⁻³ ; NH_4^+ = 0.06 mol dm⁻³ ; OH^- = 0.7 mol dm⁻³. State its units. [2]

194. Feasibility of a reaction may be established using standard electrode potentials of half-cells.

$$Fe(OH)_3(s) + e^- \rightleftharpoons Fe(OH)_2(s) + OH^-(aq) \quad E^\ominus = -0.56 \text{ V}$$

$$O_2(g) + 2H_2O(l) + 4e^- \rightleftharpoons 4OH^-(aq) \quad E^\ominus = +0.40 \text{ V}$$

a. Determine if oxygen can oxidise iron (II) hydroxide to iron (III) hydroxide. [3]

b. Write a balanced equation for the overall reaction. [3]

195. This question is about reaction rates.

Calculate the rate constant when:

i. $E_a = -5.00$ J mol^{-1} ; $A = 30.2$; $T = 315$ K ii. $E_a = 6.00$ J mol^{-1} ; $A = 24.5$; $T = 25$ °C

iii. $E_a = 8.03 \times 10^{-3}$ kJ mol^{-1} ; $A = 25.5$; $T = 302$ K iv. $E_a = -2.03$ J mol^{-1} ; $A = 29.6$; $T = 27$ °C [4]

196. A reaction will only happen if the activation energy barrier is crossed.

Calculate the activation energy, in kJ mol^{-1}, when:

i. $k = 78$; $A = 30.2$; $T = 15$ °C ii. $k = 58$; $A = 16.5$; $T = 270$ K [2]

197. The value of the rate constant changes with changes in temperature.

a. Calculate the temperature in °C, when $k = 60$, $A = 17.0$ and $E_a = -3.05$ J mol^{-1}. [2]

b. Calculate the temperature in °C, when $k = 60$, $A = 35.0$ and $E_a = -2.50$ kJ mol^{-1}. [2]

c. Calculate the temperature in K, when $k = 26.3$, $A = 29.6$ and $E_a = 6.02$ J mol^{-1}. [2]

198. A student was given an unknown metal and a solution. The metal was reacted with the solution forming a pale blue solution of **compound W**.

The student divided the solution of **compound W** into two portions and carried out two chemical tests. When aqueous sodium hydroxide was added, the pale blue solution turned into a pale blue precipitate. Then, diluted hydrochloric acid was added to the other portion, followed by barium chloride solution. A white precipitate was observed.

a. State the chemical formula of **compound W**. [1]

b. i. Write a balanced equation for the reaction between the metal and the solution. [1]

ii. State any other observations apart from the formation of the pale blue solution. [1]

199. 30.0 cm^3 of iron (II) sulfate solution of concentration 0.0200 mol dm^{-3} was acidified with an excess of sulfuric acid. It was then titrated with potassium manganate (VII) solution. 25.30 cm^3 of potassium manganate (VII) solution was required to reach the end-point of the titration. Determine the concentration of potassium manganate (VII) solution in mol dm^{-3}. [5]

200. A chemist investigated the effect of temperature on the rate constant, k, of a reaction. The following table illustrates the results of the investigation:

k (x 10^{12})	T / °C
1.963	-73.0
2.102	-123

QUESTIONS
Physical Chemistry and Transition Elements

2.409	-173
3.627	-223

$R = 8.31$ J mol^{-1} K^{-1} $A = 1.60 \times 10^{12}$

a. Plot a graph of ln k vs 1/T, determine the gradient and thus the activation energy of the reaction. [7]

b. Calculate the rate constant when the temperature is 25 °C. [2]

Physical Chemistry and Transition Elements

QUESTIONS 5

5 QUESTIONS
Physical Chemistry and Transition Elements

Multiple Choice Questions

1. What is the best definition for order of a reactant, as found in the rate equation?

    A   The time taken for the concentration of that reactant to halve.

    B   It is the power to which the concentration of the reactant is raised in the rate equation.

    C   It is a constant that links the rate of reaction with the concentration of the reactants.

    D   It is the power to which the concentration of the products is raised in the rate equation.

2. Concentration-time graphs may be drawn for reactants of different orders. Choose the statement which best explains the concentration-time graph of a zero order reactant:

    A   Half-life is constant; concentration decreases rapidly.

    B   Half-life decreases with time; concentration decreases at a constant rate.

    C   Half-life decreases with time; concentration halves in equal time intervals.

    D   Half-life increases with time; concentration stays the same.

3. Which of the following statements is true?

    A   Change in concentration of a zero order reactant has no effect on the reaction rate.

    B   Change in concentration of a first order reactant has no effect on the reaction rate.

    C   Change in concentration of a second order reactant has no effect on the reaction rate.

    D   Doubling the concentration of a zero order reactant doubles the reaction rate.

4. Which of the following statements is true?

    A   A first order reactant does not always have to be included in the rate equation.

    B   A zero order reactant is always included in the rate equation.

    C   A zero order reactant is not included in the rate equation.

    D   The reaction rate will depend on the concentration of a zero order reactant.

5. K_c changes:

    A   With pressure and the concentration of reactants.

    B   Only with temperature.

    C   Only with pressure.

    D   With concentration of reactants and temperature.

Physical Chemistry and Transition Elements — QUESTIONS 5

6. How many cm^3 are there in one dm^3?

    A   100

    B   10,000

    C   10

    D   1000.

7. What is the simplified balanced ionic equation for the reaction between an acid and an alkali?

    A $Na^+ + Cl^- \rightarrow NaCl$

    B $H^+ + 2OH^- \rightarrow H_2O$

    C $HCl + NaOH \rightarrow NaCl + H_2O$

    D $H^+ + OH^- \rightarrow H_2O$

8. A conjugate acid-base pair is a pair of two:

    A   Species which transform into each other by the gain or loss of a proton (H^+).

    B   Acids and bases that react to form salt and water.

    C   Species which always exist together while a reaction is in progress.

    D   Species that speed up a reaction by lowering the activation energy.

9. A proton is a subatomic particle found inside the nucleus of an atom. A H^+ ion is sometimes referred to as a proton, because:

    A   During the formation of the hydrogen ion, all the neutrons and electrons are donated leaving it only with one proton.

    B   An atom can donate its neutrons easily.

    C   A hydrogen atom only consists of a proton.

    D   A hydrogen ion consists of only one proton. Removal of one electron from a hydrogen atom leaves it with only one proton.

10. pH =

    A $\log [H^+(aq)]$

    B $10^{-[H^+(aq)]}$

    C $- \log [H^+(aq)]$

    D $6.02 \times 10^{23} [H^+(aq)]$

11. State the units of K_a.

QUESTIONS
Physical Chemistry and Transition Elements

    A   mol dm^{-3}

    B   mol

    C   dm^3

    D   Depends on the expression of K_a.

12. A strong acid, HA, dissociates completely in solution, whereas a weak acid only dissociates partially in solution. For a strong monobasic acid, the concentration of hydrogen ions ([H$^+$]) is equal to:

    A   The concentration of the acid.

    B   The pH of the acid.

    C   The K_a of the acid.

    D   The volume of the acid.

13. The ionic product of water,

    K_w = [H$^+$(aq)] [OH$^-$(aq)]

    What is the ionic product of water at 25°C?

    A   1.02 x 10^{-14} mol^2 dm^{-6}

    B   1.02 x 10^{-14} mol dm^{-3}

    C   1.00 x 10^{-14} mol^2 dm^{-6}

    D   1.00 x 10^{-14} mol dm^{-3}

14. Which of the following statements best explains the action of a buffer solution?

    A   A buffer solution absorbs hydrogen ions preventing the pH from fluctuating.

    B   Acids or bases do not show their effect in the presence of a buffer solution.

    C   A buffer solution prevents any change in pH when small amounts of acid or base is added.

    D   A buffer solution minimises changes in the pH when small amounts of acid or base is added.

15. Exothermic reactions "give out" heat during the course of the reaction, whereas endothermic reactions "take in" heat.

    What type of energy changes are involved in first ionisation energy and second ionisation energy?

	First ionisation energy	Second ionisation energy
A	Endothermic	Endothermic
B	Endothermic	Exothermic

Physical Chemistry and Transition Elements

QUESTIONS 5

| C | Exothermic | Exothermic |
| D | Exothermic | Endothermic |

16. Lattice enthalpy is the energy change associated with the formation of one mole of an ionic compound from its gaseous ions under standard conditions.

    Which of the following combinations result in the highest lattice enthalpy?

	Ionic size	Ionic charge
A	Large	High
B	Small	High
C	Small	Low
D	Large	Low

17. Entropy, S is always a positive number. Which of the following statements best explains the reason for this?

    A   The particles in any substance are never stationary. Therefore there is always some disorder in a system. Entropy measures the degree of disorder.

    B   Entropy is another measure for temperature. A temperature can never have a negative value.

    C   The particles in any substance are always moving. A substance whose particles are ordered will have a high entropy.

    D   Entropy is equal to free energy. Free energy will always be a positive number as energy cannot be negative.

18. Ions form when an atom gains or loses electrons. They may be negative or positive.

    A 2+ ion will form when:

    A   Two atoms lose one electron.

    B   An atom loses two electrons.

    C   Two atoms gain one electron.

    D   An atom gains two electrons.

19. Which of the following is a half-cell?

    A   An acid and a base in solution.

    B   Two metal rods in different solutions connected with a wire. The two solutions connected with a salt bridge.

C   An iron rod placed in an aqueous solution of iron ions.

D   An iron rod placed in an aqueous solution of zinc ions.

20. The following is a diagram of an electrochemical cell:

What is the arrow pointing to?

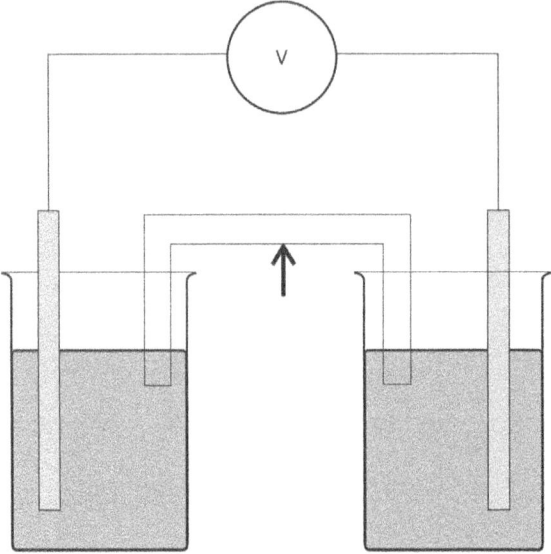

A   Metal strip

B   Pressure equalizer

C   Wire

D   Salt bridge

21. A transition element:

A   Is a d block element.

B   forms ions with only one oxidation state.

C   Forms an ion with an incomplete d sub-shell.

D   Is a non-metal.

22. Electrons are arranged in shells around the nucleus of an atom. These shells consist of sub-shells. The 1s sub-shell has lower energy than the 2s sub-shell.

Which of the following statements about sub-shells is NOT true?

A   The 2s sub-shell has lower energy than the 3s sub-shell.

B   In fluorine, the highest energy electron is not in a p sub-shell.

C   The p sub-shell can hold a maximum of six electrons.

D   In both nitrogen and oxygen, the highest energy electrons are in a 2p sub-shell.

23. Which of the following is true about the electron configuration of chromium and copper?

   A   In both chromium and copper, the 4s sub-shell is half-filled.

   B   In both chromium and copper, 4s sub-shell fills completely before the 3d sub-shell is filled.

   C   Chromium and copper do not have 4s sub-shells.

   D   The 4s sub-shells of either chromium and copper can hold a maximum of four electrons.

24. Element X has the following properties:

   - It is a metal.
   - It has variable oxidation states.
   - Compounds of X are coloured.

   What could element X be?

   A   An alkali earth metal

   B   A halogen

   C   A transition element

   D   Magnesium.

25. Which of the following statements best describes a precipitation reaction?

   A   A reaction which gives off gas.

   B   Two solutions react together forming an insoluble compound.

   C   A metal reacts with a solution forming a metal oxide and carbon dioxide.

   D   A metal oxide reacts with a base forming a salt and water.

26. Copper (II) oxide reacts with aqueous sodium hydroxide, forming copper (II) hydroxide.

   Copper (II) hydroxide is a:

   A   Pale blue precipitate.

   B   Pale blue solution.

   C   Rusty brown precipitate.

   D   Rusty brown solution.

QUESTIONS
Physical Chemistry and Transition Elements

27. Which of the following best explains the diagram shown below?

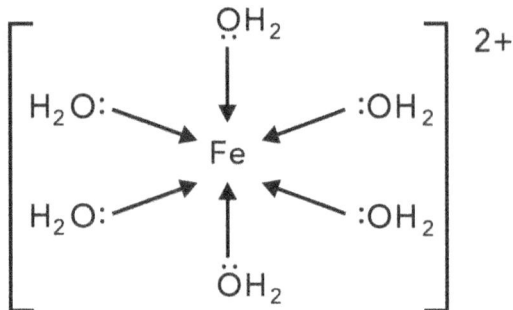

A   It is a complex ion with six H_2O ligands and a coordination number of six. Each water ligand donates a pair of electrons to the iron ion.

B   It is a complex ion with six H_2O ligands and a coordination number of twelve. Each water ligand donates a pair of electrons to the iron ion.

C   It is a complex ion with six H_2O ligands and a coordination number of six. Each water ligand donates an electron to the iron ion.

D   It is a complex ion with six H_2O ligands, each with a 2+ charge. Each water ligand donates a pair of electrons to the iron ion.

28. Cis-platin is a very effective drug against cancer. Drugs acting in the body do so by various mechanisms.

How does cis-platin exert its effects?

A   It helps in the destruction of cancer cells by acting as a cellular toxin.

B   Cancer cells cannot reproduce in the presence of cis-platin because it deprives the cells of oxygen and glucose.

C   Once cis-platin enters a cancer cell, it will lose 90% of its water resulting in dehydration and this prevents reproduction of the cell.

D   It binds to the DNA of cancer cells preventing them from reproducing.

29. A ligand substitution reaction is one where a ligand is replaced by another ligand.

These reactions can lead to the formation of coloured products.

Which of the following correctly describes the colour and state of these complex ions?

	$[CuCl_4]^{2-}$(aq)	$[CoCl_4]^{2-}$(aq)	$[Cu(NH_3)_4(H_2O)_2]^{2+}$(aq)
A	Yellow solution	Blue precipitate	Deep blue solution

B	Yellow solution	Blue solution	Deep blue solution
C	Blue solution	Yellow solution	Light blue solution
D	Deep blue solution	Blue solution	Yellow solution

30. The equation shown is that of a ligand substitution reaction:

    $[Cu(H_2O)_6]^{2+}$ (aq) + $4NH_3$ (aq) $\rightleftharpoons$ $[Cu(NH_3)_4(H_2O)_2]^{2+}$(aq) + $4H_2O$ (l)

    Which two of the following statements are **not** true?

    1: $[Cu(NH_3)_4(H_2O)_2]^{2+}$ is an octahedral complex ion.

    2: On addition of small amounts of ammonia, a pale blue precipitate is observed. When excess ammonia is added, the pale blue precipitate dissolves forming a deep blue solution.

    3: On addition of small amounts of ammonia, a deep blue colour is observed. When excess ammonia is added, a pale blue precipitate forms.

    A   2 and 3

    B   1 and 3

    C   3 only.

    D   1 only.

31. A solution of iron (II) ions may be oxidised using MnO_4^- ions. Which of the following statements is not true?

    A   The oxidation state of Mn in MnO_4^- is +7. At the end point of the reaction, a permanent pink colour will be observed.

    B $KMnO_4$ is usually contained in the burette during titrations.

    C   Iron (II) solution is usually placed in the conical flask during titrations.

    D   Initially, as MnO_4^- ions are added from the burette to the conical flask containing the iron (II) solution, an orange colour will be observed.

32. Balance the following equation:

    MnO_4^- (aq) + aH^+ (aq) + bFe^{2+} (aq) $\rightarrow$ Mn^{2+} (aq) + cFe^{3+} (aq) + dH_2O (l)

    A   a = 8, b = 5, c=2, d=4

    B   a = 4, b = 5, c=5, d=4

    C   a = 8, b = 2, c=2, d=4

    D   a = 8, b = 5, c=5, d=4

5 QUESTIONS — Physical Chemistry and Transition Elements

33. Which of the following correctly shows the formula of a thiosulfate ion?

    A $S_2O_3^{2-}$

    B $TlSO_4^{2-}$

    C $TiSO_4^{2-}$

    D $NaSO_4$

34. The following sentences are about haemoglobin and oxygen transportation

    1:   Fe^{2+} in haemoglobin forms a coordinate bond with an oxygen molecule facilitating the transport of oxygen.

    2:   Haemoglobin is found in blood.

    3:   Oxygen binds more strongly than carbon monoxide to haemoglobin.

    Which of the above are true?

	Statement 1	Statement 2	Statement 3
a	True	True	False
b	True	True	True
c	False	True	False
d	False	True	True

35. Hexadentate ligands consist of:

    A   Six pairs of electrons.

    B   Six lone pairs of electrons.

    C   Six metal ions.

    D   Six oxygen atoms.

36. One of the ligands is often referred to as "en". Which of the following about this ligand is not true?

    A   It is a bidentate ligand.

    B   It always donates only two electrons to a metal ion.

    C   The chemical name of en is ethane-1,2-diamine.

    D   En donates two lone pairs of electrons to a metal ion.

37. What is the best interpretation of $[Cu(H_2O)_6]^{2+}$?

A   A compound consisting of copper and water.

B   Concentration of $Cu(H_2O)_6$.

C $[Cu(H_2O)_6]^{2+}$ is a complex ion.

D   Concentration of $[Cu(H_2O)_6]^{2+}$.

38. Methanol may be used instead of hydrogen in fuel cell vehicles. What is an advantage of using methanol instead of hydrogen?

    A   It does not contain carbon and therefore does not produce carbon dioxide.

    B   Methanol in liquid form is easier to store than hydrogen.

    C   Methanol is easier to transport due to its lower weight.

    D   Methanol powered vehicles always generate more power than hydrogen powered vehicles.

39. Which of the following statements are true?

    1:     Adsorption involves molecules lodging inside a material.

    2:     Adsorption involves molecules attaching to the surface of a material.

    3:     Some catalysts work by adsorbing the reactants.

    A   2 and 3

    B   1 and 3

    C   Only 2

    D   Only 1

40. Which option best describes the apparatus shown?

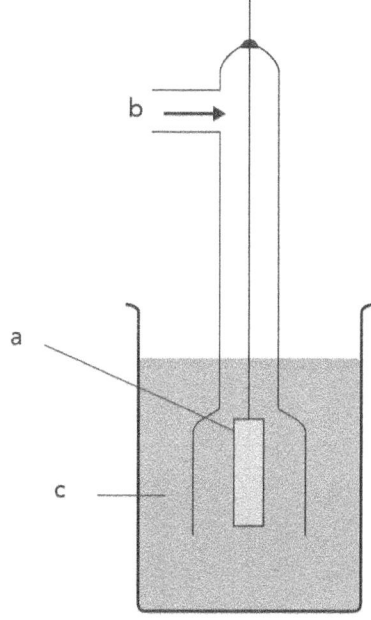

Choose a label for a, b, c and the best description for the above diagram

	The diagram shown is	a	b	c
A	A hydrogen half-cell	Salt bridge	Water in	Alkaline solution
B	A hydrogen half-cell	Magnesium electrode	Hydrogen gas in	Solution of H^+ ions
C	A hydrogen half-cell	Magnesium metal	Water in	Salt solution
D	A hydrogen half-cell	Platinum electrode	Hydrogen gas in	Solution of H^+ ions

41. The Born-Haber cycle for a compound is shown below:

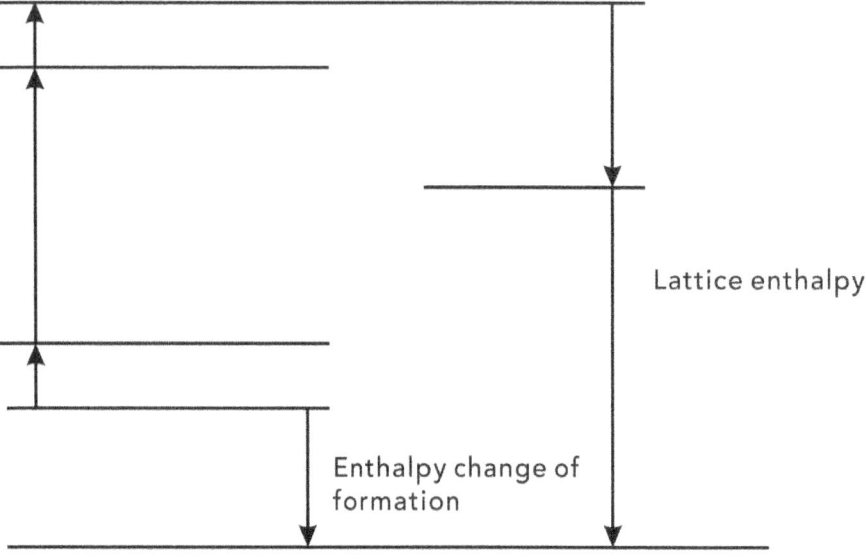

The arrows shown represent the enthalpy change of formation, standard enthalpy changes of atomisation, first ionisation energy, first electron affinity and lattice enthalpy.

Physical Chemistry and Transition Elements
QUESTIONS 5

What might this compound be?

A calcium chloride

B magnesium oxide

C lithium fluoride

D lithium oxide.

42. One of the statements given below is true. Select the statement that is true:

A A chloride ion has a more negative enthalpy change of hydration than a bromide ion.

B Lattice enthalpy may be exothermic or endothermic.

C Standard enthalpy change of hydration is the energy change that takes place when a mole of solid ions is dissolved in water.

D Entropy may take a positive or negative value.

43. Name the following (enthalpy) changes

1: $Ca(s) \rightarrow Ca(g)$ $\Delta H = +178$ kJmol^{-1}

2: $Fe^{2+} \rightarrow Fe^{3+} + e^-$

3: $Al^{3+}(g) + aq \rightarrow Al^{3+}(aq)$ $\Delta H = -4613$ kJmol^{-1}

	1	2	3
A	Enthalpy change of formation	Oxidation	Enthalpy change of hydration
B	Enthalpy change of atomisation	Oxidation	Enthalpy change of hydration
C	Oxidation	Reduction	Enthalpy change of hydration
D	Enthalpy change of atomisation	Reduction	Enthalpy change of solution

QUESTIONS
Physical Chemistry and Transition Elements

44. Titration curves of acid-base reactions can take different shapes depending on the strength of the acid and the base

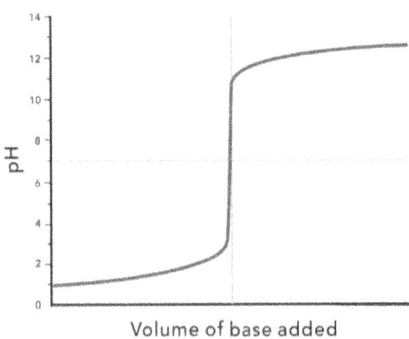

Volume of base added

Which option best describes the graph shown above?

A   Concentrated HCl - concentrated NaOH titration curve

B   Concentrated H_2SO_4 - aqueous NH_3 titration curve

C   Concentrated CH_3CH_2COOH - concentrated $Ca(OH)_2$ titration curve

D   Concentrated CH_3COOH - aqueous NH_3 titration curve

45. Rate-concentration and concentration-time graphs for zero order, first order and second order reactants take different forms.

1:
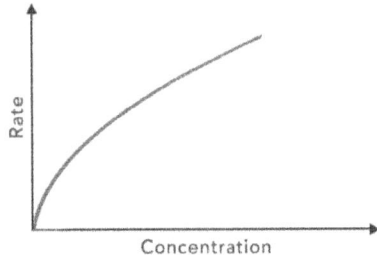

2:
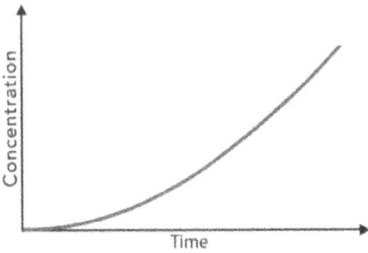

3:
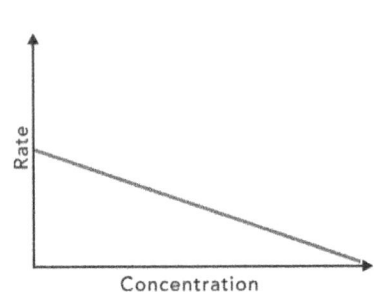

Which of the above graphs aren't true for either zero, first or second order reactants?

A   Only 1

B   Only 2

C   Only 3

D   1,2 and 3

46. A sample of hydrochloric acid has a concentration of 1.35×10^{-3} mol dm^{-3}. What is its pH?

A   2.57

B   6.61

C   2.87

D   0.13

47. A solution of Ca(OH)$_2$ has a concentration of 6.50×10^{-2} mol dm^{-3}. What is its pH?

A   13.11

B   12.80

C   1.19

D   12.50

48. A student carried out a reaction at different temperatures and plotted a graph of ln k (k / s^{-1}) against (1 / T)/ K^{-1}.

Using the graph, what is the activation energy for the reaction?

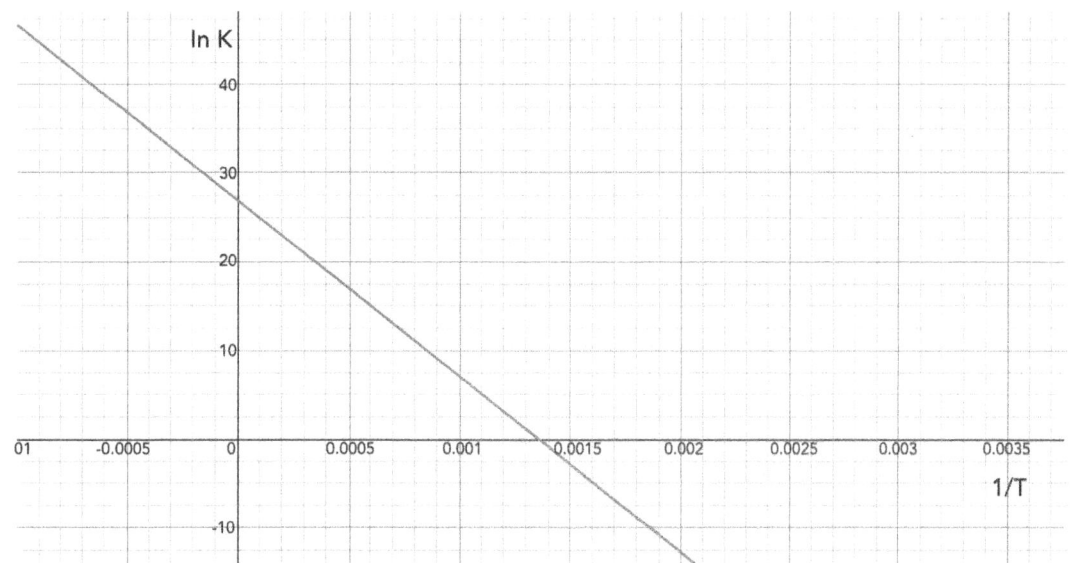

QUESTIONS
Physical Chemistry and Transition Elements

A 160 kJ mol⁻¹

B 90 kJ mol⁻¹

C 250 kJ mol⁻¹

D 300 kJ mol⁻¹

49. Which reaction contains the smallest change in oxidation state, that is not zero, for phosphorus?

 A $P_4O_{10} + 6H_2O \rightarrow 4H_3PO_4$

 B $2Ca_3(PO_4)_2 + 6SiO_2 + 10C \rightarrow P_4 + 10CO + 6CaSiO_3$

 C $P_4 + 10Cl_2 \rightarrow 4PCl_5$

 D $P_4 + 6Br_2 \rightarrow 4PBr_3$

50. This question is about the feasibility of reactions.

    The following data relates to three reactions that were carried out. Please note that the entropy here has been given in kJ K⁻¹ mol⁻¹. It would usually be given in J K⁻¹ mol⁻¹.

    Which of the following reactions **cannot** happen spontaneously?

	ΔH / kJ mol⁻¹	T / K	ΔS / kJ K⁻¹ mol⁻¹
1	-65.0	305	-2.43 x 10⁻¹
2	200	296	3.86 x 10⁻¹
3	-300	340	-8.89 x 10⁻¹

 A 1 only

 B 1 and 3 only

 C 2 only

 D 1, 2 and 3

51. The following apparatus was used to determine the enthalpy change of combustion of an alcohol.

QUESTIONS 5

Physical Chemistry and Transition Elements

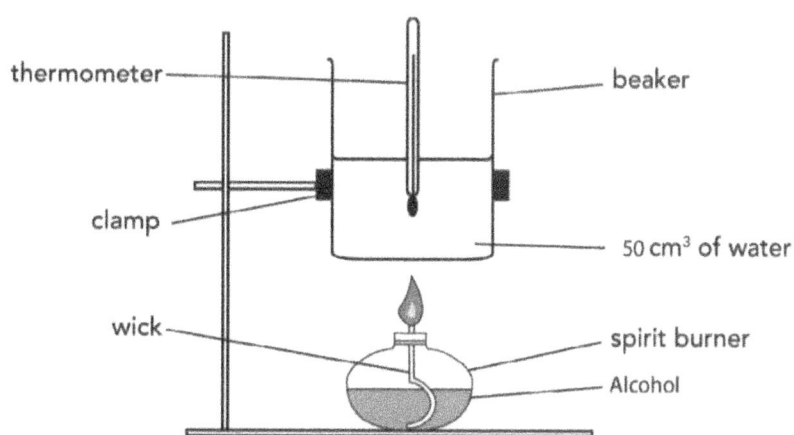

0.0200 moles of alcohol was burnt to heat up the water. Which of the following is a not possible value for the enthalpy change of combustion of the alcohol?

A   -836 kJ mol^{-1}

B   -941 kJ mol^{-1}

C   -1150 kJ mol^{-1}

D   -209 kJ mol^{-1}

52. Scandium hydroxide, Sc(OH)$_3$ is a strong base. A student was given a sample of concentration 3.50 x 10^{-2} mol dm^{-3}. What is its pH?

A   12.54

B   13.02

C   12.85

D   0.98

53. The following rate-determining step was proposed for a reaction:

$$2A + B \rightarrow E + 2F$$

Which of the following statement(s) is / are **not** true?

1:   The overall reaction could only have A and B as the reactants.

2:   The overall order of the reaction is 3.

3:   The rate equation of the reaction is rate = k2[A][B]

A   Only 1

B   Only 2

C   Only 3

D   1 and 3

54. A student used the expression $[H^+(aq)] = \sqrt{K_a \times [HA(aq)]}$ to calculate the pH of a "stronger" weak acid. Which of the following reasons explain why his answer may not be valid?

    A   The expression - $[HA]_{equilibrium} \sim [HA]_{undissociated}$ - may not be true

    B   The acid was a monobasic acid

    C   "Stronger" weak acids always fully dissociate in water

    D K_a values do not apply for "stronger" weak acids.

55. Which of the following combinations may not result in a buffer solution?

    A   25 cm^3 of 0.092 mol dm^{-3} ethanoic acid and 20 cm^3 of 0.070 mol dm^{-3} NaOH

    B   25 cm^3 of 0.050 mol dm^{-3} ethanoic acid and 20 cm^3 of 0.070 mol dm^{-3} sodium ethanoate

    C   25 cm^3 of 0.050 mol dm^{-3} ethanoic acid and 20 cm^3 of 0.070 mol dm^{-3} NaOH

    D   20 cm^3 of 0.070 mol dm^{-3} ethanoic acid and 25 cm^3 of 0.050 mol dm^{-3} sodium ethanoate

56. A student carried out two tests on two samples of un unknown **compound X**.

    Test 1: Warm the mixture with aqueous NaOH. The gas produced turned damp red litmus paper blue.

    Test 2: Add dilute nitric acid, followed by silver nitrate solution. A yellow precipitate was observed.

    Which of the following correctly identifies **compound X**?

    A NH_4Br

    B NH_4I

    C $CuBr_2$

    D CuI_2

57. Which of the following show *cis-trans* isomerism?

    1:    $[Pt(en)_2]^{2+}$

    2:    $[Co(NH_3)_4Cl_2]^+$

    3:    $Br(CH_3)C=CBr_2$

    A   Only 1

B   Only 2

C   Only 3

D   1 and 3

58. Which of the following statement(s) is / are true?

$$Ca^{2+}(aq) + 2e^- \rightleftharpoons Ca(s) \quad E^\ominus = -2.76\ V \qquad Br_2(l) + 2e^- \rightleftharpoons 2Br^-(aq) \quad E^\ominus = 1.07\ V$$

1:      $Ca + Br_2 \rightarrow Ca^{2+} + 2Br^-$: this reaction is feasible

2:      $2Br^- + Ca^{2+} \rightarrow Ca + Br_2$: this reaction is feasible

3:      If an electrochemical cell was made using these two half-cells, the Ca/Ca^{2+} half-cell would be oxidised.

A   Only 1

B   Only 2

C   Only 3

D   1 and 3

59. The following gaseous equilibrium was allowed to reach equilibrium in a closed container, so there were about equal amounts of reactants and products and the temperature was raised to 720 K. K_c is 52.1 at 720 K.

$$H_2\ (g) + I_2\ (g) \rightleftharpoons 2HI\ (g)$$

Which of the following statement is true?

A   The concentration of HI decreased

B   The concentration of H_2 increased

C   The concentration of HI increased

D   Changes in temperature does not affect pressure

60. The 1st to 8th successive ionisation energies, in kJ mol^{-1}, of an element are:

    780   1600   3200   4500   16100   19900   24000   29000

What is the element?

A   Al

B   P

C   Be

D   Si

QUESTIONS

61. The rate equation of a reaction is : rate = k [Br_2]. The half-life of Br_2 was found to be 25 seconds. What is the rate of reaction when the concentration of Br_2 is 0.05 mol dm^{-3}?

    A   5.99 x 10^{-3}

    B   17.3

    C   1.25

    D   1.39 x 10^{-3}

62. Which of the following statements about titrations are **not** true?

    1:    An indicator with a wide pH range is often used for weak acid / weak base titration.

    2:    The indicator is considered a weak base in titrations.

    3:    0.0630 mol dm^{-3} of HCl of 20 cm^3 will react exactly with 0.0315 mol dm^{-3} of NaOH of 40 cm^3.

    A   Only 1

    B   Only 2

    C   1 and 2

    D   2 and 3

63. A student was required to measure the pH of two separate solutions using a pH meter. The following steps were carried out:

    1:    Select the pH mode and set temperature to 25 °C.

    2:    Place the electrode in the solution containing pH 7 buffer. Allow the display to stabilize and set the display to read 7.

    3:    Place the electrode in the solution containing pH 2 buffer. Allow the display to stabilize and set the display to read 2.

    4:    Place the electrode in 1st sample to be tested and once reading is taken, place in the 2nd sample.

    The student did not follow the correct procedure. What could be done differently?

    A   Use only one buffer of either pH 7 or 2.

    B   Carry out the 3rd step before the 2nd.

    C   Rinse the electrode with deionized water between solutions and dry using tissue.

    D   Move the apparatus to a room with a lower temperature and lower the temperture on the meter.

64. You are provided with the following table.

Enthalpy change	Energy / kJ mol^{-1}
Lattice enthalpy of AgCl	-910
Enthalpy change of hydration of chlorine	-390
Enthalpy change of solution of AgCl	+45.6

Physical Chemistry and Transition Elements

QUESTIONS 5

What is the enthalpy change of hydration of silver, in kJ mol^{-1}?

A   -474

B   +474

C   -566

D   1346

65. This question is about the volatility of propan-1-ol and pentane.

    Which of the following options is correct?

A   Propan-1-ol is more volatile than pentane because its molecules can form hydrogen bonds.

B   Pentane is more volatile than propan-1-ol because pentane molecules induce dipoles on each other.

C   Propan-1ol is less volatile because its molecules can form hydrogen bonds with each other.

D   Propan-1-ol is less volatile because the molecule is shorter than pentane and therefore have more Van Der Waals forces between molecules.

66. Which of the following statement(s) is / are **not** true?

1:    A complex ion with a coordination number of four will always take a square planar shape.

2:    Complex ions can only show optical isomerism.

3:    The coordination number of [Ni(NH$_2$CH$_2$CH$_2$NH$_2$)$_3$]$^{2+}$ is 3.

A   Only 1

B   Only 2

C   Only 3

D   1, 2 and 3

67. Which of the following options could be true?

	pK$_a$		
	HNO$_3$	CH$_3$COOH	HCl
A	1.4	-4.7	6.3
B	-1.4	4.7	6.3
C	-1.4	-4.7	-6.3
D	-1.4	4.7	-6.3

68. 1.80 g of aqueous hydrated copper (II) ions were reacted with 1.70 g of aqueous sodium hydroxide. Which of the following statements is true?

A   There are 0.0283 moles of aqueous hydrated copper (II) ions.

B   0.860 g of sodium hydroxide will stay unreacted.

C   1.28 g of sodium hydroxide will stay unreacted.

D   0.0425 moles of sodium hydroxide react completely with the copper (II) ions.

69. 5.30 g of aqueous hydrated chromium (III) ions were reacted with an excess of aqueous ammonia until no further colour changes were observed. Which of the following statements are true?

    1:     3.38 g of aqueous ammonia is required to completely react with the chromium ions.

    2:     A green precipitate is formed at the end of the reaction.

    3:     The major product is a complex ion with two different ligands.

    A   Only 1

    B   Only 2

    C   2 and 3

    D   1 and 3

70. The following equilibrium is established if steam comes in contact with red hot carbon.

$$H_2O(g) + C(s) \rightleftharpoons H_2(g) + CO(g)$$

Which of the following correctly states the K_p expression for the above equilibrium?

A $$K_p = \frac{[H_2][CO]}{[H_2O][C]}$$

B $$K_p = \frac{[H_2][CO]}{[H_2O]}$$

C $$K_p = \frac{[H_2][CO]}{[C]}$$

D $$K_p = \frac{[H_2O]}{[H_2][CO]}$$

71. A student carried out a reaction at different temperatures and plotted a graph of ln k (k / s^{-1}) against (1 / T)/ K^{-1}.

Using the graph, what is the pre-exponential factor, A, for the reaction?

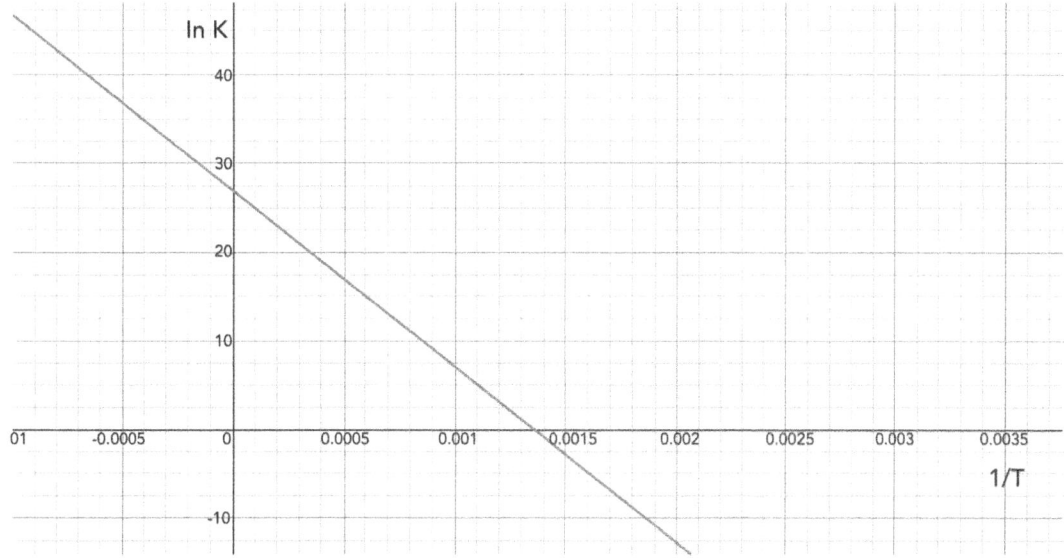

A 1.9×10^{-12}

B 5.3×10^{11}

C   3.3

D   27

MODULE 6

Organic Chemistry And Analysis

- Aromatic compounds
- Carbonyl compounds
- Carboxylic acids and esters
- Nitrogen compounds
- Polymers
- Organic synthesis
- Chromatography and Spectroscopy (NMR)

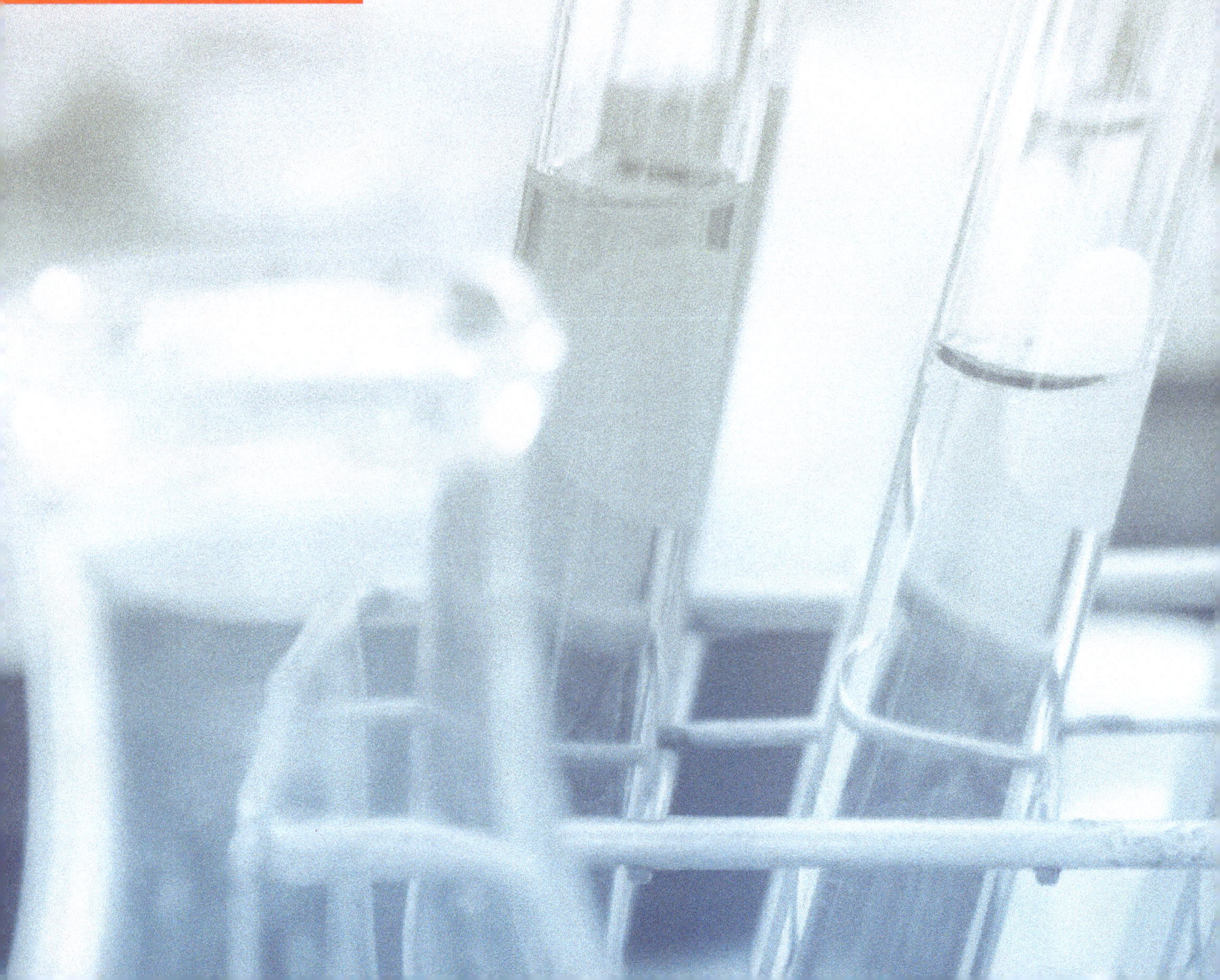

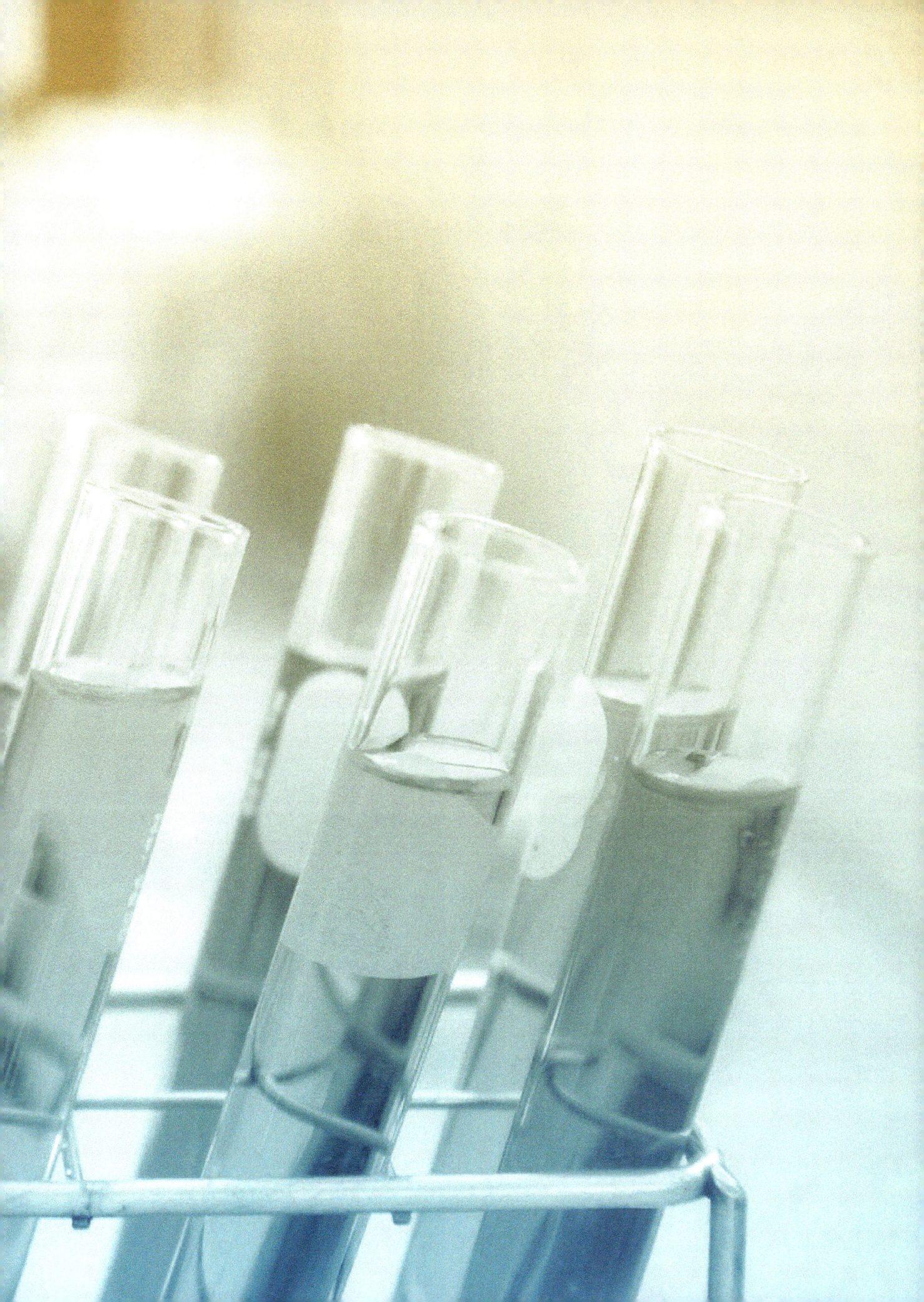

Revision Guide — Organic Chemistry and Analysis

Aliphatic and aromatic compounds

Aromatic compounds consist of a delocalised electron ring structure such as benzene, whereas aliphatic compounds do not.

Benzene and phenol are examples of aromatic compounds.

How many double bonds in benzene?

Benzene is represented as shown in figure 6.0.
There are no strict double bonds in benzene, as such. This is because the electron in p-orbital of each carbon atom overlaps with the electron in p-orbitals of other carbon atoms. These electrons are "delocalised" in that they do not belong to any one carbon; they form a cloud of delocalised electrons above and below the plane of the ring of carbon atoms.

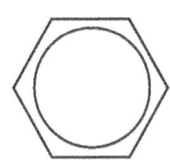

Figure 6.0 The delocalised conjugated structure of Benzene

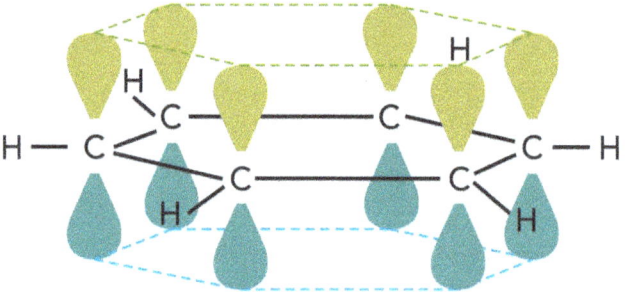

Figure 6.1 A diagram showing the p orbitals of the C atoms in Benzene and their overlap

Bonding around the C in benzene

Carbon has four electrons in its outer shell. In benzene, three out of the four form single bonds with the two other adjacent C atoms and the H atom. This leaves one electron unbonded in a 2p orbital above and below the plane of the carbon atoms. It is this electron that overlaps with other similar 2p orbital electrons of the adjacent C atoms.

Reactions of benzene

Benzene undergoes **electrophilic substitution reactions** as they **maintain the stability of the benzene ring** unlike addition reactions.

Benzene can be nitrated using a mixture of concentrated nitric acid and concentrated sulfuric acid as a catalyst (at a temperature of 50 °C). The reaction mechanism for the reaction is given below.

The equation for the formation of the nitronium ion (NO_2^+) is : $HNO_3 + H_2SO_4 \rightarrow NO_2^+ + HSO_4^- + H_2O$

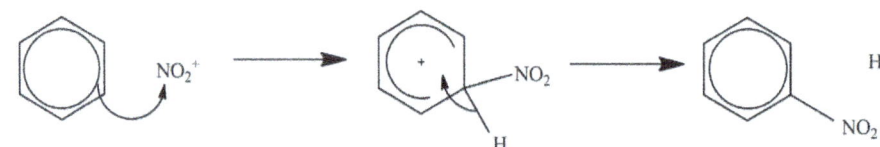

Organic Chemistry and Analysis — Revision Guide 6

The H⁺ formed at the end of the mechanism above reacts with HSO_4^- from the initial step to reform H_2SO_4. This illustrates the action of H_2SO_4 as a catalyst.

Benzene reacts with halogens in the presence of a halogen carrier in a similar way. The reaction mechanism is the same, except the NO_2^+ ion is replaced by a halogen ion such as Br^+.

Alkylation of benzene (Friedel-Crafts alkylation)

Benzene reacts at room temperature with a haloalkane (such as chloromethane) in the presence of a metal halide as a catalyst.

$$C_6H_6 + CH_3Cl \longrightarrow C_6H_5CH_3 + HCl$$

(Benzene also reacts with acyl chlorides in the presence of a catalyst and heat - Friedel-Crafts acylation)

Benzene

- A halogen carrier generates a powerful electrophile. This can then attract π electrons from benzene
- Halogenation: Reacts with halogens in the presence of a halogen carrier. Eg. $FeBr_3$
- Nitration: Benzene reacts with a mixture of conc. HNO_3 and conc. H_2SO_4, at 50°C
- Normally take part in substitution reactions and not addition
- Region of high electron density of benzene attracts electrophiles
- Benzene does not react like other alkenes
- preserves the delocalisation and stability of benzene ring
- Phenols consist of a benzene ring with an OH group attached directly to it

Delocalised model
- Cyclic hydrocarbon
- C_6H_6
- Shape around each C: trigonal planar (bond angle: 120°)
- 6 C atoms arranged in planar hexagonal ring
- Each C-C bond is the same length
- Electron in p-orbital of C atoms overlaps with electrons in p-orbitals of other C atoms = ring of delocalised electrons

- Benzene: delocalised π electrons spread over all six C atoms
  Alkenes: π electrons above and below the two C atoms in double bond
- Benzene: lower π electron density than alkenes
- There is insufficient π electron density in benzene to polarise a molecule such as bromine

Phenol

Phenols react differently to benzene due to the presence of the OH group. Remember, OH groups help molecules form hydrogen bonds with water. This may make the molecule soluble in water.

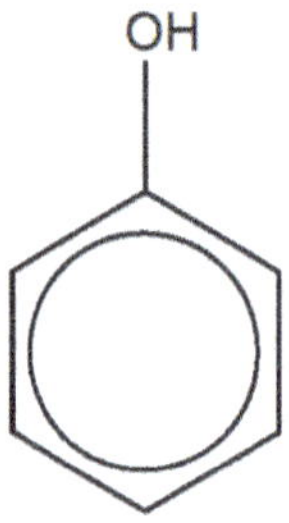

phenol

Phenol undergoes electrophilic substitution reactions relatively easily than benzene, because the lone pair of electrons on the O atom of the OH group is donated to the π-system of benzene. This increases the electron density of benzene's ring structure and enables phenol to polarise molecules relatively easily.

Phenol is weakly acid which means it will react with strong bases such as NaOH but not with carbonates.

$C_6H_5OH + NaOH \longrightarrow C_6H_5O^-Na^+ + H_2O$

Electron-donating and withdrawing groups

Phenol can also be nitrated like benzene. However, the reaction **does not require concentrated HNO_3** or the presence of a H_2SO_4 catalyst.

Phenols react with dilute nitric acid to form 2-nitrophenol.

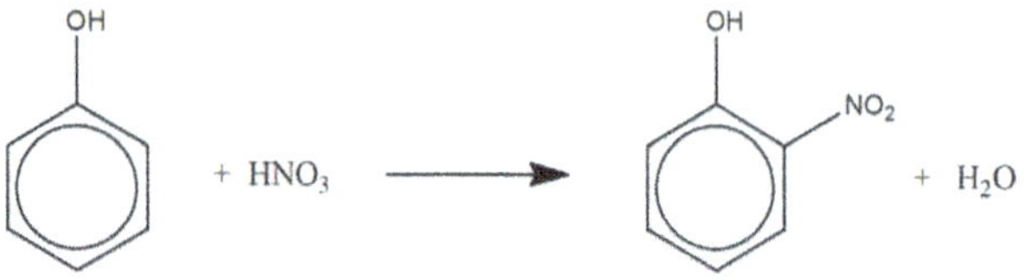

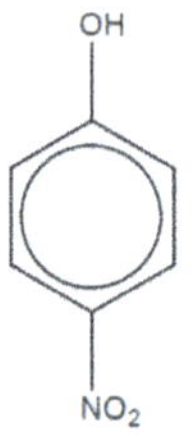

$C_6H_5OH + HNO_3 \longrightarrow$ 2-nitrophenol $+ H_2O$

4-nitrophenol

Note that the **OH** group is known as an **electron-donating group**. Electron-donating groups have a **2- and 4-directing effect**. This means the OH group directs the new group into the 2nd or the 4th position of the ring (In phenol, the carbon atom to which the OH group is attached is taken to be the 1st carbon. Nitration of phenol may also form 4-nitrophenol. Another example of an electron-donating group is **NH$_2$**.

Revision Guide 6
Organic Chemistry and Analysis

On the other hand, **NO₂** is an example of an **electron-withdrawing group**. Electron withdrawing groups have a **3-directing effect**.

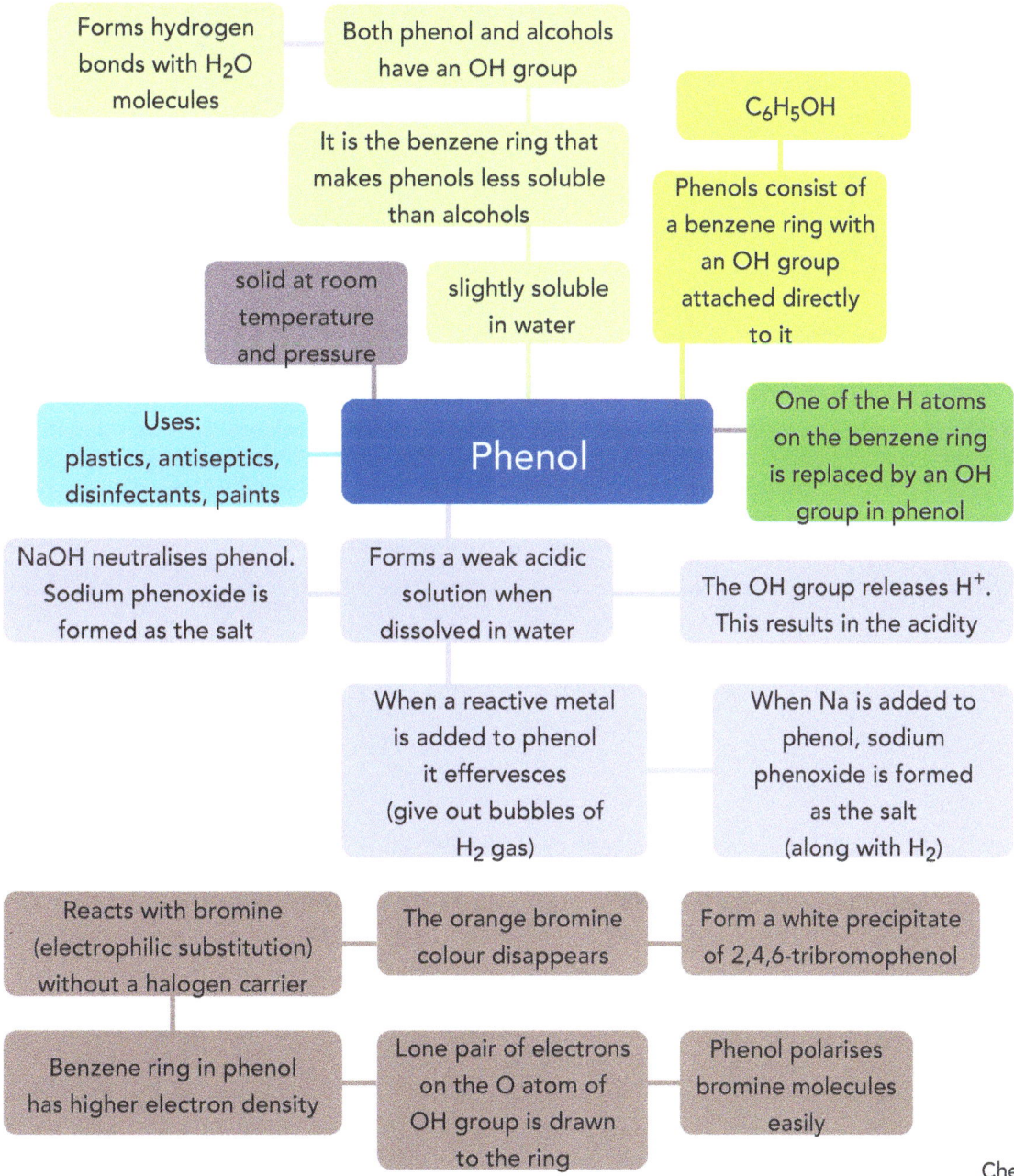

For your exam, you will not be required to know any further electron-donating or electron-withdrawing groups.

Note how the NO_2 group directs the incoming Br on to the 3rd carbon atom of the ring.

Phenol

- Forms hydrogen bonds with H_2O molecules
- Both phenol and alcohols have an OH group
- It is the benzene ring that makes phenols less soluble than alcohols
- C_6H_5OH
- Phenols consist of a benzene ring with an OH group attached directly to it
- solid at room temperature and pressure
- slightly soluble in water
- One of the H atoms on the benzene ring is replaced by an OH group in phenol
- Uses: plastics, antiseptics, disinfectants, paints
- NaOH neutralises phenol. Sodium phenoxide is formed as the salt
- Forms a weak acidic solution when dissolved in water
- The OH group releases H^+. This results in the acidity
- When a reactive metal is added to phenol it effervesces (give out bubbles of H_2 gas)
- When Na is added to phenol, sodium phenoxide is formed as the salt (along with H_2)
- Reacts with bromine (electrophilic substitution) without a halogen carrier
- The orange bromine colour disappears
- Form a white precipitate of 2,4,6-tribromophenol
- Benzene ring in phenol has higher electron density
- Lone pair of electrons on the O atom of OH group is drawn to the ring
- Phenol polarises bromine molecules easily

Revision Guide
Organic Chemistry and Analysis

Carbonyl compounds

The functional group (C=O) found in aldehydes and ketones is a carbonyl group. The name of an aldehyde end in -al (pentanal) and that of a ketone end in -one (pentanone).

Like alkanes and alkenes, aldehydes and ketones are homologous series. A homologous series is a series of organic compounds having the same functional group but with each successive member differing by CH_2.

> Study the definition of a homologous series

Oxidation of alcohols

Primary alcohols are alcohols where the OH group is bonded to a carbon atom that is bonded only to one other carbon. **Primary alcohols can be oxidised to aldehydes using acidified dichromate ions.** Acidified dichromate ions can be made from potassium dichromate, $K_2Cr_2O_7$ and sulfuric acid, H_2SO_4. The oxidising agent is shown as [O] in the following equations.

propan-1-ol + [O] → propanal + H_2O

Aldehydes can be further oxidised to carboxylic acids. Note that **when the aldehyde is being made, the aldehyde is distilled off** (removed) from the reaction mixture. This prevents it from being further oxidised to a carboxylic acid. **When making the carboxylic acid, the reaction mixture is usually heated under reflux** (this prevents any products escaping and ensures that any aldehyde produced in the primary oxidation is returned to the reaction mixture to undergo the second oxidation to a carboxylic acid).

propanal + [O] → propanoic acid

Secondary alcohols are alcohols where the OH group is bonded to a carbon atom that is bonded two other carbon atoms. **Secondary alcohols can be oxidised to ketones using acidified dichromate ions.**

Reduction of carbonyl compounds

Aldehydes are reduced to primary alcohols by sodium tetrahydridoborate(III), $NaBH_4$. The reducing agent is shown as [H] in the following equations.

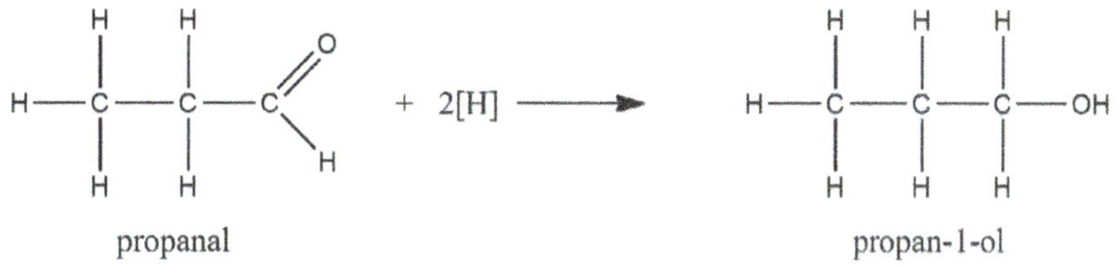

propanal + 2[H] ⟶ propan-1-ol

Ketones are reduced to secondary alcohols by $NaBH_4$.

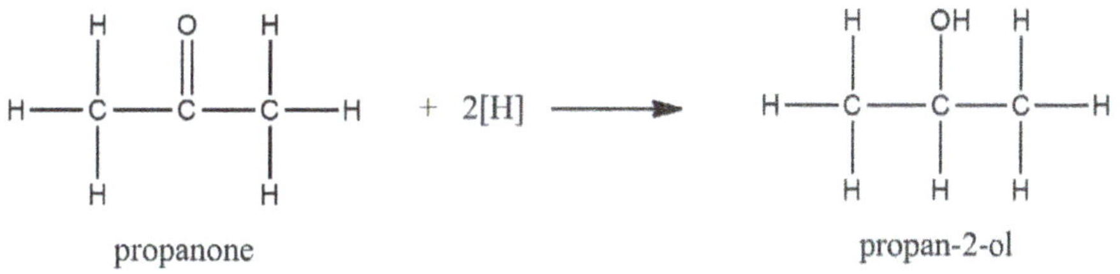

propanone + 2[H] ⟶ propan-2-ol

Aldehydes and ketones react with $NaBH_4$ in a nucleophilic addition reaction. The hydride ion, H^-, acts as the nucleophile.

The mechanism for the reduction of propanone is shown in the following figure.

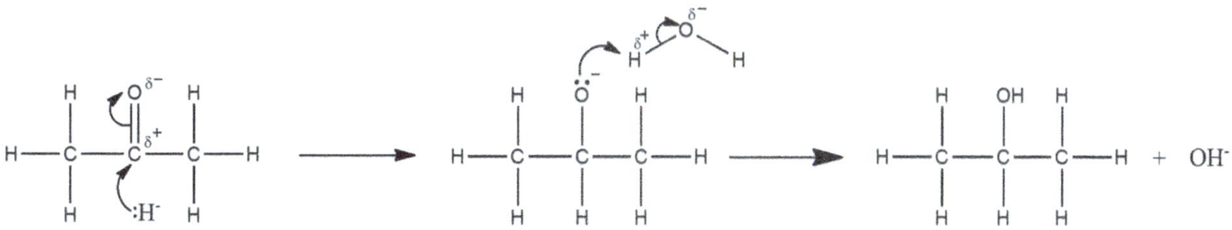

Revision Guide
Organic Chemistry and Analysis

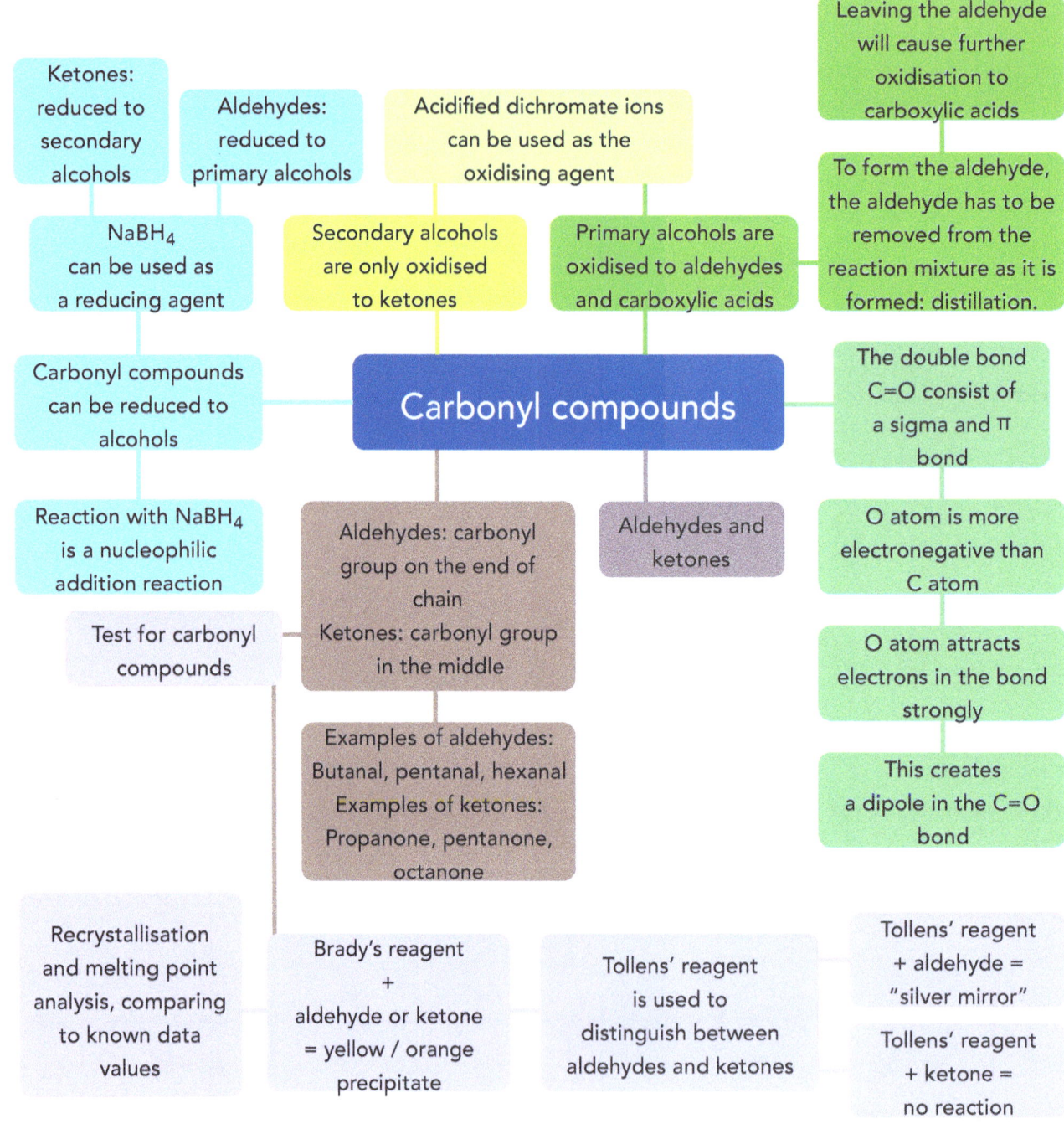

Functional groups and Chemical tests

Functional group - A group of atoms responsible for the characteristic reactions of a compound.

Alcohols have the -OH functional group. For example pentanol has the formula $C_5H_{11}OH$. It is often written this way rather than $C_5H_{12}O$, because it highlights the structure of the functional group.

Carboxylic acids have the -COOH functional group. Ethanoic acid, with the structural formula CH_3COOH, is an example. The same chemical formula may also be written as $C_2H_4O_2$. However, since this may be the same as another compound such as ethene-1,2-diol, writing the structural formula is more appropriate.

Test for carbonyl compounds

Aldehydes and ketones react with Brady's and Tollen's reagents as shown below.

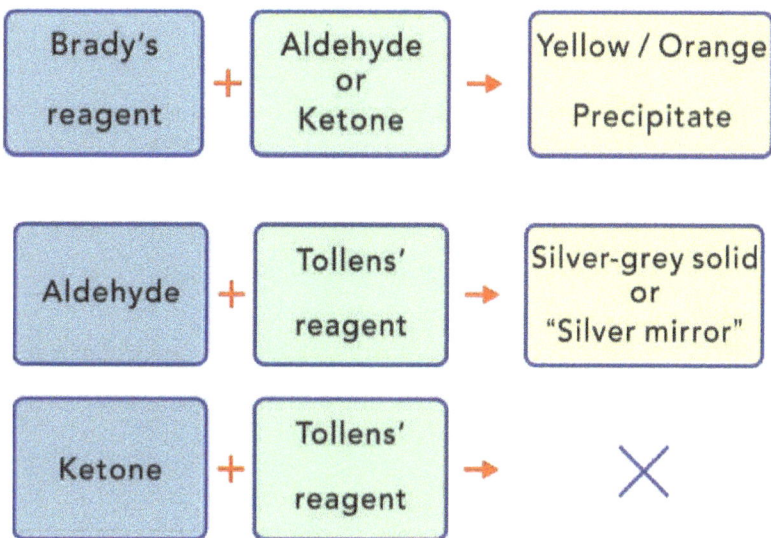

Figure 6.2 A diagram showing the reactions of Tollen's and Brady's reagents with Aldehydes and Ketones.

Test for alkenes

Shake or bubble an alkene through bromine water.
Colour changes from Orange (reddish-brown) to colourless.

Test for haloalkanes (also known as halogenoalkanes)

Warm the haloalkane with sodium hydroxide solution in a mixture of ethanol and water. Then add dilute nitric acid. Now add silver nitrate solution.

Different coloured precipitates will be seen depending on the haloalkane used. However as the colours are quite similar, identifying them can be difficult. Ammonia solution can be used to help with the identification.

Chloroalkane – White precipitate (Precipitate dissolves when dilute aqueous ammonia is added.)
Bromoalkane – Cream precipitate (Precipitate only dissolves when concentrated aqueous ammonia is added.)
Iodoalkane – Yellow precipitate (Precipitate does not dissolve with aqueous ammonia of any concentration.)

Revision Guide — Organic Chemistry and Analysis

Test for phenols

Bromine water reacts with phenols. The orange colour (of bromine) disappears and a white precipitate (2,4,6-tribromophenol) is formed.

OR

Reacts with sodium hydroxide solution resulting in a colourless solution
Does not react with sodium carbonate.

(Note that this shows that phenol is weakly acidic, but isn't acidic enough to react with carbonates. Acids such as HCl and HNO_3 would react with carbonates giving off carbon dioxide).

Test for primary alcohols, secondary alcohols and aldehydes

When acidified potassium dichromate is added to a primary alcohol, secondary alcohol or an aldehyde, the orange solution changes to a green solution.

Test for carboxylic acids

Effervescence (gas given off is CO_2) is seen when a carbonate is added to a carboxylic acid. Test the gas with limewater (turns milky).

Esterification reactions

These reactions result in the formation of an ester.

Carboxylic acids react with alcohols to form esters. When naming an ester, the first part of the name comes from the alcohol and the second part from the carboxylic acid. For example, butanoic acid reacts with ethanol to form ethyl butanoate.

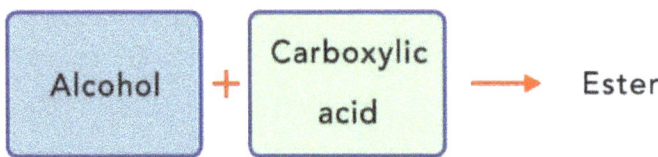

Figure 6.3 Alcohol and carboxylic acid react to form an

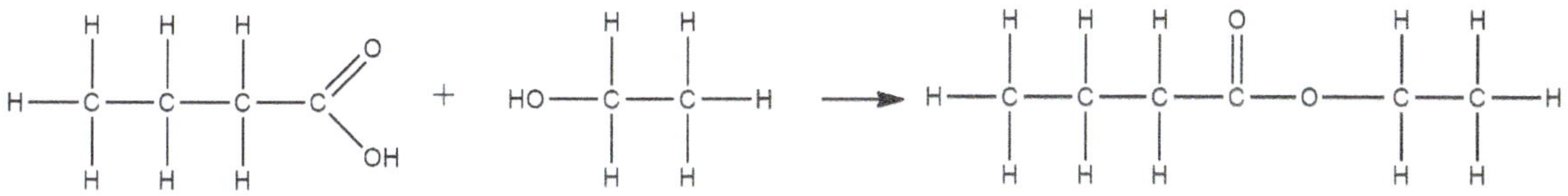

+ H_2O

Acyl chlorides react with alcohols to form esters. For example, pentanoyl chloride can react with ethanol to form ethyl pentanoate.

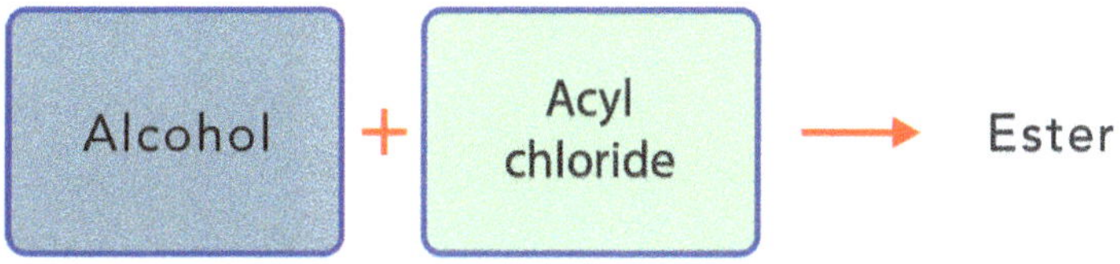

Figure 6.4 Alcohol and acyl chloride react to form an ester

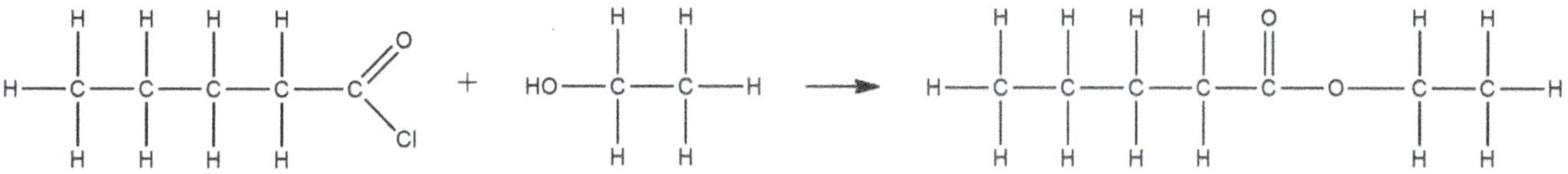

+ HCl

Phenols also react with acyl chlorides resulting in the formation of an ester. In the following example, phenol reacts with propanoyl chloride to form phenyl propanoate

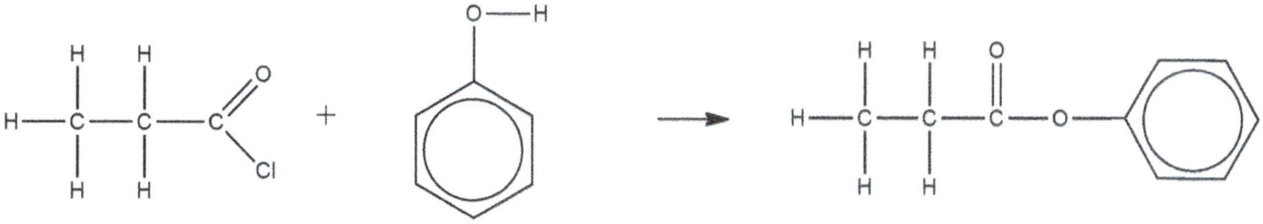

+ HCl

Esters can also be made by reacting an acid anhydride with an alcohol.

For example, butanoic anhydride reacts with propan-1-ol to form propyl butanoate as in the following example.

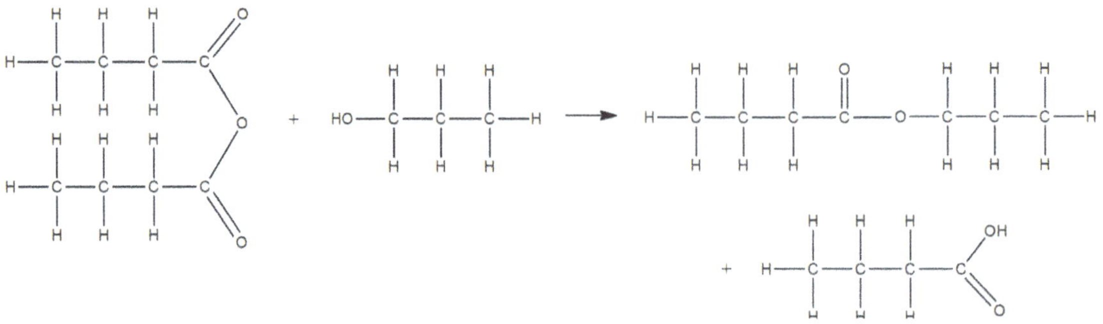

Hydrolysis of Esters

Esters can be hydrolysed by heating the ester with either an acid or an alkali. This is the opposite reaction to esterification.

Acid hydrolysis of an ester forms a carboxylic acid and an alcohol. The hydrolysis of ethyl propanoate is shown below.

$$CH_3CH_2COOCH_2CH_3(aq) + H_2O(l) \underset{}{\overset{H^+(aq)}{\rightleftharpoons}} CH_3CH_2COOH(aq) + CH_3CH_2OH(aq)$$

Base (alkali) hydrolysis of ethyl propanoate using sodium hydroxide is as follows.

$$CH_3CH_2COOCH_2CH_3(aq) + NaOH(aq) \rightarrow CH_3CH_2COONa(aq) + CH_3CH_2OH(aq)$$

Base hydrolysis of an ester forms the salt of the carboxylic acid and the alcohol.

Revision Guide 6

Organic Chemistry and Analysis

Carboxylic acids

Carboxylic acids are weak acids which means when they are added to water, they do not dissociate completely unlike strong acids.

Reacting carboxylic acids with metals and metal carbonates will result in the formation of hydrogen and carbon dioxide respectively. Effervesence can often be seen during these reactions.

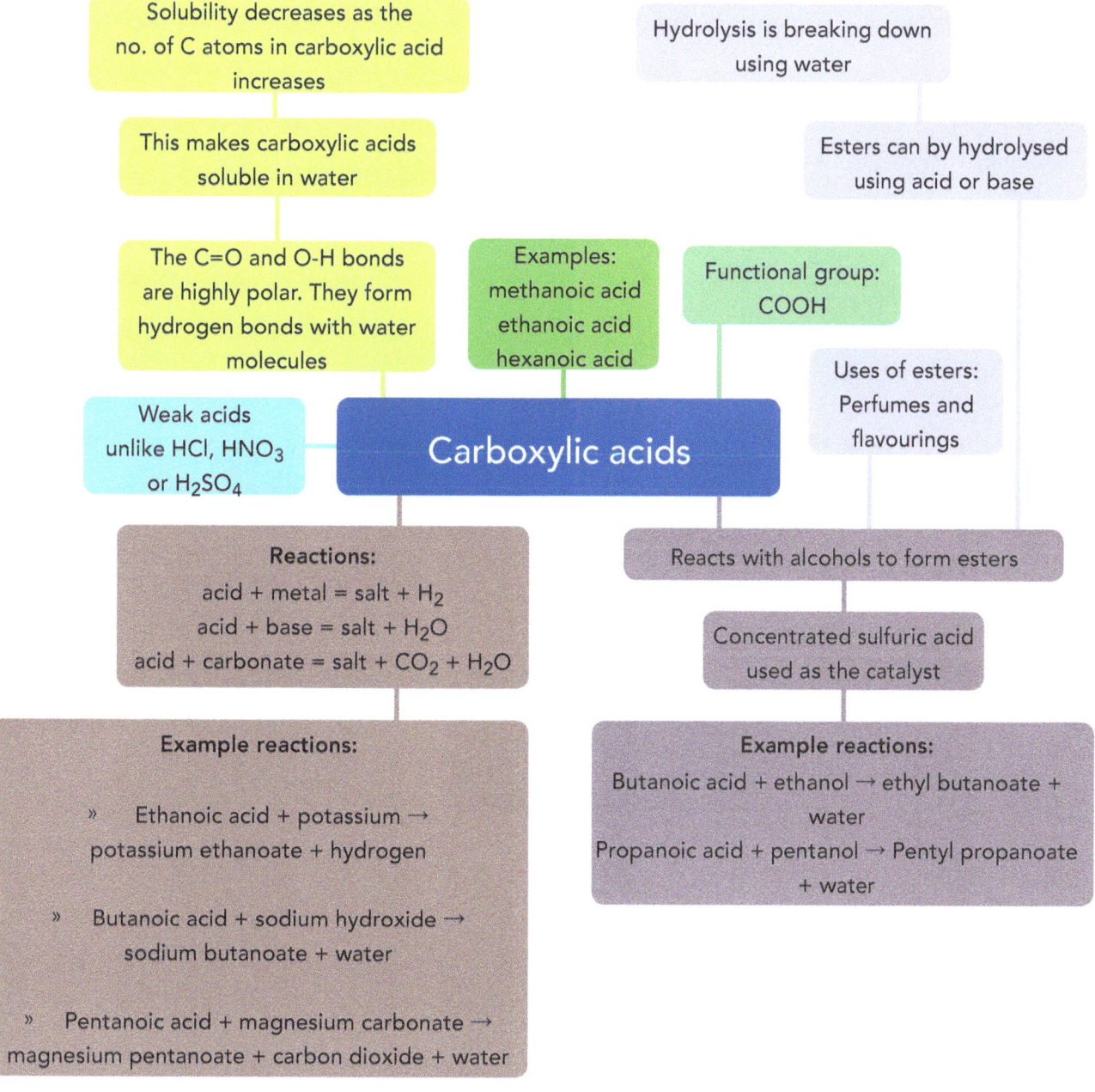

Chemistry Q & A | 127

Revision Guide — Organic Chemistry and Analysis

Amines

A **primary amine** is a molecule in which the **nitrogen atom** is **bonded** to only **one carbon** atom.
A **secondary amine** is a molecule in which the **nitrogen atom** is **bonded** to **two carbon** atoms.

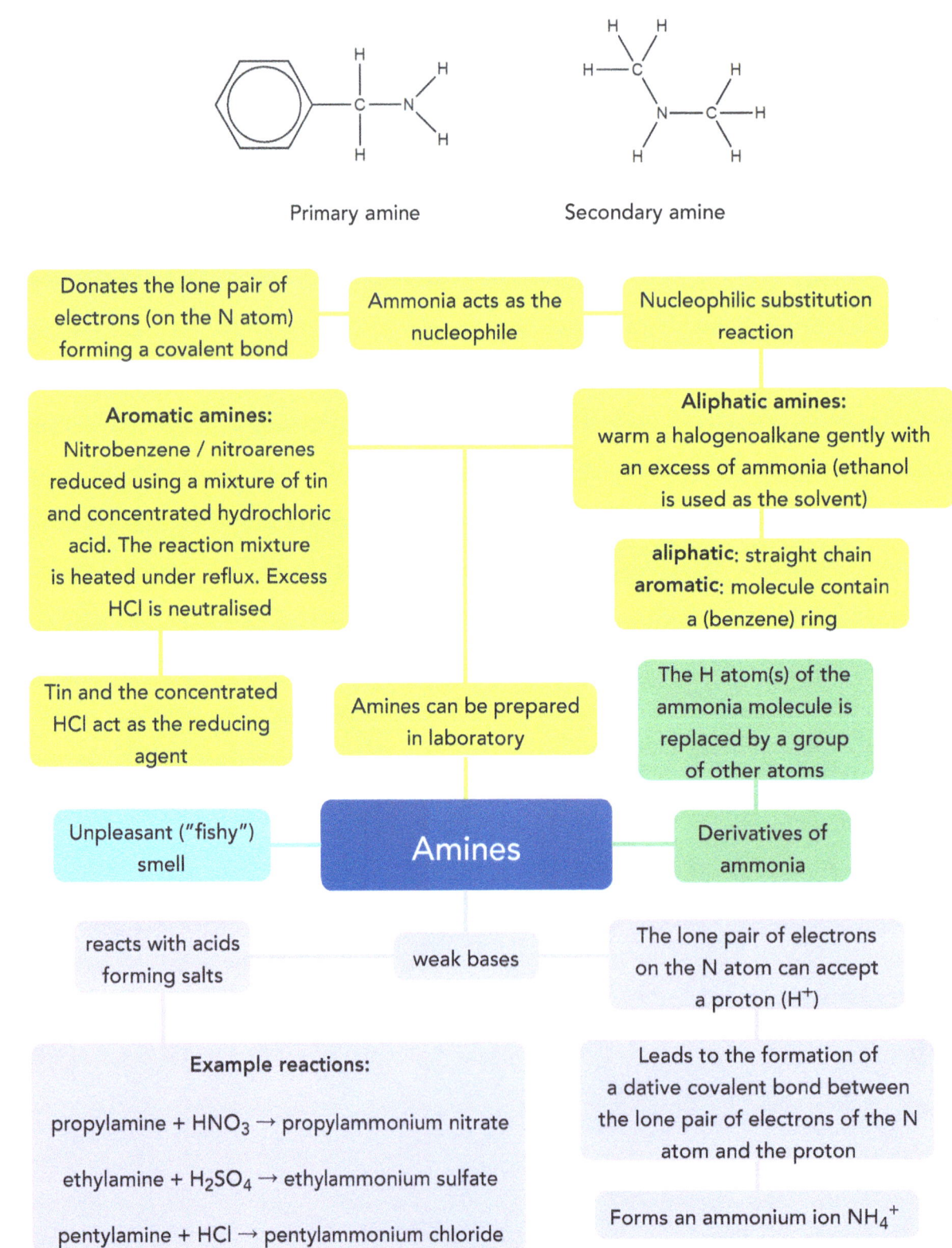

Primary amine   Secondary amine

- Donates the lone pair of electrons (on the N atom) forming a covalent bond
- Ammonia acts as the nucleophile
- Nucleophilic substitution reaction

Aromatic amines: Nitrobenzene / nitroarenes reduced using a mixture of tin and concentrated hydrochloric acid. The reaction mixture is heated under reflux. Excess HCl is neutralised

Aliphatic amines: warm a halogenoalkane gently with an excess of ammonia (ethanol is used as the solvent)

aliphatic: straight chain
aromatic: molecule contain a (benzene) ring

Tin and the concentrated HCl act as the reducing agent

Amines can be prepared in laboratory

The H atom(s) of the ammonia molecule is replaced by a group of other atoms

Unpleasant ("fishy") smell

Amines

Derivatives of ammonia

reacts with acids forming salts

weak bases

The lone pair of electrons on the N atom can accept a proton (H^+)

Leads to the formation of a dative covalent bond between the lone pair of electrons of the N atom and the proton

Forms an ammonium ion NH_4^+

Example reactions:

propylamine + HNO_3 → propylammonium nitrate

ethylamine + H_2SO_4 → ethylammonium sulfate

pentylamine + HCl → pentylammonium chloride

Organic Chemistry and Analysis — Revision Guide 6

Amino acids

α-amino acids have the general formula $RCH(NH_2)COOH$. The amino acid shown in figure 6.6 is Serine, with its R group as $CH_2(OH)$.

The carboxylic acid group of an amino acid can react with alkalis, whereas the amine group can react with acids. The reactions of the amino acid Alanine (whose R group is CH_3) is shown in figure 6.7.

Figure 6.5 A diagram showing the different functional groups of amino acids

Figure 6.6 Equations showing the reactions of the amino acid Alanine with a base and an acid

A **chiral carbon** is a carbon atom attached to four different atoms or groups of atoms. The central carbon atom in alpha amino acids, like Serine shown below, is chiral. Compounds with chiral centres show optical isomerism. **Optical isomers are non-superimposable mirror images about a chiral centre.** The two optical isomers of Serine is shown below.

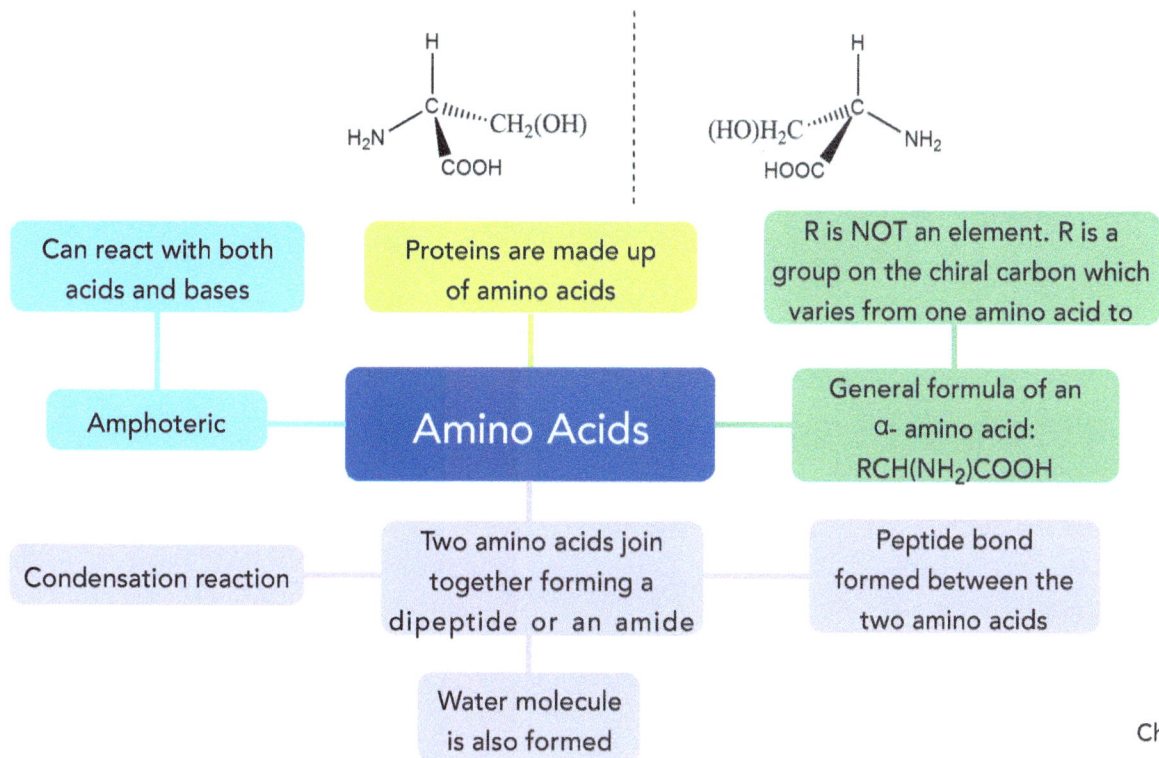

- Can react with both acids and bases
- Amphoteric
- Condensation reaction
- Proteins are made up of amino acids
- **Amino Acids**
- Two amino acids join together forming a dipeptide or an amide
- Water molecule is also formed
- R is NOT an element. R is a group on the chiral carbon which varies from one amino acid to
- General formula of an α- amino acid: $RCH(NH_2)COOH$
- Peptide bond formed between the two amino acids

Chemistry Q & A

Revision Guide — Organic Chemistry and Analysis

Polymers

Polyesters and polyamides

Polyamides are formed by **condensation polymerisation** when carboxylic acids / dicarboxylic acids react with amines / diamines. (A dicarboxylic acid has two carboxylic acid groups and a diamine has two amine groups as in the following example).

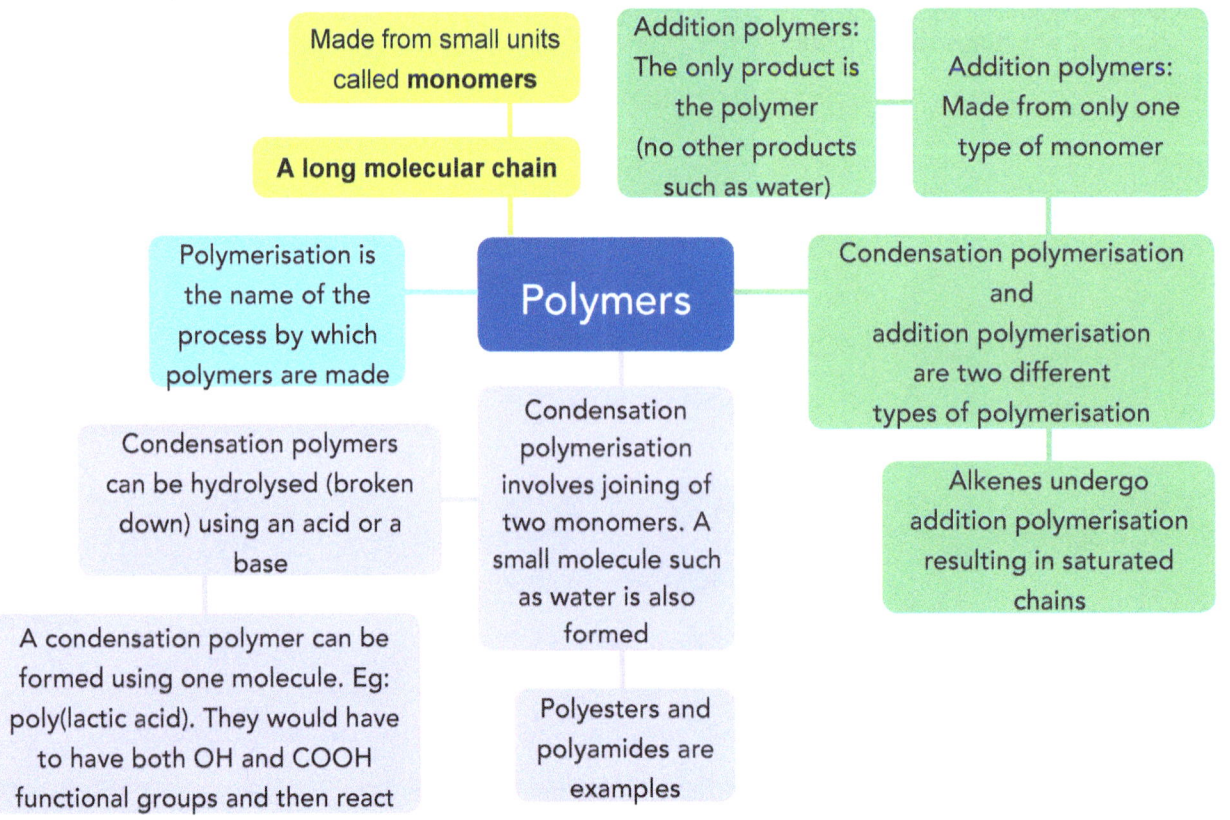

Polyesters are formed by **condensation polymerisation** when carboxylic acids / dicarboxylic acids react with alcohols / diols.

A photodegradable polymer breaks down when exposed to light for a long period of time. It can be made by adding C=O double bonds into the backbone of the polymer. C=O bonds absorb light and break, resulting in the breaking of the polymer chain.

- Made from small units called **monomers**
- A long molecular chain
- Polymerisation is the name of the process by which polymers are made
- Condensation polymers can be hydrolysed (broken down) using an acid or a base
- A condensation polymer can be formed using one molecule. Eg: poly(lactic acid). They would have to have both OH and COOH functional groups and then react
- Condensation polymerisation involves joining of two monomers. A small molecule such as water is also formed
- Polyesters and polyamides are examples
- Addition polymers: The only product is the polymer (no other products such as water)
- Addition polymers: Made from only one type of monomer
- Condensation polymerisation and addition polymerisation are two different types of polymerisation
- Alkenes undergo addition polymerisation resulting in saturated chains

Revision Guide 6

Organic Chemistry and Analysis

Gas chromatography

Gas chromatography is used to separate and identify volatile (easily evaporates) components in a mixture. A gas chromatogram gives two key pieces of information:

- Identity of each component - from the retention time.
- Amount of a compound in the sample - this is proportional to the area under each peak.

The retention time is dependent on the attraction of the compounds to the stationary phase. In the following chromatogram, the numbers next to the peaks, give the peak integration values. These indicate the peak areas. Using these integration values, the exact concentrations of the compounds in the sample can be determined. To do this, a calibration graph has to be plotted.

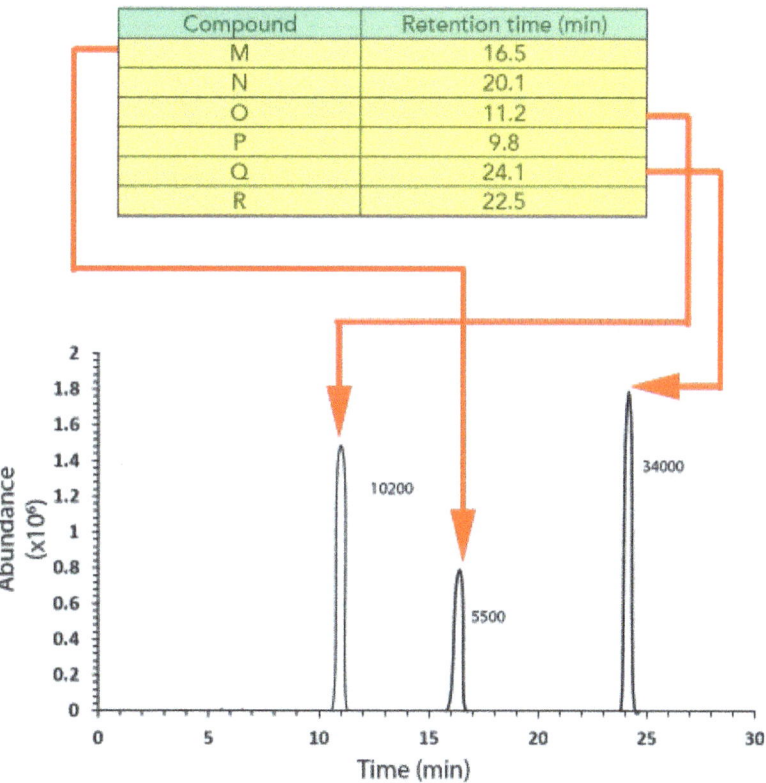

Figure 6.7 Gas chromatography

Limitations of Thin-Layer Chromatography and Gas Chromatography

Similar compounds can often have similar R_f values or retention times
Unknown compounds does not have R_f values or retention times which can be compared
Not all the components in the mixture will be separated

Revision Guide — Organic Chemistry and Analysis

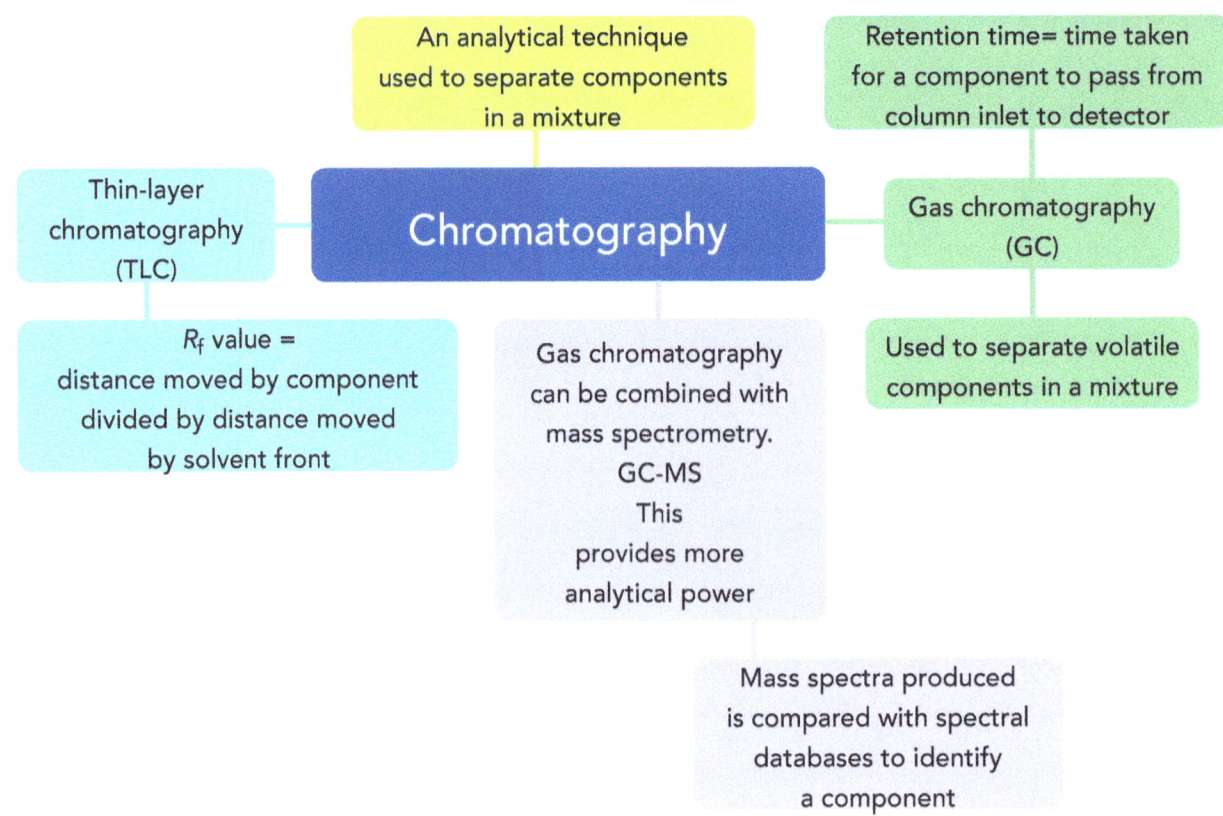

Proton NMR Spectroscopy

Unlike carbon-13 NMR spectra, proton NMR spectra may show spin-spin coupling patterns. When you observe one of these spectra, you will notice that each peak may consist of sub-peaks.

A proton NMR spectrum gives us four keys pieces of information:

- Number of different proton environments - from the number of peaks
- Type of proton environments in the compound - from the chemical shifts
- Proportions of protons in each environment - from the peak areas
- Number of protons bonded to the adjacent carbon atoms - from the spin-spin coupling patterns

The following figures show the proton NMR spectra of two compounds. The number next to each peak represents the integration values of the peak. The integration values indicate the area under the peaks which in turn tells us the relative ratio of protons in each environment. **Both of the following spectra were obtained with D_2O present. Therefore, any peaks due to OH protons will not be present.**

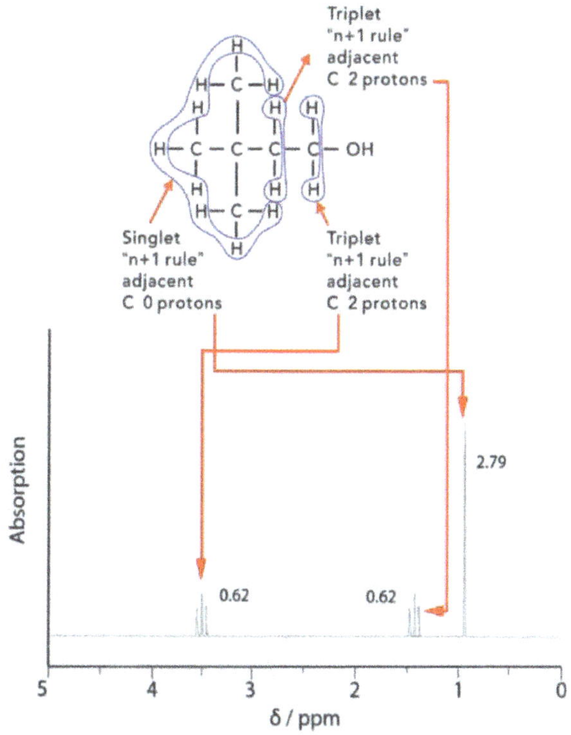

Organic Chemistry and Analysis — Revision Guide 6

Figure 6.8 Proton NMR of 3,3-dimethylbutan-1-ol

The integration values in this spectrum are:

0.62 : 0.62 : 2.79

The numbers can then be divided by the smallest number to give:

1 : 1 : 4.5

Multiplying the new set of numbers by two gives:

2 : 2 : 9

Note that, as these are ratios 0.62 : 0.62 : 2.79 = 1 : 1 : 4.5 = 2 : 2 : 9. If you multiply or divide numbers in a ratio by the same number, its meaning will not change.

In this example, these numbers are equal to the number of protons in each environment.

The integration value of the triplet at δ = 3.5 ppm is **0.62**. This was equal to **2**. This is because there are 2 protons in the proton environment responsible for the triplet.

The fact that peak is a triplet tells us that there are two protons bonded to the adjacent carbon.

The chemical shifts of the peaks also gives us valuable information about the structure of the molecule. Again, lets take the triplet at δ = 3.5 ppm as an example. If you look at the data sheet, 3.5 ppm corresponds to HC-O protons. This makes sense as the carbon atom to which those protons are bonded are attached to an OH group.

Note that the integration values in the spectrum are as follows:

0.52 : 0.52 : 0.26 : **1.56**

We can convert them to whole numbers by dividing by the smallest number which in this case is 0.26. This gives us:

2 : 2 : 1 : **6**

In this example, these numbers are equal to the number of protons in each environment. However, this is not always the case.

The integration value of the doublet at δ = 1.1 ppm is **1.56**. When all the numbers were divided by 0.26, this was equal to **6**. This is because there are 6 protons in the proton environment responsible for the doublet. The fact that peak is a doublet tells us that there is only one proton bonded to the adjacent carbon.

The peak at δ = 2.6 ppm is a multiplet (7 small peaks). This is because the adjacent carbons are bonded to 6 equivalent protons. This peak is due to the proton of the $HC(CH_3)_2$ group.

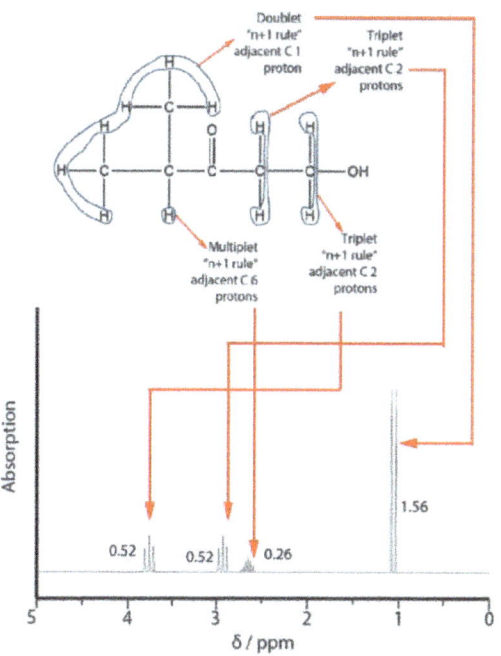

Figure 6.9 Proton NMR of 1-hydroxy,4-methylpentan-3-one

Chemistry Q & A | 133

Revision Guide
Organic Chemistry and Analysis

Proton environments

Figure 6.9 has three proton environments. Even though there are thirteen protons (hydrogen) in total, nine of them belong to the same environment, two to another and finally two to another different environment. We ignore the proton of the OH group as this proton would not produce a distinctive peak in a proton NMR spectrum. (As the proton NMR spectrum in figure 6.9 has been produced with D_2O, any peaks due to OH protons will not be present).

In figure 7.0, there are four different proton environments. There are eleven protons in total, excluding the one from the OH group. Six of these protons are in the same environment. Any given proton in this environment is bonded to a carbon atom that is bonded to $C(CH_3)H$. Two of the protons in the compound are in a different environment as they are bonded to a carbon that is bonded to OH and CH_2. Another two are bonded to a carbon that is bonded to C=O and CH_2 which places them in another environment. The final proton exists by itself in a separate environment with the peak at 2.6 ppm.

You will often get questions in the exam asking you to identify a peak or to come up with a structure for a compound. Having a structure to your answer will often help with these questions. Using the first peak in Figure 7.0 as an example:

1. State the location of the peak
2. Comment on the splitting pattern of the peak
3. Refer to the number of associated protons
4. Refer to the number of adjacent protons
5. Identify the peak shift from the data table and use all the information to determine its likely identity

The peak at 1.1 ppm is a doublet with an integral of 6. Therefore there are six protons in this environment and one proton adjacent. The shift of the peak corresponds to a CH-CH group which is likely to be a $CH(CH_3)_2$ group as this has 6 associated protons in the same environment $(CH_3)_2$ and one adjacent proton (CH).

Carbon-13 NMR Spectroscopy

A carbon-13 NMR spectrum gives the following information about a compound.

- Number of different carbon environments - from the number of peaks.
- Type of carbon environments - from the chemical shifts.

Organic Chemistry and Analysis
Revision Guide 6

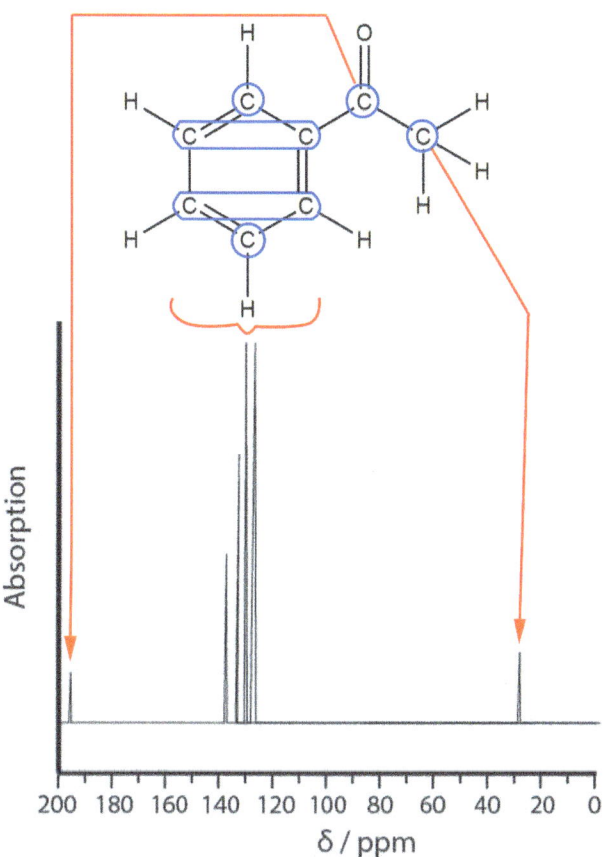

Figure 6.10

This compound has a total of six different carbon environments and therefore there are six peaks in the spectrum.

Four of these are within the benzene ring which corresponds to chemical shift values between 120 and 140 ppm.

One of the peaks is at $\delta = 197$ ppm. This peak is due to the carbon atom of the C=O group.

The other peak is at $\delta = 29$ ppm. The carbon atom from the CH_3 group is responsible for this peak.

Note that there is no splitting in carbon-13 NMR spectra, as carbon-13 has a low abundance.

Revision Guide
Organic Chemistry and Analysis

Mass spectroscopy

Mass spectrometry tells us the molecular weight of a compound and also provides information about the percentage by mass of each element present. It allows us to determine the empirical formula. Using the molecular weight of the compound and the empirical formula, the molecular formula can be found.

The mass spectrum of pentan-1-ol is shown in the following figure. The *m/z* values of other peaks gives us hints about the structure of the compound. These peaks are known as fragmentation peaks as they are formed from fragments of the main molecule.

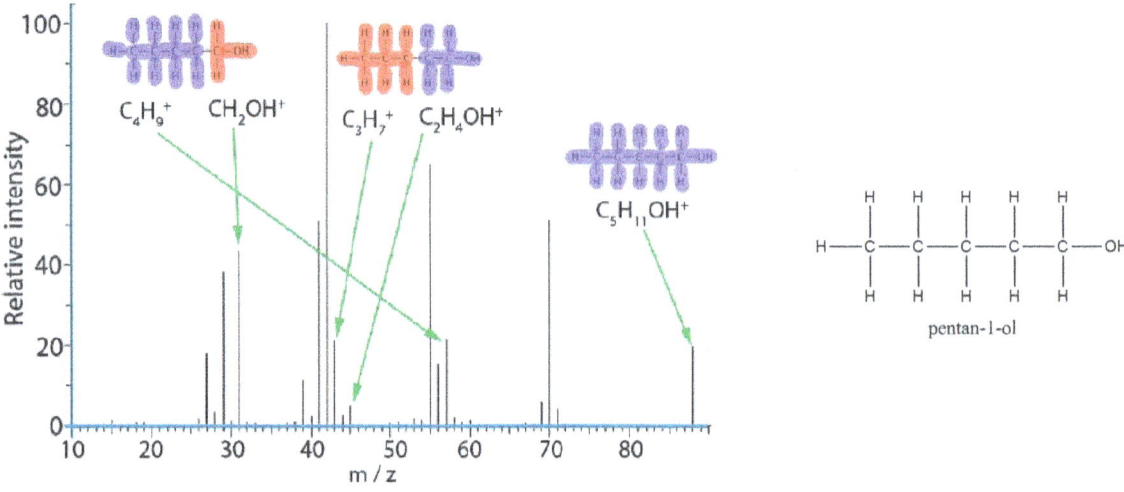

Figure 6.11 Mass spectrum of pentan-1-ol

For example, a fragment of CH_3 has a mass of 15. So, if the difference between a large fragment and the m/z of the molecular ion is 15, then a CH_3 has fragmented. Knowing the masses of some common fragments can make it easy to construct a good estimate of the molecular structure.

Some common fragment masses:

OH	17
CH_3	15
CH_2CH_3	27
$CH_2CH_2CH_3$	42
CH_2OH	31

You may be asked to write equations to explain fragmentation peaks. Two example equations are given below.

$C_5H_{12}O \rightarrow C_5H_{12}O^+ + e^-$ ✓   m/z of position ion formed = 88   (molecular ion peak).

$C_5H_{12}O^+ \rightarrow CH_2OH^+ + C_4H_9\cdot$ ✓ m/z of position ion formed = 31

Revision Guide 6

Organic Chemistry and Analysis

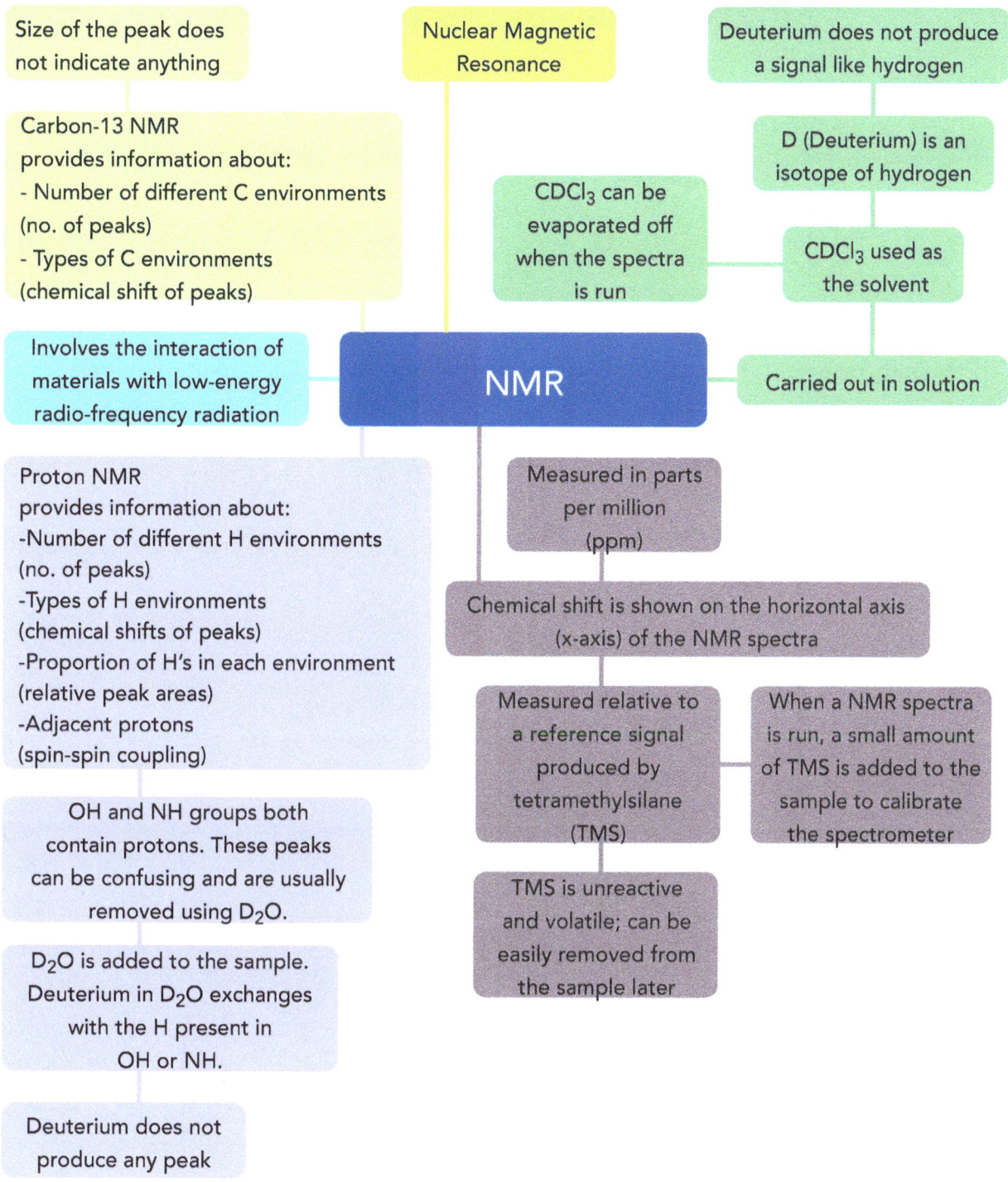

Reactions of hydrogen cyanide

When a haloalkane is heated under reflux with a solution of sodium or potassium cyanide in ethanol, a nitrile is formed. This is a nucleophilic substitution reaction.

The reaction mechanism for the reaction between chloropropane and the cyanide ion (CN⁻) is shown in the following figure. The resulting compound is butanenitrile. Note that the reaction mixture has to be **heated under reflux**.

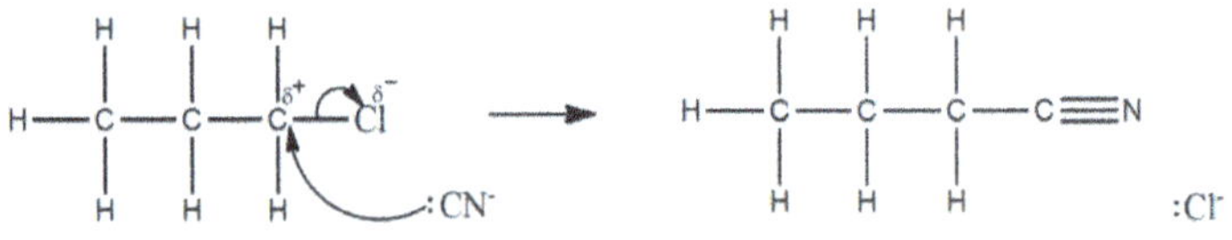

Carbonyl compounds react with HCN to form hydroxynitriles in a nucleophilic addition reaction.

The carbonyl compound is reacted with a mixture of NaCN or KCN and HCl. HCN is formed as follows:

$$NaCN + HCl \rightarrow HCN + NaCl$$

The following mechanism shows the reaction between ethanal and HCN to form 2-hydroxypropanenitrile.

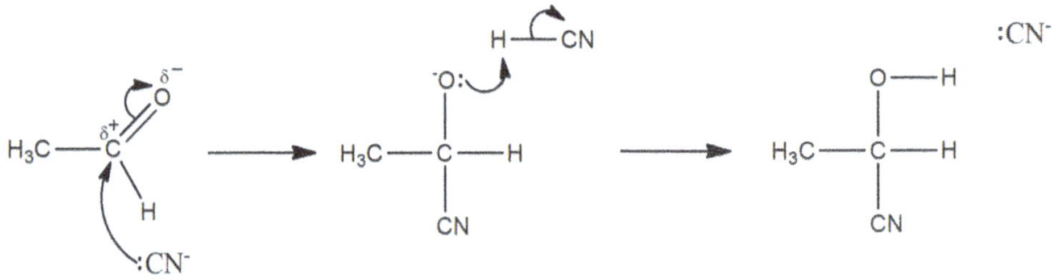

Reactions of nitriles

Nitriles can be reduced to form amines by reacting with hydrogen and using nickel as a catalyst

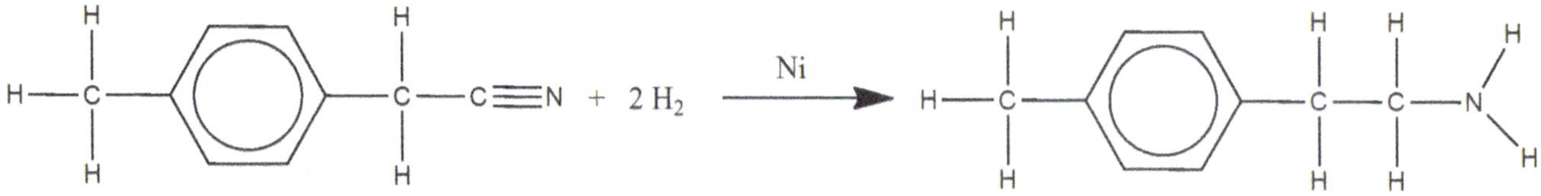

Nitriles can be hydrolysed using hot aqueous acid to form carboyxlic acids

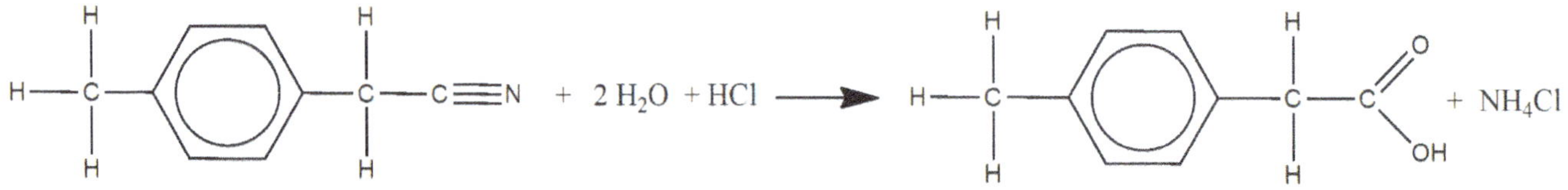

Revision Guide 6

Organic Chemistry and Analysis

Acyl Chlorides and their Reactions

Acyl chlorides are derivatives of carboxylic acids - the OH group of the carboxyl functional group has been replaced by a chlorine atom.

Carboxylic acids react with sulphur dichloride oxide (thionyl chloride) to form an acyl chloride, sulphur dioxide and hydrogen chloride. The equation for the reaction between ethanoic acid and sulphur dichloride oxide is shown below:

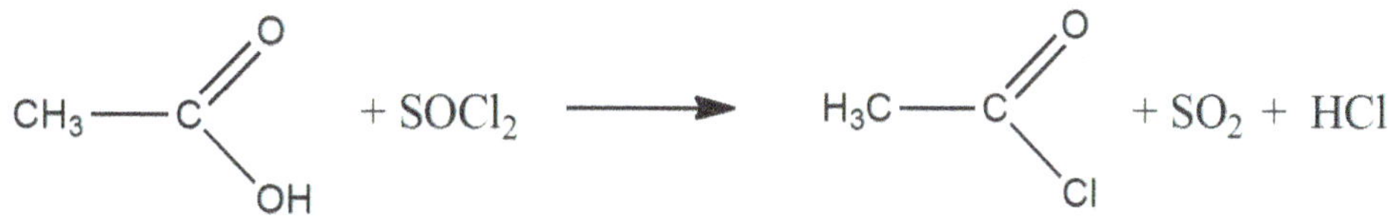

$$CH_3-COOH + SOCl_2 \longrightarrow H_3C-COCl + SO_2 + HCl$$

Acyl chlorides are named as follows

Carboxylic acid name	Formula of carboxylic acid	Acyl chloride name	Formula of acyl chloride
Ethanoic acid	CH_3COOH	Ethanoyl chloride	CH_3COCl
Propanoic acid	CH_3CH_2COOH	Propanoyl chloride	CH_3CH_2COCl
Pentanoic acid	$CH_3(CH_2)_3COOH$	Pentanoyl chloride	$CH_3(CH_2)_3COCl$

Acylation: Benzene reacts with an acyl chloride as shown in the following reaction:

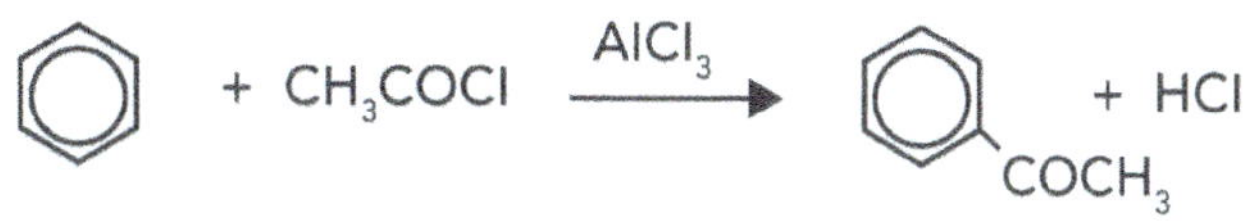

$$C_6H_6 + CH_3COCl \xrightarrow{AlCl_3} C_6H_5COCH_3 + HCl$$

A halogen carrier is required for the reaction to take place. The reaction mixture has to be heated at 60 °C for about 30 minutes. This reaction is known as **Friedel-Crafts acylation** and is an electrophilic substitution reaction.

Acyl chlorides react with ammonia to form primary amides and hydrogen chloride as shown in the following equation:

$$CH_3COCl + NH_3 \rightarrow CH_3CONH_2 + HCl$$

The hydrogen chloride will react with any excess ammonia to form ammonium chloride as shown next.

$$NH_3 + HCl \rightarrow NH_4Cl$$

Combining these two equations gives us:

$$CH_3COCl + 2NH_3 \rightarrow CH_3CONH_2 + NH_4Cl$$

Acyl chlorides react with primary amines to form secondary amides as shown in the following equation:

$$CH_3COCl + CH_3NH_2 \rightarrow CH_3CONHCH_3 + HCl$$

Acyl chlorides react with alcohols to form esters as shown below:

$$CH_3COCl + CH_3CH_2CH_2OH \rightarrow CH_3COOCH_2CH_2CH_3 + HCl$$

$$H_3C-\overset{\displaystyle O}{\underset{\displaystyle Cl}{C}} + CH_3CH_2CH_2OH \longrightarrow H_3C-\overset{\displaystyle O}{\underset{\displaystyle O-CH_2CH_2CH_3}{C}} + HCl$$

Acyl chlorides are hydrolysed in a nucleophilic addition reaction to form carboxylic acids and hydrogen chloride as shown below:

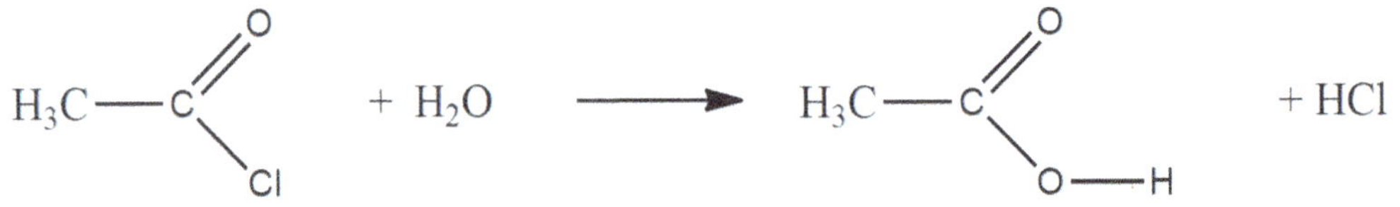

$$H_3C-\overset{\displaystyle O}{\underset{\displaystyle Cl}{C}} + H_2O \longrightarrow H_3C-\overset{\displaystyle O}{\underset{\displaystyle O-H}{C}} + HCl$$

Practical Procedures

This section outlines some of the common practical procedures you would be expected to know.

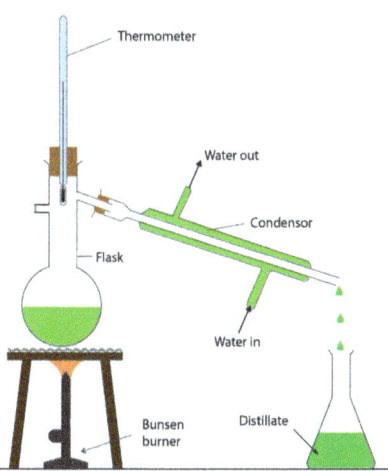

Distillation

Used to separate components liquids from a mixture based on their boiling points.

Example use: In the preparation of aldehydes using primary alcohols.

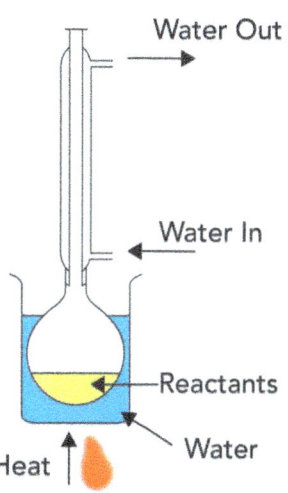

Heating under reflux

Ensures that no reactants or products escape the reaction mixture.

Example use: In the preparation of carboxylic acids using primary alcohols.

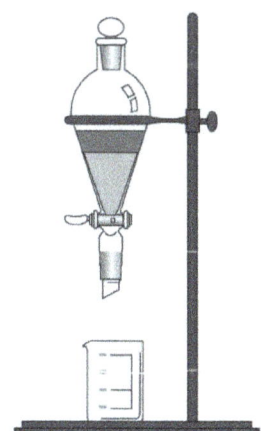

Separating funnel

Used to separate immiscible liquids.

Example use: Separating an organic solvent from an aqueous solution.

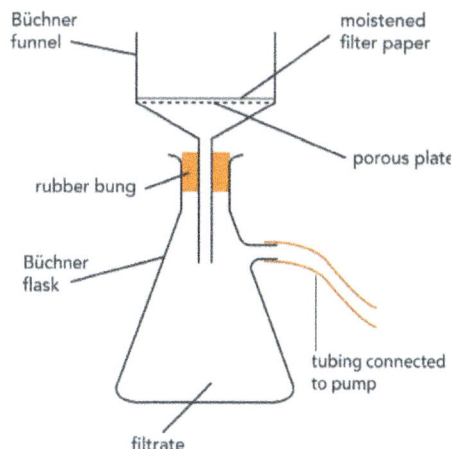

Filtration under reduced pressure

Used to filter a liquid.

Example use: Purification of an organic solid. Allows for recrystallisation to take place.

Revision Guide — Practical Procedures (Practical Endorsement)

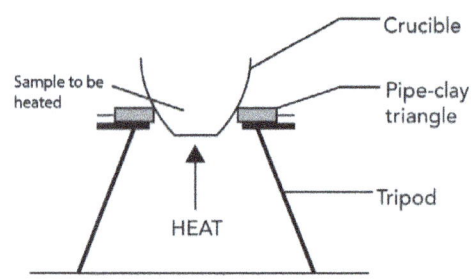

Evaporating the water

Can be used to evaporate water from a sample.

Example use: Finding the water of crystallisation.

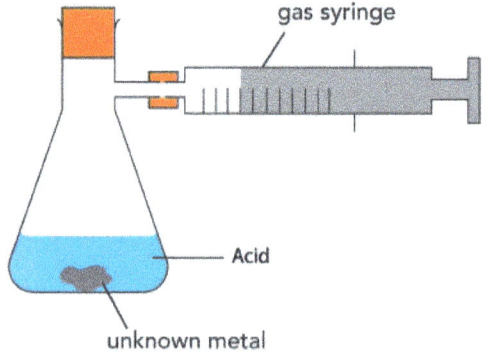

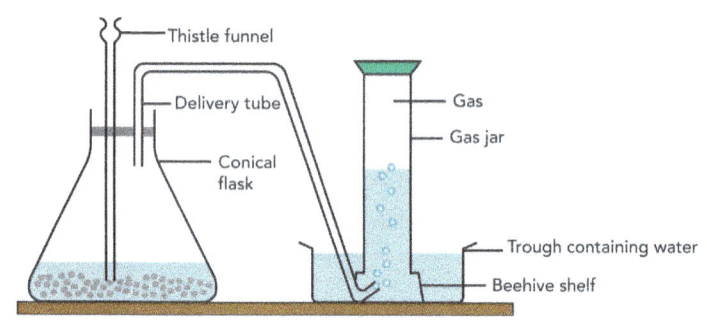

Measuring volume of gas formed

Gas given off during the reaction collects in the gas syringe pushing the plunger across the syringe.

Example use: Measuring the volume of carbon dioxide produced during the reaction between a metal carbonate and an acid. Can be used to work out the amount, in mol, of a reactant.

Collecting a gas formed during a reaction

Reactants are placed in the conical flask. The gas formed during the reaction collects in the gas jar.

Example use: The gas formed can be analysed. This can be used to identify a reactant.

Titrations

A volumetric flask can be used to make up a standard solution (a solution of known concentration).

In an acid-base titration, burette holds the acid. The volumetric pipette is used to add a known volume of the alkali to the conical flask. The acid is then added drop by drop using the tap of the burette.

Example use: Determining the concentration of a sample.

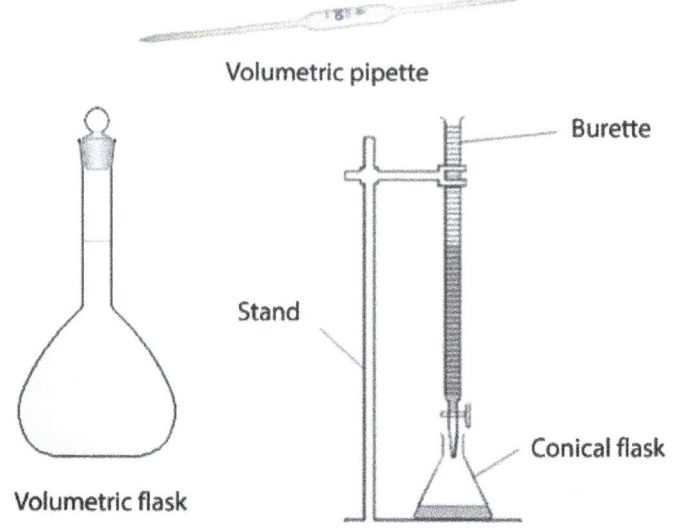

Practical Procedures (Practical Endorsement)

Revision Guide

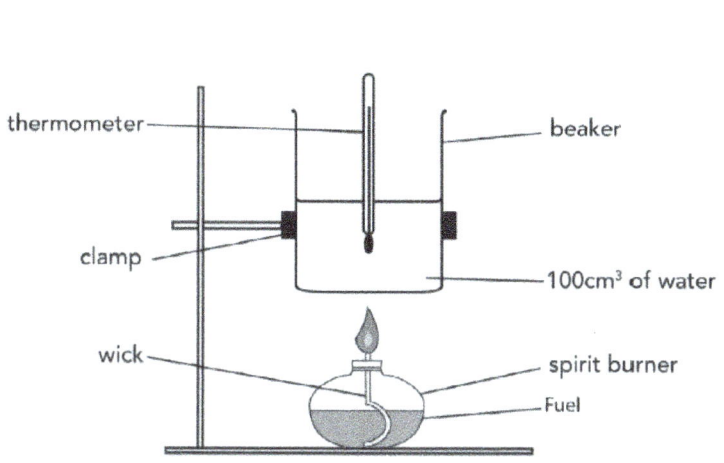

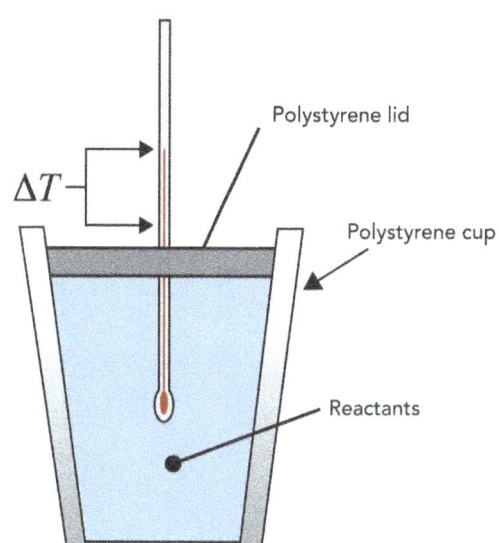

Combustion of a compound

The compound / fuel is placed in the spirit burner and combusted. The energy given out is proportional to the rise in temperature of the water.

Example use: To find the enthalpy change of combustion.

The formula $Q = mc\Delta T$ is useful in calculations involving this apparatus.

Calorimeter

Enthalpy changes involving solutions can be determined. The solutions are placed in the polystyrene cup and allowed to react resulting in a change in temperature.

Example use: Determining enthalpy change of neutralisation.

Melting point apparatus

A melting point apparatus is used to measure the melting point of a sample.

Example use: Determining the purity of an organic solid.

Most organic solids formed as part of chemical reactions are not 100% pure. Measuring the melting point of the organic product and comparing it with the melting point of a pure sample helps in determine the purity of the product.

pH meter

This is an electronic device consisting of a special measuring proble (a glass electrode). The glass electrode is dipped into the solution whose pH needs to be measured. Before use, a pH meter has to be calibrated. To do this, rinse the pH probe with deionised water and shake gently to remove any excess water.

Then place the probe in three standard buffer solutions (whose pH are known) that span the range of the pH values to be measured and measure their pH(despite the fact that you know their pH). Before each measurement is taken, the probe has to be rinsed again with deionised water. Record your pH readings of the buffer solutions in a table. These results will be used to adjust any readings you may take with the pH meter.

Example use: Determining the concentration of an acid.

Revision Guide — Practical Procedures (Practical Endorsement)

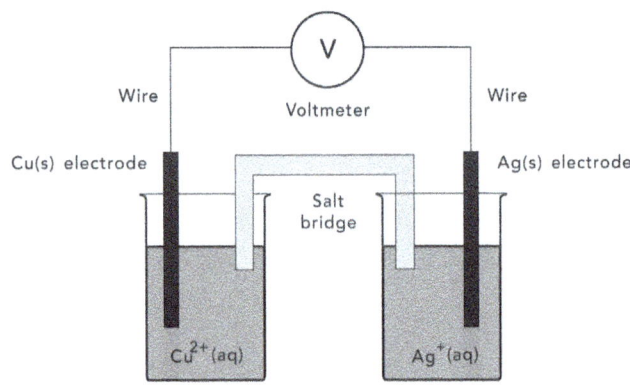

Electrochemical cell

Two half-cells can be connected as shown and the reading on the voltmeter will be equal to the cell potential, $E^\ominus$, of the electrochemical cell.

$E^\ominus_{(cell)} = E^\ominus$ (positive terminal) $- E^\ominus$ (negative terminal)

Example use: Determining the effect of concentration on the cell potential of an electrochemical cell.

Note: You can measure the standard electrode potential of a half-cell by connecting it to a standard hydrogen half-cell.

Thin-layer chromatography

A pencil line is drawn at the bottom of the plate and a small drop of the mixture to be analysed is placed on it. After the drop has dried, the plate is stood in a layer of solvent in a covered beaker. The level of the solvent has to be below the pencil line.

R_f values are calculated by dividing the distance between the component mark of interest and the pencil line by the distance between the solvent front and the pencil line. R_f value is then compared with data table values to help identification.

Example use: Identifying components in a mixture.

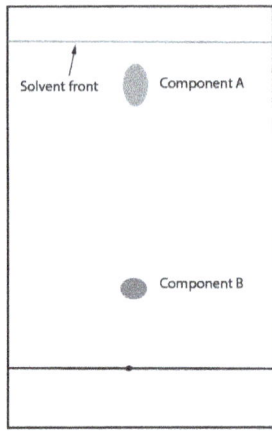

Colorimetry

A colorimeter is a device used to measure concentration of colored compounds in solution. It measures the amount of light that is absorbed as it passes through the solution. The results are recorded as the absorbance of the solution.

Example use: Determining reaction rates.

Purifying a product

When a solid is formed as part of a reaction along with a liquid, the solid product can be purified by filtering the solution and recrystallising.

When an organic liquid is formed along with an aqueous liquid such as water, the organic liquid can be extracted and purified using the following steps: shake and leave the mixture to settle in a separating funnel. Separate the layers by tapping off. Add some anhydrous $MgSO_4$ or $CaCl_2$ to the organic layer in a dry conical flask (this absorbs any water from the organic layer). Distil the oganic layer.

When a mixture of liquids are formed from a reaction, the liquids can be separated by distilling the mixture. The different liquids will be distilled off at different temperatures based on their boiling points.

Practical Procedures (Practical Endorsement)

Revision Guide

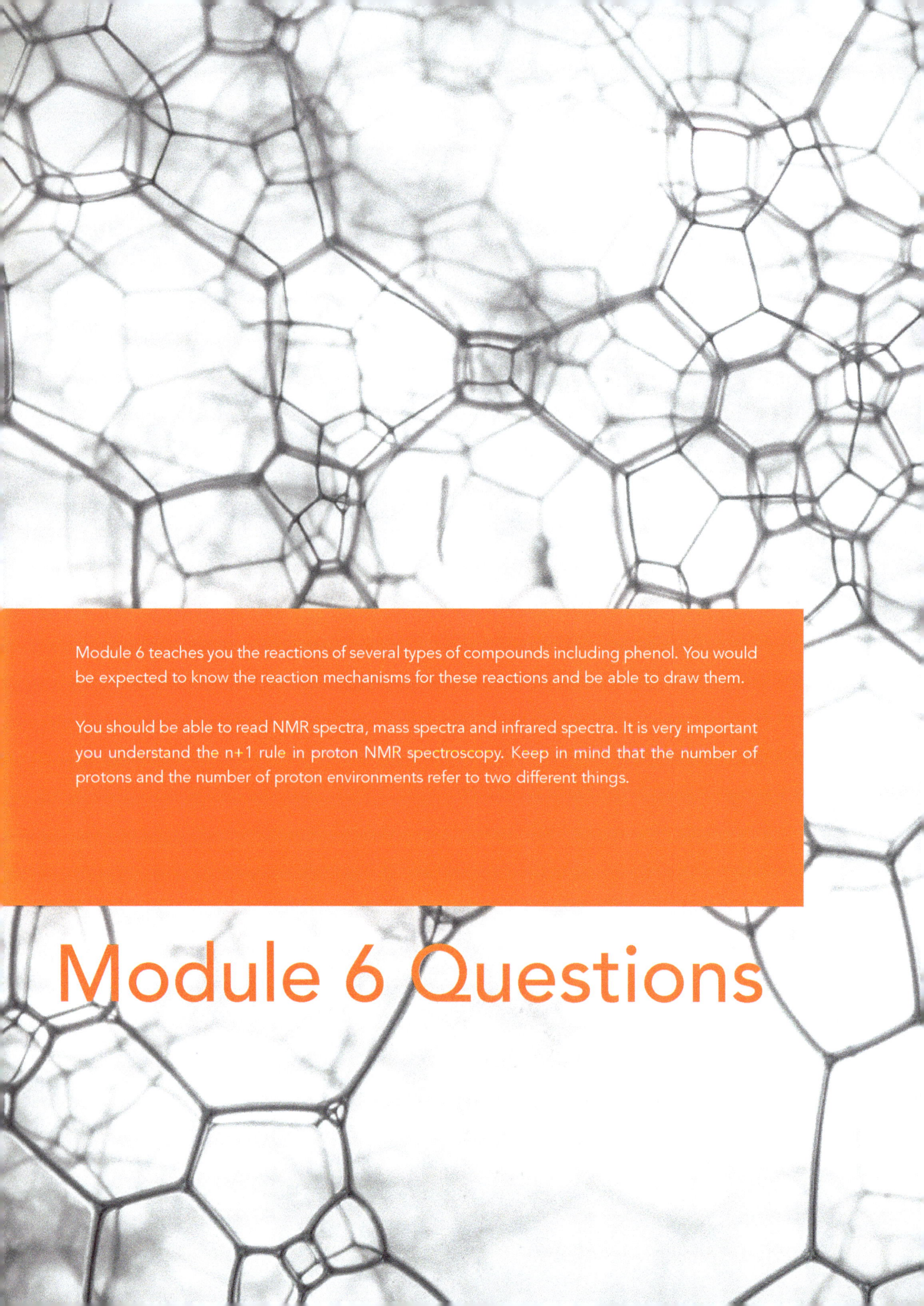

Module 6 teaches you the reactions of several types of compounds including phenol. You would be expected to know the reaction mechanisms for these reactions and be able to draw them.

You should be able to read NMR spectra, mass spectra and infrared spectra. It is very important you understand the n+1 rule in proton NMR spectroscopy. Keep in mind that the number of protons and the number of proton environments refer to two different things.

Module 6 Questions

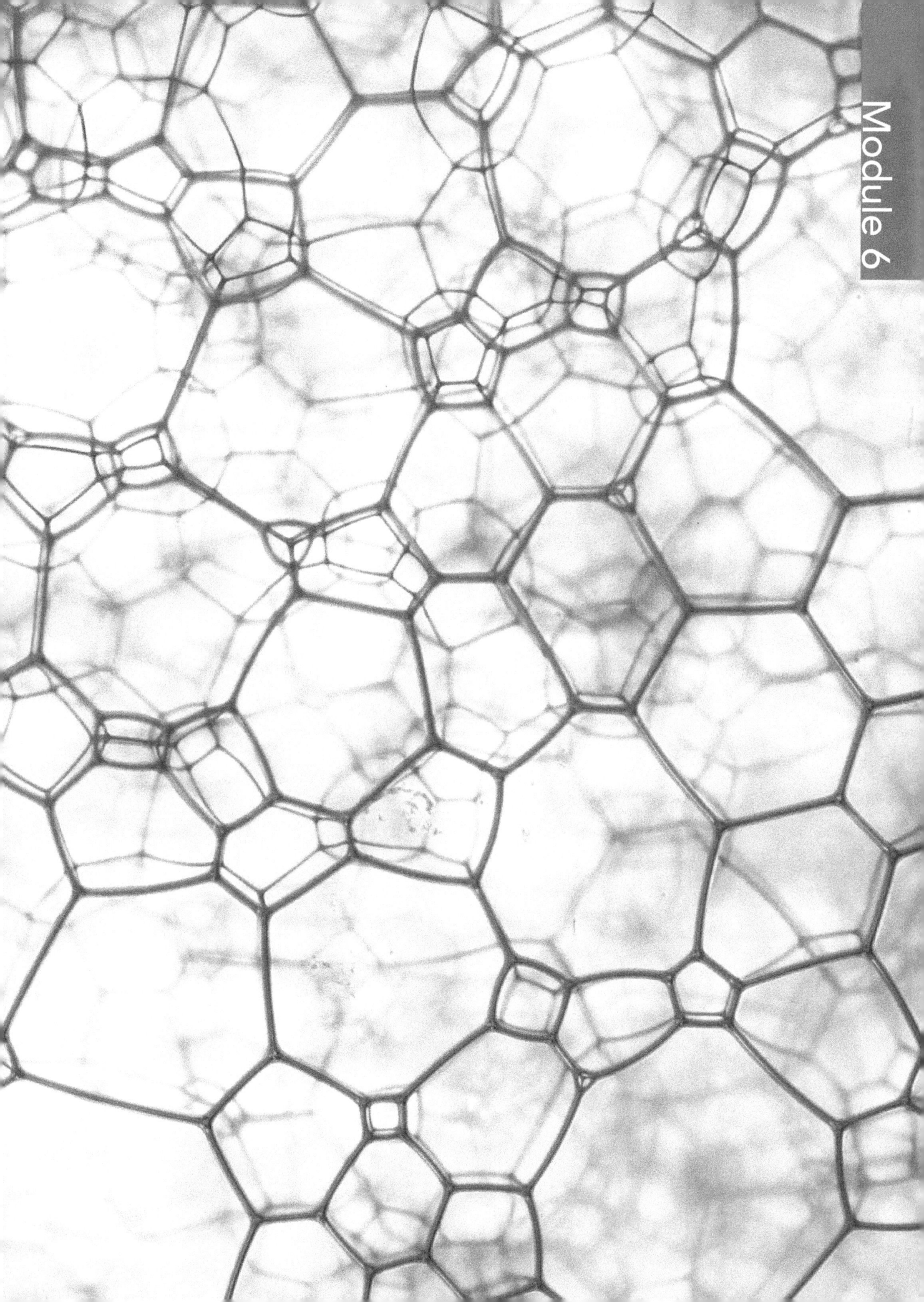

Module 6

6 QUESTIONS — Organic chemistry and analysis

1. Scientific theories change over time. Friedrich Kekulé von Stradonitz discovered the ring structure of benzene in 1865. However, since then new discoveries have taught us that the structure of benzene is in fact different to the one suggested by Kekulé.

    a. What evidence suggested that Kekulé's structure of benzene was incorrect? [3]

    b. State the delocalisation or resonance energy of benzene. Explain. [2]

    c. Describe the delocalised model for the structure of benzene. [5]

2. Nitrobenzene is an important chemical in the synthesis of many organic compounds. It may be prepared in the laboratory using standard laboratory equipment.

    a. Using curly arrows show the reaction mechanism for the nitration of benzene. [4]

    b. Write equations to show how sulfuric acid acts as a catalyst in the nitration of benzene. [2]

    c. Outline the practical method for the synthesis of nitrobenzene starting from benzene. [4]

3. The structure of compound Q is shown below.

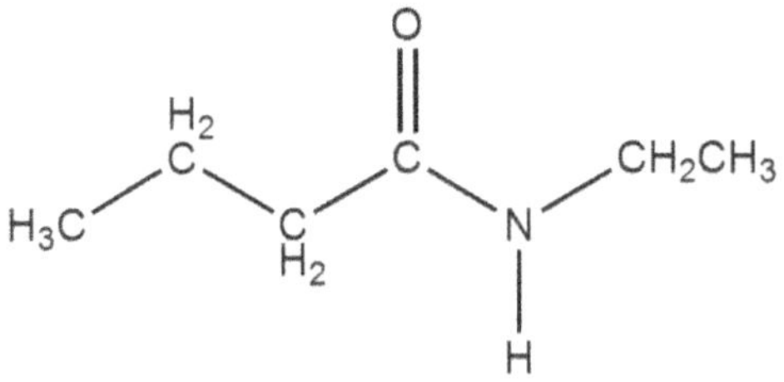

    Q

    a. State the name of the functional group of the compound shown above. [1]

    b. These compounds may further be classified based on the bonding around the nitrogen atom. State the type of **compound Q**. [1]

    c. A student reacted **compound Q** with dilute hydrochloric acid. State the molecular formula of the product(s) formed. [2]

    d. A student was asked to prepare a sample of **compound Q** in the laboratory. Devise a one-step synthesis. Give details of appropriate reagents. Write an equation for the reaction. [4]

    e. This compound is soluble in water. Explain. [1]

4. But-2-ene, C_4H_8, is used in the production of gasoline and is produced via catalytic cracking of crude oil. The displayed formula of but-2-ene is shown below.

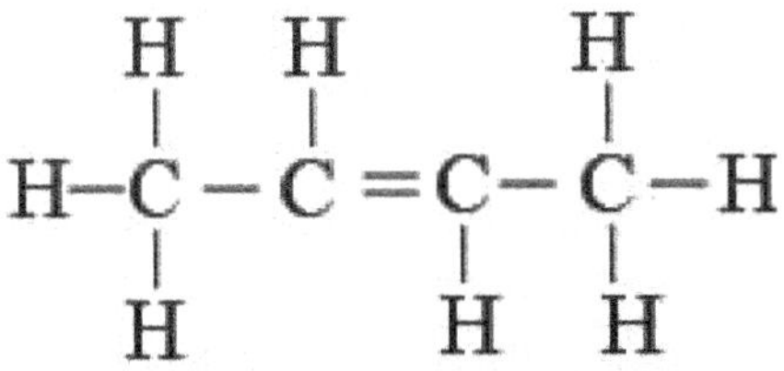

a. This compound has stereoisomers.

　i. What is meant by a stereoisomer? [2]

　ii. Name the type of stereoisomerism shown by this compound. [1]

　iii. Name a different type of stereoisomerism.

　　Explain the characteristics of this type of stereoisomerism. [2]

b. But-2-ene may be hydrogenated to form an alkane.

　i. State the name of the alkane formed and a suitable metal catalyst for this reaction. [2]

　ii. Write the electronic configuration of this metal catalyst. [1]

　iii. What type of a metal is this? [1]

c. But-2-ene may be used to produce butan-2-ol.

　Infrared spectroscopy of but-2-ene and butan-2-ol was carried out. Describe how their spectra would differ. [2]

5. Benzene is a colourless liquid with a sweet smell and is highly flammable. It occurs naturally in crude oil.

　a. Benzene may react with bromine in a substitution reaction.

　　i. Write an equation for the reaction using chemical formulae. [1]

　　ii. Using curly arrows show the reaction mechanism for the bromination of benzene and write an equation to show how the halogen carrier acts as a catalyst. [6]

　　iii. State the formulae of two halogen carriers that may be used for the bromination of benzene. [1]

　　iv. State the formulae of two halogen carriers that may be used for the chlorination of benzene. Describe the need for a halogen carrier. [3]

　b. Describe substitution reaction. [1]

　c. Explain why benzene take part in substitution reactions but not addition reactions. [2]

6. Thin layer chromatography is used to analyze a sample containing 2 components, A and B. Component A has moved a distance of 1.3 cm, component B has moved a distance of 2.7 cm and the solvent front has moved a distance of 6.2 cm. Calculate the R_f values of each component. [2]

7. An unknown sample may be analysed using various techniques, including mass spectrometry, chromatography, NMR spectroscopy and infrared spectroscopy.

A student analysed a sample of butan-2-ol.

　a. Determine the m/z value of the molecular ion peak in the mass spectrum of this compound. [1]

　b. State the number of peaks that would be present in the carbon-13 NMR spectrum of this compound. [1]

　c. The sample was shaken with D_2O. Proton NMR spectroscopy was carried out on the compound. State the number of peaks you would see in the spectrum. [1]

QUESTIONS
Organic chemistry and analysis

8. Halogens are found in the group 7 of the periodic table. Their reactivity decreases as you go down the group. Bromine and chlorine reacts with benzene if a halogen carrier is present.

   The structure of benzene is as follows.

   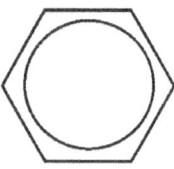

   a. Explain the reactivity of halogens. [3]

   b. State the shape and bond angle around each carbon in benzene. [2]

   c. Write an equation for the formation of the bromonium ion using Br_2 and $FeBr_3$. [1]

   d. Explain with an equation how bromobenzene can be formed starting with benzene. [2]

   e. State the type of reaction involved in the halogenation of benzenes. [1]

9. On a thin layer chromatogram, the solvent front has moved a distance of 4.0 cm. Component X has moved a distance of 0.8 cm, component Y has moved a distance of 1.2 cm and component Z has moved a distance of 2.0 cm. What are the R_f values of each component? [3]

10. Kekulé was an important chemist in Europe who suggested a structure for benzene. However, his theory about the structure of benzene was proved to be wrong.

    a. State the chemical name for the Kekulé structure of benzene. [1]

    Benzene may be nitrated using nitric acid and sulfuric acid.

    b. State the name of the mechanism in nitration of benzene. [1]

    c. Why is it important for the temperature to be at 50 °C in the formation of nitrobenzene? [1]

    d. Write an equation for the formation of the nitronium ion using nitric acid and sulfuric acid. [1]

11. H^+ is an example of an electrophile. Electrophiles are an important part of many chemical reactions.

    a. Define an electrophile. [1]

    b. Describe an electrophilic substitution reaction. [1]

12. Bromobenzene is a toxic compound and may cause damage to the nervous system if inhaled. This compound may be produced by reacting benzene with bromine. The mechanism for this reaction is shown below.

    Step 1: $Br_2 + FeBr_3 \rightarrow Br^+ + FeBr_4^-$

    Step 2:

    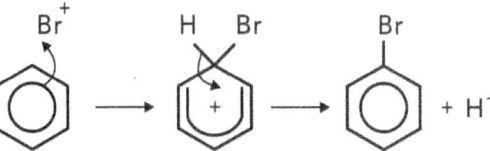

Step 3: $H^+ + FeBr_4^- \rightarrow FeBr_3 + HBr$

a.

   i. State the type of reaction given above. [1]

   ii. Reaction mechanisms use curly arrows to describe some events occurring in the process. State what a curly arrow represents. [1]

   iii. Br^+ is acting as an electrophile in this reaction. State why Br^+ is described as an electrophile. [1]

   iv. How do we know that $FeBr_3$ is acting as a catalyst? [1]

b. Bromine reacts with cyclohexene without the need for a catalyst. State the type of reaction occurring between bromine and cyclohexene and the colour changes observed during the reaction. [1]

13. Benzene undergoes substitution reactions but not addition reactions. In the bromination of benzene, a halogen carrier is used to generate the bromonium ion.

    a. Explain why benzene reacts with the bromonium ion but not with bromine. [2]

    b. Describe an addition reaction. [1]

    c. Explain why bromine react with cyclohexene but not with benzene. [4]

14. Phenol is useful in the production of many products including adhesives, paint, rubber and toys.

    a. Describe phenols. [1]

    b. i. Write an equation for the reaction of phenol with bromine water. State any observations. [2]

       ii. Explain why phenol react with bromine water relatively easily. [2]

    c. Explain the solubility of phenol in water. [2]

    d. Write an equation for the reaction of phenol with aqueous sodium hydroxide. [1]

15. Use curly arrows to show the reaction mechanism for the reaction between cyclohexene and bromine. [4]

16. Butan-1-ol can be oxidised to form butanal. It is important to distil the aldehyde from the reaction mixture as it is formed.

    Distilling the aldehyde prevents it from being oxidised to a **compound W**.

    a. State the name and molecular formula of **compound W**. [2]

    b. Draw a labelled diagram to show an apparatus that could be used in the distillation process. [2]

    c. Suggest the name of two compounds that can be used to form a mixture that can oxidise butan-1-ol to butanal. [1]

17. Write an equation for the formation of the carboxylic acid from propanol. [2]

18. Describe electronegativity. [1]

QUESTIONS — Organic chemistry and analysis

19. Describe how you could prepare Tollens' reagent in the laboratory. [2]

20. Write equations to show how iron(III) bromide is acting as a catalyst in the bromination of benzene. [2]

21. Give chemical explanations for the following statements:

    a. Phenol has a higher electron density in the ring than benzene. [1]

    b. An aldehyde has to be distilled off from the reaction mixture as it is formed from the oxidation of an alcohol. [1]

22. Alcohols are organic compounds in which a hydroxyl functional group is bonded to a saturated carbon atom. Alcohols may be oxidised to form other products.

    a. Name a suitable oxidising agent that can oxidise primary and secondary alcohols. [1]

    b. Write an equation for the oxidation of butan1-ol to butanal. [2]

    c. Why is it necessary to distil aldehydes from the reaction mixture in the oxidation of primary alcohols to aldehydes? [1]

    d. Describe a simple chemical test that can be used to identify a carbonyl group (as found in aldehydes and ketones). [2]

23. Some alcohols may be oxidised using an oxidising agent. The product formed will depend on the type of alcohol used. Alcohols may be categorized as primary, secondary and tertiary.

    In a reaction, butan-2-ol was oxidised to form product Y.

    a. Name Y and state the name of the family of compounds to which Y belongs. [2]

    b. Write a balanced equation for the oxidation of butan-2-ol. You may represent the oxidising agent as [O]. [2]

    c. State the type of the alcohol, butan-2-ol. [1]

    d. State the type of alcohol which cannot be oxidised. [1]

24. Carbonyl compounds contain the C=O functional group. Carbonyl compounds may be reduced.

    a. Suggest a suitable reducing agent for the reduction of carbonyl compounds. [1]

    b. Write an equation to show the reduction of butanal. [2]

    c. Write an equation to show the reduction of butan-2-one. [2]

    d. State the type of mechanism involved in the reduction of carbonyl compounds. [1]

25. Formaldehyde is an aldehyde used in the production of resins. It is produced in very large quantities around the world.

    You are given two samples A and B, one of which is formaldehyde and the other a ketone. Describe various methods that can be used to confirm the presence of carbonyl bonds in these compounds, identify the ketone and confirm the identity of formaldehyde. Your answer should include details of any chemical tests and physical measurements.

    [8]

QUESTIONS 6

Organic chemistry and analysis

26. Define nucleophile. [1]

27. Butanal, also known as butyraldehyde is a colourless flammable liquid. Butanal can be oxidised to form **Compound X**.

    a. Write an equation for the oxidation of butanal. [2]

    b. State a suitable oxidising agent and any observations. [2]

    c. Name **compound X**. [1]

    132g of **compound X** was dissolved in 2000 cm^3 of water to form a solution.

    d. Calculate the concentration of this solution in mol dm^{-3}. [2]

28. Write an equation for the reaction of sodium with phenol. [2]

29. A carbonyl compound could be identified using its 2,4-dinitrophenylhydrazone derivative. Describe the stages in this process. [3]

30. Write an equation to show the reaction between sodium and propanoic acid. Name the products formed. [2]

31. Describe how you could prepare an aromatic amine in the laboratory. Name the reagents used. [3]

32. Esters are widely found in nature and are also made industrially across the world. Esters are responsible for the aroma of many fruits including apples and bananas.

    a. State some of the uses of esters. [1]

    b. Briefly describe three ways in which an ester may be formed. [3]

    c. Name the raw chemicals that may be used to form the following esters:

    i. Methyl butanoate [2]

    ii. Propyl methanoate [2]

    d. State the functional group in an ester. [1]

33. Describe how pentyl propanoate can be made using an acid anhydride and an alcohol. [2]

34. Describe the term "hydrolysis". [1]

35. Sodium tetrahydridoborate(III) reacts with aldehydes and ketones differently. The student was given an unlabelled bottle of a chemical which is suspected to be an aldehyde.

    a. State the chemical formula of sodium tetrahydridoborate(III). [1]

    b. Describe a series of chemical tests which can be used to test whether the sample is an aldehyde. Include any observations. [2]

    Following some tests, the compound was identified to be 4-bromopentanal.

QUESTIONS
Organic chemistry and analysis

c. i. Name the product formed from the reaction between sodium tetrahydridoborate(III) and 4-bromopentanal. [1]

ii. State the name of the reaction mechanism. [1]

36. Esters may be hydrolysed using acid or alkali.

   a. Describe the acid hydrolysis of esters. [2]

   b. Describe the alkaline hydrolysis of esters. [2]

37. Infrared spectroscopy may be used to identify unknown compounds. Infrared spectra of two compounds, ethanol and ethanoic acid, along with the two unidentified samples, were presented to a chemical analyst. The first of these spectra had a broad absorption peak from 2480 - 3100 cm^{-1}.

   a. State which compound is responsible for the first spectrum. [1]

   The chemist wanted to identify the ethanoic acid.

   b. Suggest a test that could be carried out to test for carboxylic acids. State any observations. [2]

   c. Write a balanced equation to show the reaction between lithium carbonate and ethanoic acid. Name the product formed. [3]

38. Ethylamine is a colourless gas and is often used in the chemical industry. It is a weak base like other amines. It has a pK$_a$ of 10.7.

   Ethylamine is produced in large scales by reacting ethanol and ammonia.

   a. What are amines? [1]

   b. Explain why amines can act as weak bases. [1]

   c. Write an equation for the formation of ethylamine from ethanol and ammonia. [2]

   d. Calculate the K_a of ethylamine. [2]

   e. Explain how the K_a of a stronger acid would differ from that of a weaker acid. [1]

39. Describe the difference between saturated and unsaturated fatty acids. [1]

40. Describe how ethyl butanoate can be made using an alcohol and a carboxylic acid. [2]

41. 4-hydroxybenzoic acid is a white crystalline solid which can be used in the preparation of esters. Its structure is show below. It is slightly soluble in water.

   a. Explain why 4-hydroxybenzoic acid is soluble in water.
   Use a labelled diagram to support your answer. [4]

b. Name the functional groups in the compound. [1]

c. Predict the number of peaks in the carbon-13 NMR spectrum of 4-hydroxybenzoic acid. [1]

42. Describe a simple chemical test to distinguish between aldehydes and ketones. [2]

43. Explain the solubility of carboxylic acids in water. [2]

44. Write an equation to show the reaction between propylamine and nitric acid. Name the product(s) formed. [2]

45. Aliphatic amines do not contain a ring structure such as benzene. They made be prepared in the laboratory using halogenoalkanes.

    a. Describe how an aliphatic amine may be prepared using halogenoalkanes. You may use 1-bromobutane as the halogenoalkane. [3]

    b. What type of a reaction is this? [1]

    c. Write a balanced equation for the reaction that is taking place. [1]

    d. Name the product(s) formed. [1]

46. Carboxylic acids react with bases in a neutralisation reaction.

    a. Write an equation to show the reaction between butanoic acid and sodium hydroxide. Name the products formed. [3]

    A student measured the boiling points of some alcohols and alkanes of similar sizes and concluded that alcohols have higher boiling points than the alkanes.

    b. Explain the student's findings. [2]

    A sample of liquid pentane was heated changing it into a gas. The student said, "as the liquid changes into a gas, the covalent bonds within the molecules are broken, so the atoms are free to move".

    c. Comment on this statement and explain the changes that take place when the liquid changes into a gas. [2]

47. State the general formula for an α-amino acid. [1]

48. A student carried out mass spectrometry on an unknown sample to distinguish the following polymer from another compound.

    a. State the type of polymer shown above. [1]

    b. Draw the structure of the monomer that could be used to make this compound. [1]

QUESTIONS

Organic chemistry and analysis

The student was given two unknown samples of carboxylic acids by a teacher.
Thin-layer chromatography may be used to identify these compounds.

c. Explain how the student could analyse the results from chromatography to identify the compounds. [2]

d. Several similar compounds have similar properties. Suggest why this could be a problem when it comes to identifying them using chromatography. [1]

49. Amino acids are important building blocks of enzymes and hormones. They can act as both acids and bases and are hence termed amphoteric.

   a. An amino acid may gain or lose a proton resulting in different electric charges. The isoelectric point of an amino acid is the pH at which it is neutral and forms its zwitterion. Suggest a factor which might affect the isolectric point of an alpha amino acid. [1]

   b. Explain why amino acids are amphoteric. [2]

50. Pent-2-ene may undergo polymerisation to form poly(pent-2-ene).

   a. Draw the polymer showing two repeating units. [2]

   b. i. Describe addition polymerisation. [1]

      ii. Name a family of compounds whose species may show addition polymerisation. [1]

   c. Describe condensation polymerisation. [1]

   d. Compare condensation and addition polymerisation. [2]

   Pent-2-ene forms stereoisomers.

   e. Draw the skeletal formulae of the two isomers and name each of them. State the type of isomerism. [5]

51. Many biological molecules show optical isomerism. They include naturally occurring amino acids.

   a. What can be said about optical isomers found in nature as opposed to artificially made ones, in terms of their interaction with enzymes? [1]

   b. Describe optical isomerism. [1]

   c. Explain how the chemical reactions of one optical isomer differ from the other. [1]

52. Describe the acid and alkaline hydrolysis of polypeptides and proteins. [2]

53. $(CH_3)CH(NH_2)COOH$ reacts with $(CH(CH_3)_2)CH(NH_2)COOH$ to form a dipeptide:

   a. Draw an equation showing the structural formula of the reactants and products formed. [2]

   b. State the name of the type of reaction shown in your equation. [1]

54. Bromobenzene may be formed by reacting benzene with bromine in the presence of $AlBr_3$ which acts as a catalyst.

a. i. The equation for the reaction is shown below:

   $C_6H_6 + Br_2 \rightarrow C_6H_5Br + HBr$

   Outline the mechanism for the formation of bromobenzene.

   Show how $AlBr_3$ behaves as a catalyst. [5]

   ii. State what type of reaction this is. [1]

b. Phenol reacts more readily with bromine than benzene reacts with bromine. Explain why. [2]

55. Cyclobutane and but-1-ene are examples of isomers. There are various different types of isomerism. Stereoisomerism is one of these types.

    a. Draw the skeletal formulae of cyclobutane and but-1-ene and state the name of the isomerism. [3]

    b. i. Describe stereoisomers. [2]

       ii. Name two types of steroisomerism. [1]

    c. i. What is a chiral carbon? [1]

       ii. State the type of isomerism shown by a compound with a chiral carbon. [1]

56. Organic compounds called Toluidines can be used in the preparation of dyes. There are three structural isomers of this compound.

    a. i. One of these isomers is 2-amino-1-methylbenzene($C_6H_4CH_3NH_2$). Draw the displayed formula of this isomer. [1]

       ii. Name two other isomers of Toluidine. [2]

    Because of the amino group, Toluidine is weakly basic. It can react with acids forming salts.

    b. i. Explain why amines can behave as bases. [1]

       ii. Write the formula of the salt that would be formed when an excess of $C_6H_4CH_3NH_2$ reacts with sulfuric acid. [1]

57. In the manufacture of pharmaceuticals often only one optical isomer is formed. Having both optical isomers can be dangerous for various reasons.

    a. Describe optical isomers. [1]

    b. Why can it be an advantage to produce pharmaceuticals that are a single optical isomer? [2]

    c. Draw the opitcal isomer of the following compound: [1]

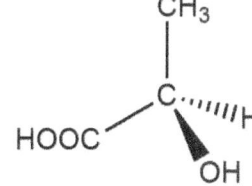

QUESTIONS
Organic chemistry and analysis

58. Describe a peptide linkage. [1]

59. The isoelectric point of an amino acid is the pH at which it is neutral, with both a positive and negative charge. Glycine has an isoelectric point of 5.97. Describe what happens to glycine at a pH of 1 and 13. [2]

60. Amino acids are the building blocks of proteins. Proteins are important biological molecules. Haemoglobin found in the red blood cells of humans, is a protein. It is important in carrying oxygen around the body.

    a. State the general formula of an α-amino acid. [1]

    b. The structure of the amino acid alanine is shown below:

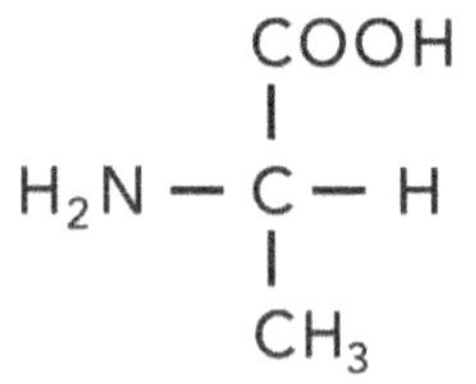

    $$H_2N-\underset{\underset{CH_3}{|}}{\overset{\overset{COOH}{|}}{C}}-H$$

    i. Draw the ions formed by alanine at a pH of 1 and 12. The isoelectric point of alanine is 6.11 [2]
    ii. Two alanine units can bond together forming a dipeptide. Draw the structure of this dipeptide, stating the name of the bond linking the two amino acids. [2]

61. Describe polyamides. [1]

62. This question is about the following polymer.

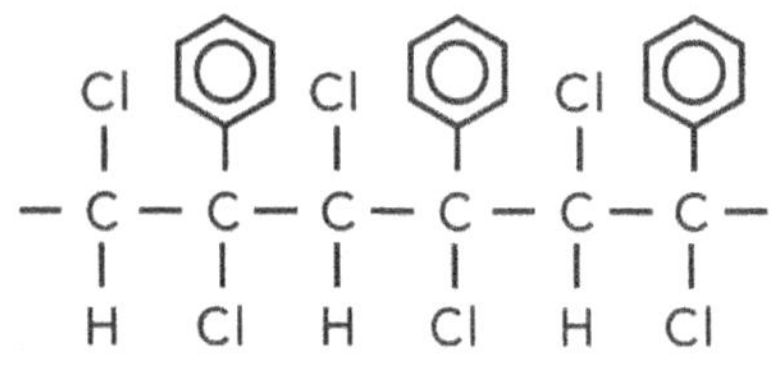

    $$-\underset{\underset{H}{|}}{\overset{\overset{Cl}{|}}{C}}-\underset{\underset{Cl}{|}}{\overset{\overset{\phi}{|}}{C}}-\underset{\underset{H}{|}}{\overset{\overset{Cl}{|}}{C}}-\underset{\underset{Cl}{|}}{\overset{\overset{\phi}{|}}{C}}-\underset{\underset{H}{|}}{\overset{\overset{Cl}{|}}{C}}-\underset{\underset{Cl}{|}}{\overset{\overset{\phi}{|}}{C}}-$$

    a. Identify the monomer from which the following polymer may be made: [1]

    b. State the name of the polymerisation involved in its formation. [1]

63. Enzymes are biological catalysts which work by binding to its substrate. The substrate will have a complementary shape to the active site of the enzyme. Different optical isomers of the same compound may bind differently to its enzyme. Optical isomers are also known as enantiomers.

    a. Suggest why the following compound has an optical isomer [1]

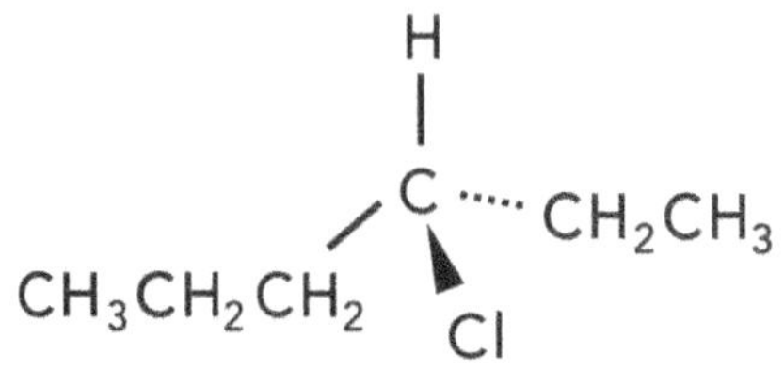

    Structure: central C bonded to H, CH₂CH₃, Cl, and CH₃CH₂CH₂

b. Draw the optical isomer of this compound [1]

c. Enzymes are biological catalysts. Explain how a catalyst can increase the rate of a reaction. [1]

64. The following compound Z, is an unstable colourless solid. Due to its instability, it is not usually available for purchase and is prepared when required.

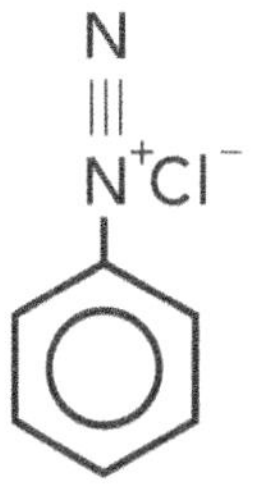

0.57g of Z was formed in the laboratory. A student worked out that based on the amounts of the reactants used, the yield should be 0.70g.

a. Calculate the amount, in mol, of the compound formed. Give your answer to a suitable number of significant figures in the standard form. [4]

b. Work out the percentage yield of Z. [2]

65. DNA contains genetic information in humans and is an example of a polymer. Polymers exist naturally in nature. However, some polymers are made by man. One such polymer is shown below.

a. State the type of polymerisation involved in the formation of this polymer. [1]

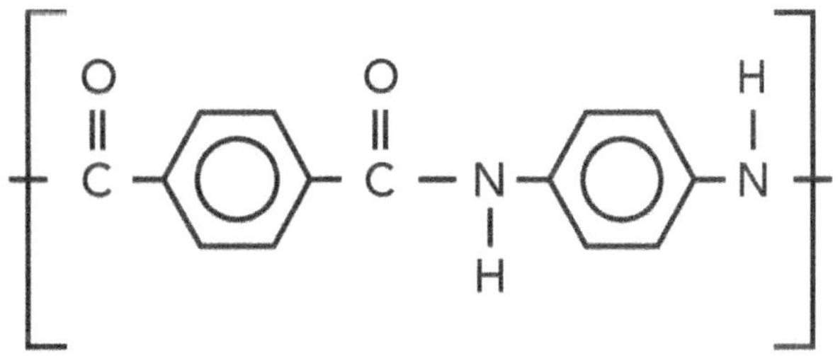

b. Suggest another type of polymer that exists in nature. [1]

66. Describe the acid and base hydrolysis of polyesters. [2]

67. Polyvinyl chloride, PVC, is one of the most commonly produced synthetic polymers. It has a wide range of uses including the production of pipes, cables and clothing.

A short section of the PVC polymer is shown below.

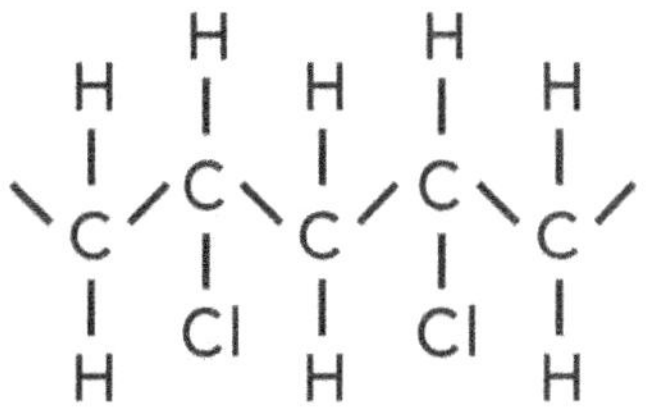

a. Identify the monomer from which PVC is obtained. [1]

QUESTIONS
Organic chemistry and analysis

b. State the type of polymerisation involved in the production of this polymer. [1]

A student says, "The monomer from which PVC is produced does not show E/Z isomerism. Although it meets one requirement for E/Z isomerism, it does not meet another."

c. Comment on the student's statement. [2]

68. Phenol is toxic to the human nervous system and could cause coma and ultimately death. However, it is also widely used throughout the world.

   a. A student was given a test tube of an unknown compound which could be phenol. The student was told that it is NOT an alkene. Describe a test tube test which can be carried out to find out whether the sample is in fact phenol. State the name of the compound responsible for the colour which would be observed if the sample is phenol. [3]

   0.800 moles of phenol was dissolved in 2300 cm^3 of water.

   (Molar mass of phenol: 94.1 gmol^{-1}.)

   b. Calculate the concentration of the solution in gdm^{-3}. [2]

   c. Explain why phenol is soluble in water. [2]

69. Describe the acid and base hydrolysis of polyamides. [2]

70. Describe the term "photodegradable". [1]

71. Compounds containing benzene rings are referred to as aromatic. Many years ago, chemists believed that benzene's structure was more like that of alkenes. However, over time, benzene was found to have a delocalised electronic structure.

    The following compound contains a benzene ring.

    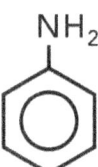

    a. Describe what happens when benzene and an alkene are added to bromine water. [1]

    b. Explain what is meant by the delocalised electron structure of benzene. [2]

    c. Name the compound shown above. [1]

    d. Write the structural formula of this compound. [1]

    e. Describe how this compound may be produced starting from nitrobenzene. State the type of reaction involved. [3]

72. Carbon-13 and proton NMR may be used to identify unknown compounds. Carbon-13 NMR was carried

out on the following **compound Q**.

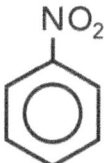

a. Write the molecular formula of this compound. [1]

b. State the number of peaks you would see in the carbon-13 NMR spectrum of **Q**. Explain your reasoning. Identify the different carbon environments by labelling them on the formula of **compound Q**. [3]

73. **Compound X**, shown here is used in the production for paracetamol. Paracetamol is an analgesic which means it is used to achieve relief from pain. It is thought that an overdose of this drug could cause skin conditions which may be fatal.

a. Name **compound X**. [1]

b. Write the structural formula of **compound X**. [1]

c. Describe how phenylamine may be produced starting from this compound. [3]

Compound X may be formed using nitric acid. Nitric acid is an example of a strong acid.

d. State the formulae of two other strong acids. [2]

e. State the formula of the electrophile involved in the formation of compound X. [1]

74. The following compound is corrosive and may cause irritation. When handling this chemical, personal protective equipment may need to be worn.

a. Name this compound. [1]

b. Write its structural formula. [1]

c. State the number of chiral centres in this compound and the atoms / groups of atoms attached to each chiral centre you have identified. [2]

d. Draw the optical isomers of this compound. [2]

e. When a compound with optical isomers are manufactured in a laboratory, the resulting product may contain a mixture of both compounds. However, sometimes it is necessary to form a single isomer of a compound. Suggest a reason as to why one may want to produce a single isomer of a compound. [1]

75. Chromatography is a very useful tool that is often used by students as well as chemists. There are different types of chromatography that may be used depending on the nature of the sample in question.

   a. Describe the term chromatography. [1]

   b. State two different types of chromatography. [2]

   c. Describe the term "mobile phase" and "stationary phase" [2]

76. This question is about thin-layer chromatography.

   a. Describe how thin-layer chromatography may be done. [3]

   b. What is the stationary and mobile phase in thin-layer chromatography? [2]

   c. How does separation take place in the solid stationary phase of thin-layer chromatography? [1]

77. Carbon-13 NMR spectra of two aromatic compounds are given below.

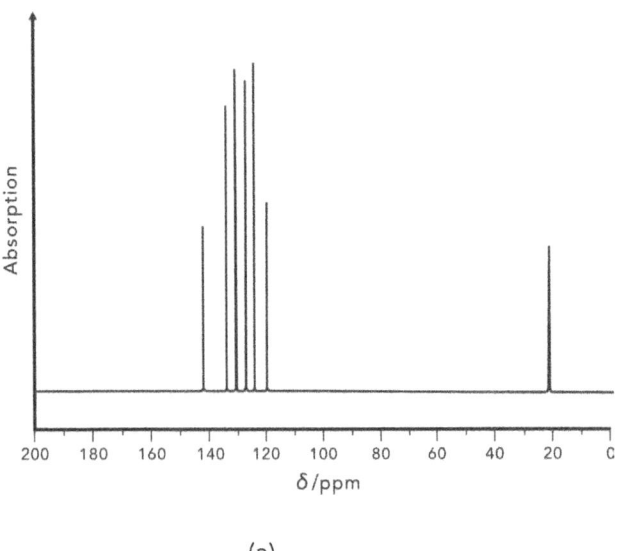

(a)

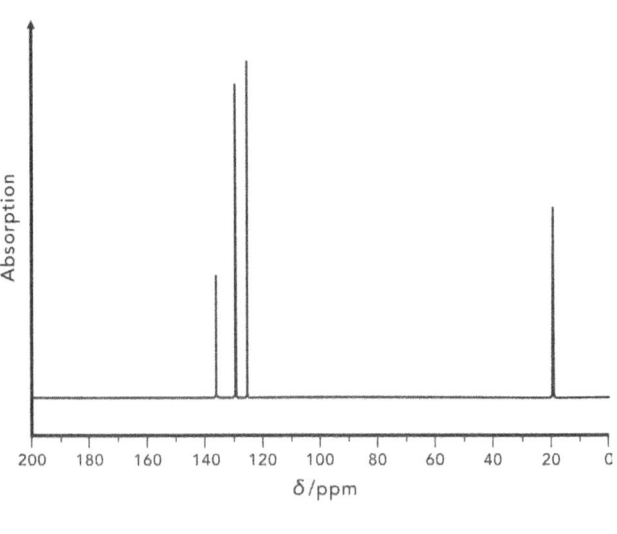

(b)

One of these spectra is derived from 1-bromo-3-methylbenzene and the other from 1,2-dimethylbenzene.

   a. Draw the displayed formulae of the two compounds. [2]

   b. On your diagrams from part a, label the carbon atoms that are in the same enviornment for both compounds. [2]

c. Match each of the carbon-13 NMR spectra to its compound. Explain your reasoning. [3]

d. State the number of proton environments in 1-bromo-3-methylbenzene. [1]

78. Describe the term R_f value. [1]

79. A student was given one sample each of compounds, $CH_3CHBrCOOH$, $CH_3CHBrCH_3$ and CH_3CH_2COOH. The samples were not labelled. The student carried out proton NMR spectroscopy on one of the samples. Its spectrum is shown below. Any peaks due to the proton of the OH group have been removed from the spectrum using D_2O. Identify the compound, explain your reasoning. [4]

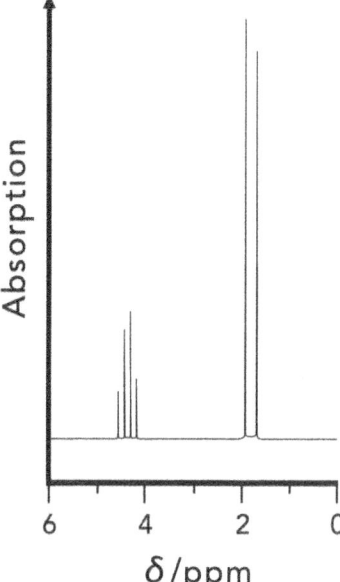

80. Fritz Prior developed solid state gas chromatography in 1947. Liquid-gas chromatography was later produced in 1950. Chromatography enables the identification of components in a mixture and is often used in forensics and in airport security.

   a. Describe the term "retention time". [1]

   b. How can gas chromatography be combined with mass spectrometry in analysis? [2]

81. Name the following compound: [1]

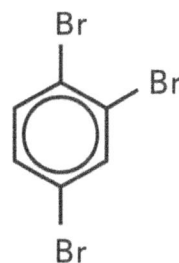

82. Butan-2-ol and Butan-1-ol cannot be easily identified using their carbon-13 NMR spectra. Explain why. [1]

83. NMR spectroscopy may be based on carbon-13 atoms or protons. TMS is often used in NMR spectroscopy. The chemical shifts of the peaks in a NMR spectrum provides information about the atoms or groups in the compound being analysed.

QUESTIONS — Organic chemistry and analysis

    a. Describe the use of tetramethylsilane (TMS) in NMR spectroscopy. [1]

    b. Describe the term, "chemical shift". [1]

    c. State two types of information given by a carbon-13 NMR spectrum. [1]

84. NMR is a very useful analytical technique. MRI is based on NMR technology and helps in the diagnosis of certain medical conditions. State with an example, the type of solvents used in NMR spectroscopy. [2]

85. There are many structural isomers of $C_5H_{12}O$ that are alcohols including pentan-3-ol. The carbon-13 NMR of two of these isomers are given below. Draw the structural isomers of two of these isomers that may have produced the following spectra (not including pentan-3-ol). Explain your answer.

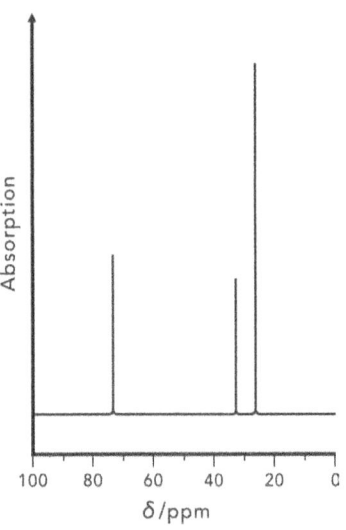
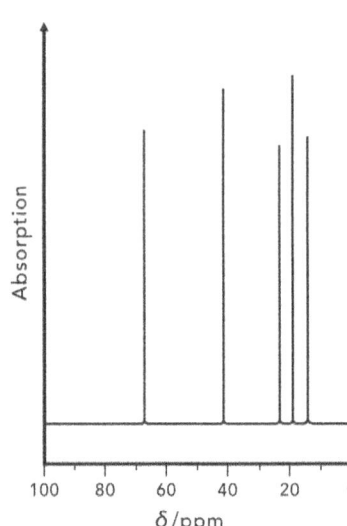

[4]

86. Esters are known for their sweet smells and can be prepared by reacting alcohols with carboxylic acids.

    They are widely used in perfumes.

    Butan-2-ol reacts with propanoic anhydride to form an ester. Using structural formulae, write an equation for this reaction. [2]

87. Predict the number of peaks and the chemical shift of those peaks in the carbon-13 NMR spectrum of $CH_3COOCH_2CH_3$. Name this compound. [2]

88. Compound X and Y reacted together to form compound Z. The rate of this reaction increased proportionally to the concentration of compound X. Compound Y has the formula $CH_3COOCH_2CH_3$.

    a. State the type of compound Y. [1]

    b. State the order with respect to compound X. Sketch a graph of rate against concentration. [2]

    Compound Y may be hydrolysed using aqueous sodium hydroxide. One of the products formed during this process is ethanol.

    c. Name the other product formed during this process and write its formula. [2]

Organic chemistry and analysis

QUESTIONS 6

89. Predict the number of peaks and the chemical shift of those peaks in the carbon-13 NMR spectrum of $CH_3CHBrCH_2CH_2CH(OH)CH_3$. Name the compound and state the name of a functional group in the compound. [4]

90. Mass spectrometry can be used to distinguish between bromobenzene and 1,3,5-tribromobenzene.

    a. Draw the structural formula of 1,3,5-tribromobenzene. [1]

    b. Describe how the mass spectrum of a compound may be used to identify that compound. [3]

91. Two isomers of $C_4H_{10}O$ are given below. Match the isomers to its carbon-13 NMR spectrum. Identify the peaks at 70 ppm in each spectrum. [2]

    $CH_3CH(OH)CH_2CH_3$
    $CH_3C(OH)(CH_3)CH_3$

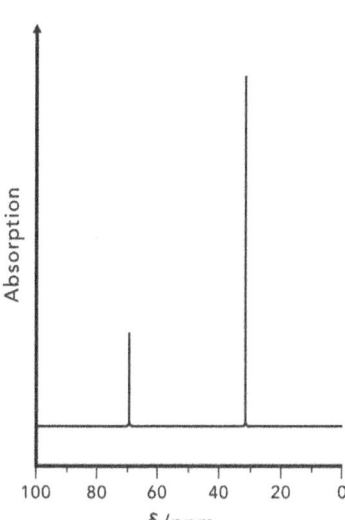
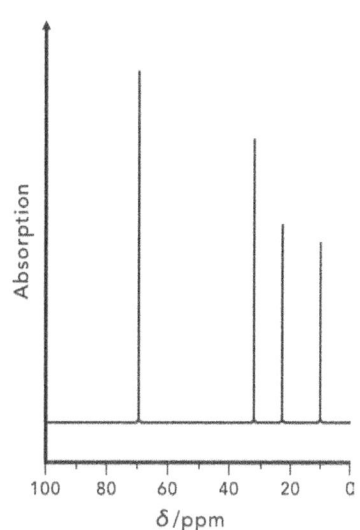

92. Draw the displayed formulae of CH_3COOH and CH_3CH_2COOH. The carbon-13 NMR spectrum of one of these compounds is given below. Identify the compound which produced this spectrum. Explain your answer. [3]

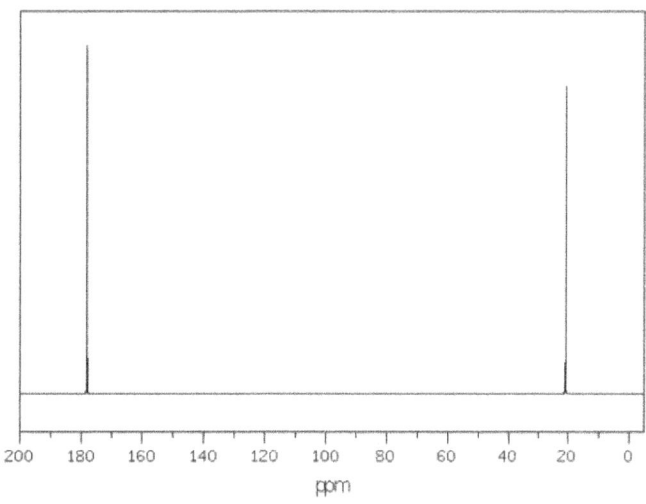

6 QUESTIONS
Organic chemistry and analysis

93. State some limitations of gas chromatography. [2]

94. Name the compound shown below. Explain whether the spectrum shown could be the carbon-13 NMR spectrum of this compound [3]

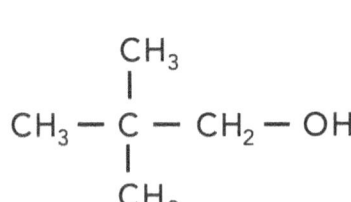

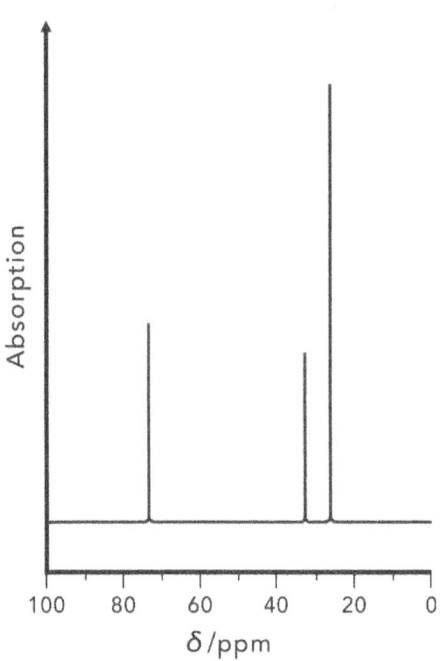

95. Draw the structural formula of 2,2-dimethylbutane. The carbon-13 NMR spectrum of this compound is one of the spectra shown below. Identify which one, explaining your reasoning. [4]

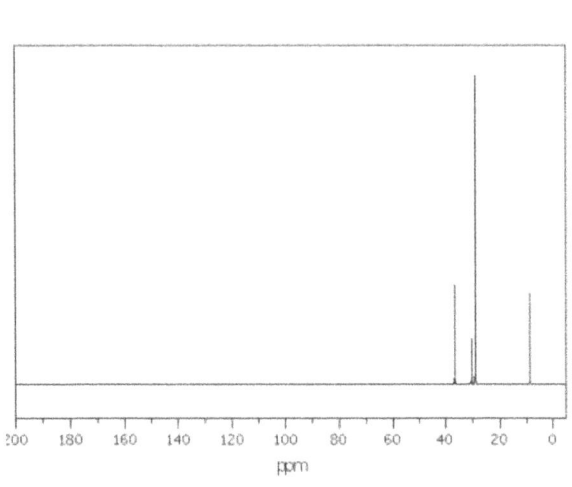

(a)

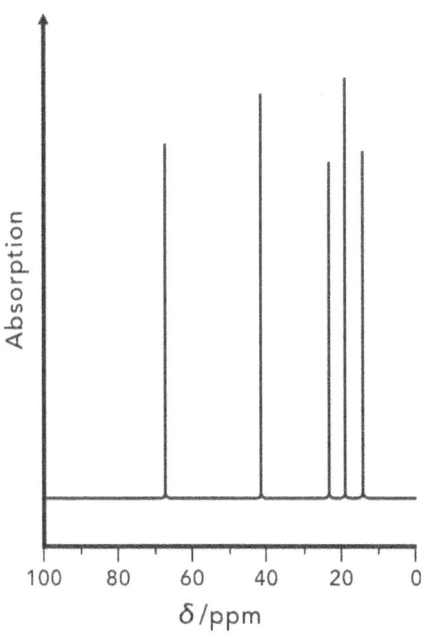

(b)

QUESTIONS 6

Organic chemistry and analysis

96. Three compounds are given below, each with their proton NMR spectra. Match the compound to its correct proton NMR spectrum, explaining your reasoning. [9]

2-methylpropane
2,2-dimethylpropanal
methyl propanoate

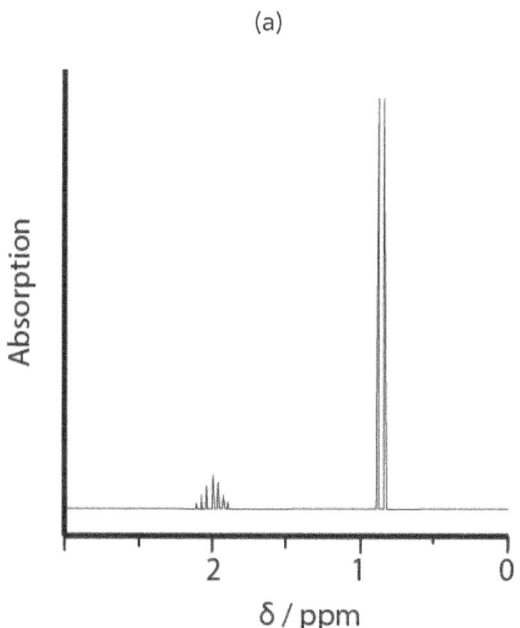
(a)

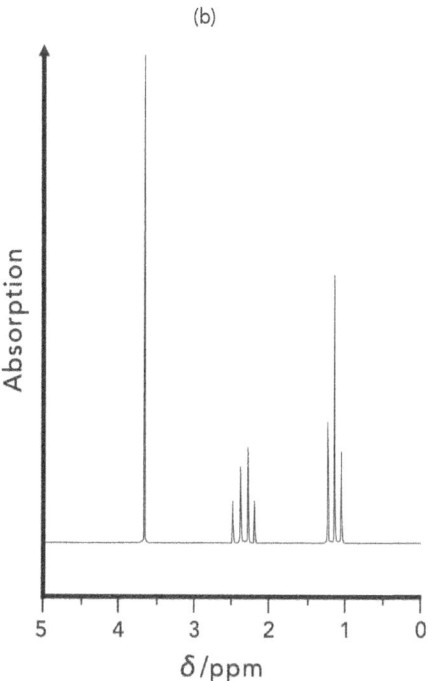
(b)

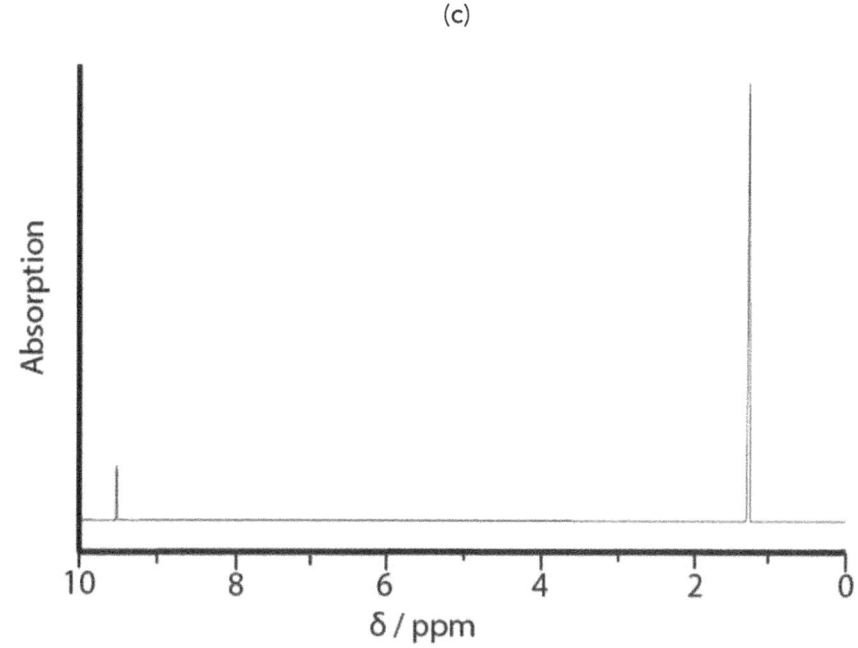
(c)

QUESTIONS
Organic chemistry and analysis

97. The carbon-13 NMR of two of the structural isomers of $C_6H_{12}O_2$, along with the two isomers are given below. Match each of these spectra to its isomer. Explain your reasoning. [2]

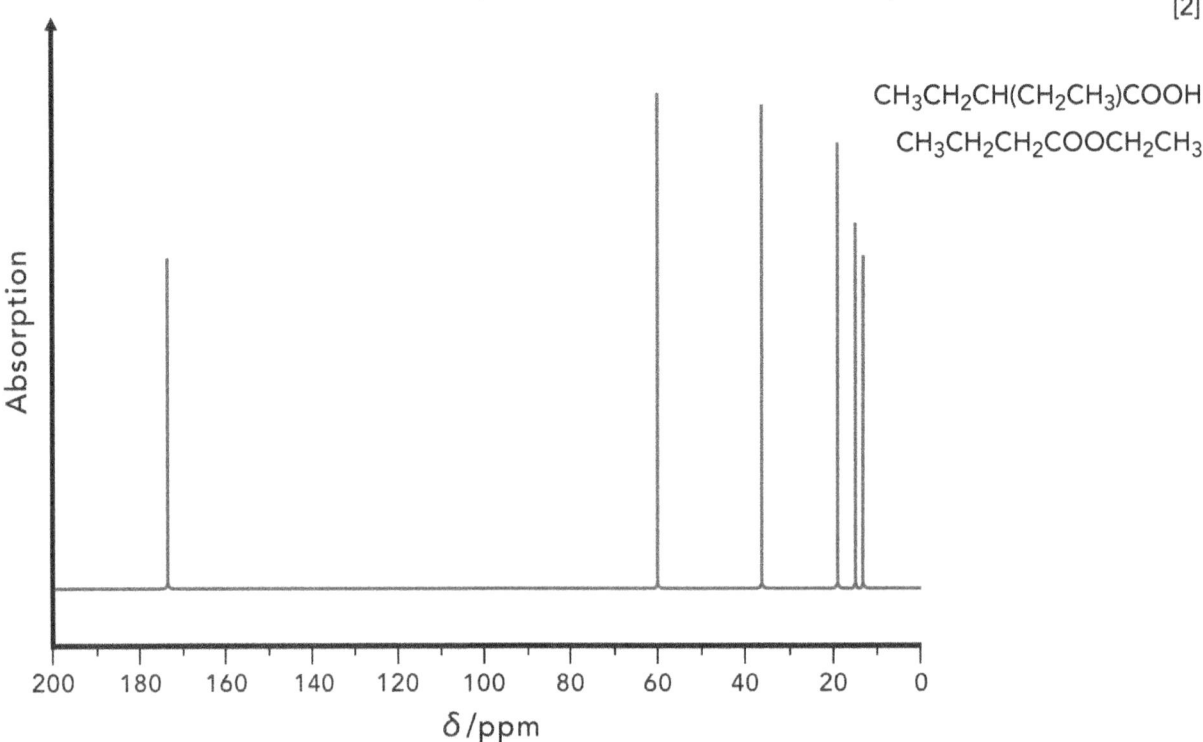

$CH_3CH_2CH(CH_2CH_3)COOH$

$CH_3CH_2CH_2COOCH_2CH_3$

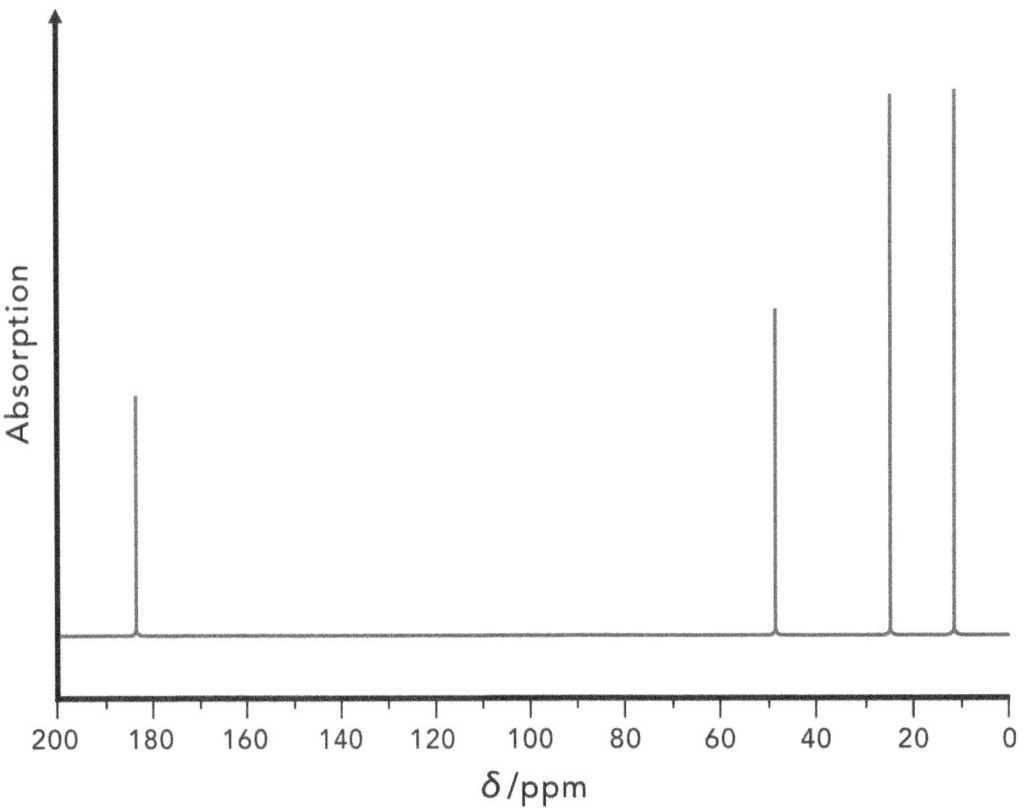

98. The following proton NMR spectrum is derived from a ketone with a molecular weight of 86. Identify this ketone, explaining your reasoning. (The ketone has the formula C_xH_yO). [5]

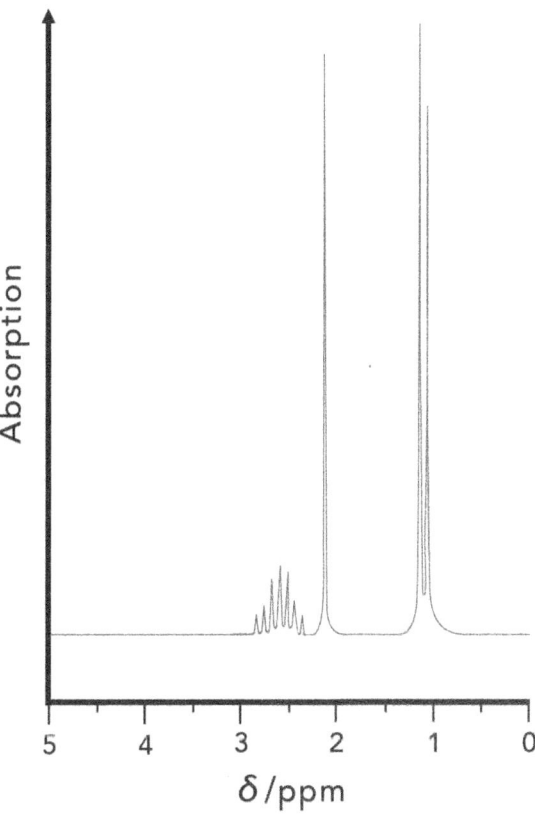

99. Proton NMR spectroscopy is based on the magnetic properties of hydrogen nuclei. It can be used to determine the structural formulae of unknown compounds.

   a. What is spin-spin coupling, as seen in a proton NMR spectrum? What information does spin-spin coupling tell you? [1]

   b. Explain how D_2O may be used to identify O-H and N-H protons in proton NMR spectroscopy? How does this work? [2]

100. Mass spectrometry may be used to determine the structure of an unknown compound.

   a. State how you would determine the molecular mass of a compound from its mass spectrum. [1]

   b. Describe the terms "molecular ion" and "molecular ion peak". [2]

101. What information can be obtained by mass spectrometry, infrared spectroscopy and NMR spectroscopy? [3]

102. Explain the spin-spin coupling patterns observed in the proton NMR spectrum of 2-iodopropane. [2]

103. ^1H spectroscopy is used in the identification of compounds in universities and in forensic medicine.

   a. Describe the "n+1 rule" of ^1H NMR spectroscopy. [1]

QUESTIONS — Organic chemistry and analysis

b. Explain the use of D_2O in ^1H NMR spectroscopy. [1]

c. State four key pieces of information provided by a ^1H NMR spectrum. [3]

104. NMR spectroscopy and gas chromatography can be used to analyse unknown compounds.

By comparing the retention time of an unknown sample with that of known compounds, a sample may be identified.

a. State what is meant by the term retention time. [1]

b. What are the limitations of using gas chromatography as an analytical tool? [2]

c. A student analysed a compound with the molecular formula C_4H_7OCl using ^1H NMR spectroscopy. Its NMR spectrum is shown below. The number next to each signal represents the number of protons responsible for that signal.

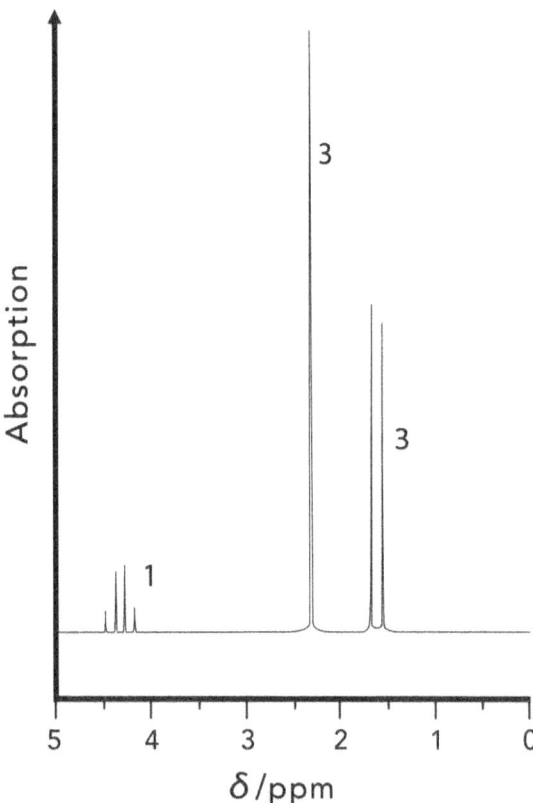

Suggest a structure for the compound. Draw its skeletal formula. Show **all** your reasoning. [4]

QUESTIONS 6

105. Explain the spin-spin coupling patterns observed in the proton NMR spectrum of 2-chloro-2-methylbutane with their relative chemical shifts. [4]

106. The three spectra given here relate to the same compound. The proton-NMR spectrum does not include any peaks resulting from the -OH protons. The number next to each signal represent the number of ^1H responsible for that signal. Use the spectra to suggest a possible structure for this compound. Draw its skeletal formula.

    The empirical formula of the compound is C_3H_6O. [6]

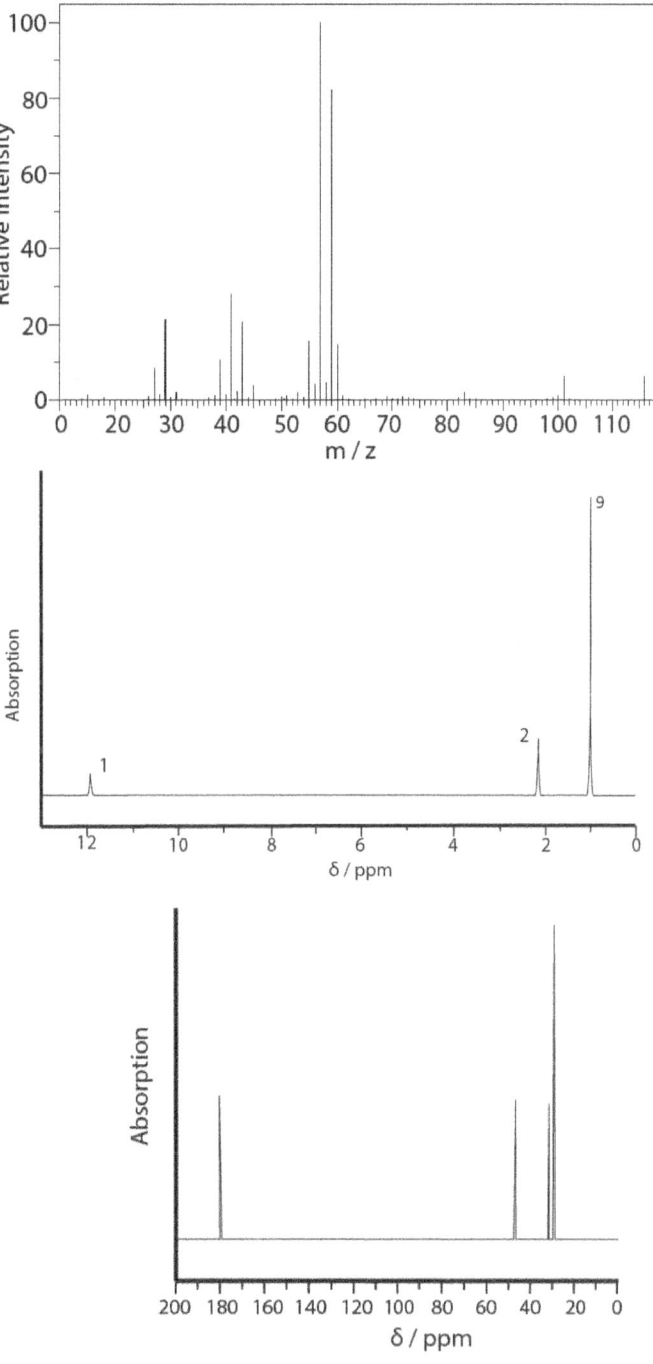

6 QUESTIONS — Organic chemistry and analysis

107. A compound with a molecular formula $C_6H_{13}Cl$ has its proton NMR and carbon-13 NMR spectra as shown below. Using this information, identify the compound. [5]

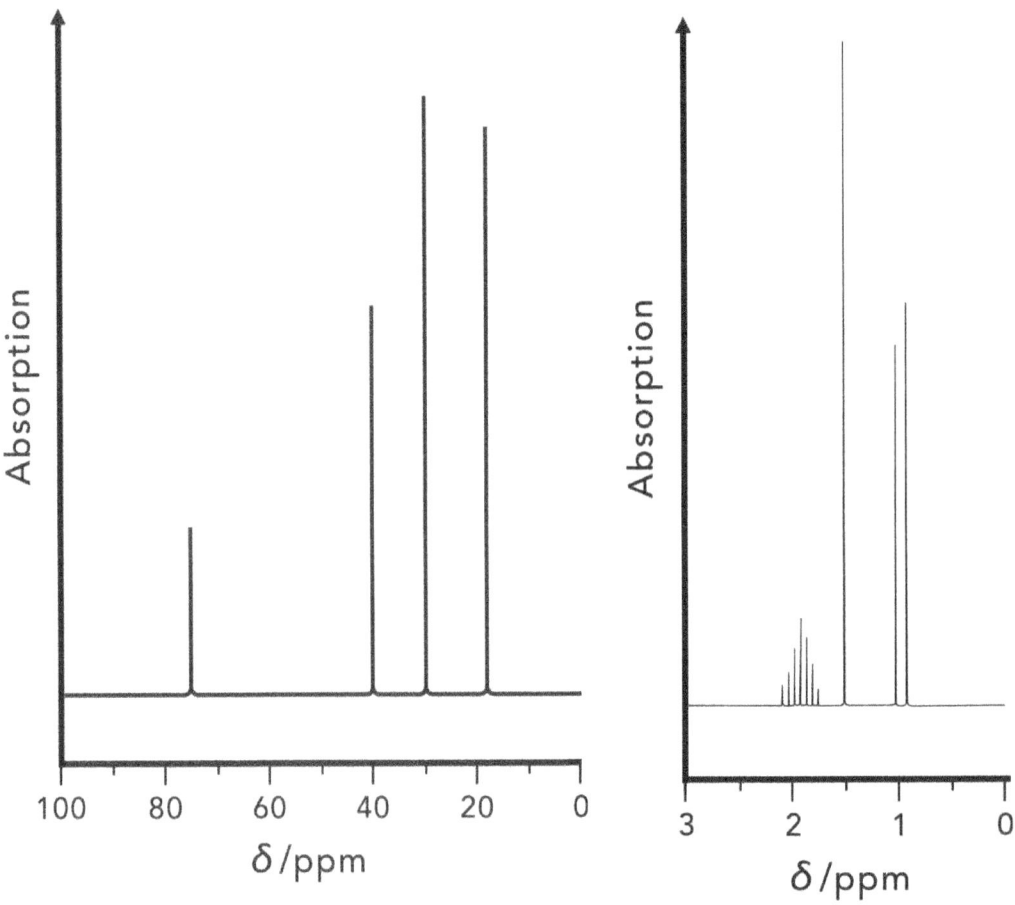

108. Carbon-13 NMR spectroscopy was carried out on 1-bromo-4-methylbenzene and 2,2-dimethylbutane. One of these spectra is given below.

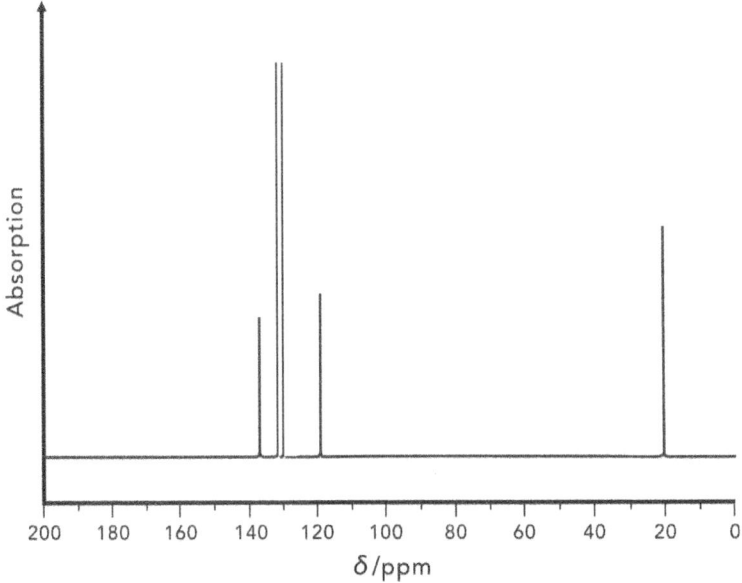

a. Draw the displayed formula of 1-bromo-4-methylbenzene. [2]

b. On the diagram you have drawn, label the carbon atoms that are in the same enviornment and state the number of peaks you would expect to see in the carbon-13 NMR spectrum of 1-bromo-4-methylbenzene. [2]

c. Identify the compound which produced the above spectrum. Explain your reasoning. [2]

109. Alkylated derivatives of **Compound X** is often used in computer hard drives and other similar equipments. In its pure form, it is extremely flammable. Its skeletal formula is shown below.

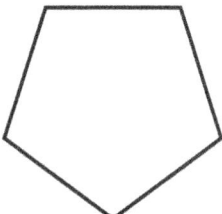

a. Name this compound. [1]

One of the following spectra is of **Compound X**.

(a)

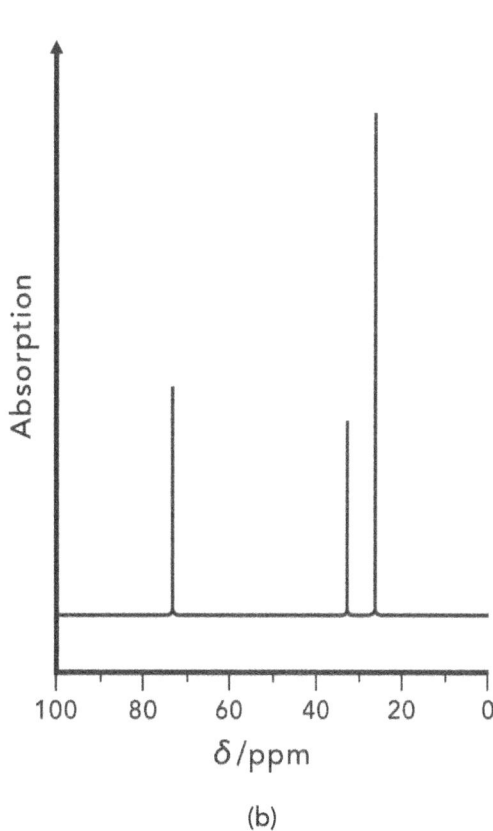

(b)

b. State which spectrum belongs to **Compound X**. Explain your reasoning. [2]

6 QUESTIONS — Organic chemistry and analysis

c. Draw the displayed formula of **Compound X** showing any other atoms present in the compound. [1]

110. Acetone is widely used as a solvent in many chemical reactions. The displayed formula of acetone is shown below.

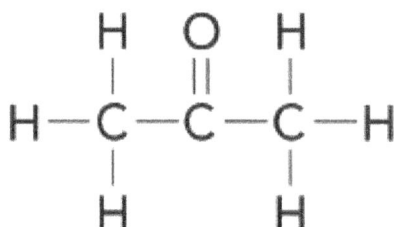

a.

i. State the systemic name of acetone. [1]

ii. Draw the skeletal formula of acetone. [2]

b. This compound reacts with HCN to form **Compound N**.

i. Draw the reaction mechanism for the reaction. [5]

ii. Name **Compound N**. [1]

iii. State the type of the above reaction. [1]

111. Testosterone is a hormone found in humans. It is important in the prevention of osteoporosis.

Enanthic acid is useful in the formation of synthetic testosterone. Its chemical formula is $CH_3(CH_2)_5COOH$.

a. State the systemic name of enanthic acid. [1]

Enanthic acid may be formed starting from an aldehyde.

b. State the name of an aldehyde that could be used. [1]

c. Using a labelled diagram of the apparatus required, describe how enanthic acid may be formed starting from the aldehyde. [5]

Aldehydes can be reduced to primary alcohols using $NaBH_4$.

d.

i. Draw the reaction mechanism for the reduction of the aldehyde you named in part (b). [4]

ii. State the type of reaction involved in this reduction. [1]

112. Hexanoic acid is found naturally in animal fats and oils. It is an oily liquid.

A student was provided with a mixture of hexanoic acid and water in a conical flask and was asked to separate the two liquids.

a. Draw the labelled diagram of a suitable apparatus which could be used to carry out the separation. [2]

b. Describe a simple chemical test which the student could use to identify the carboxylic acid. [4]

113. You are provided with the following **Compound X**.

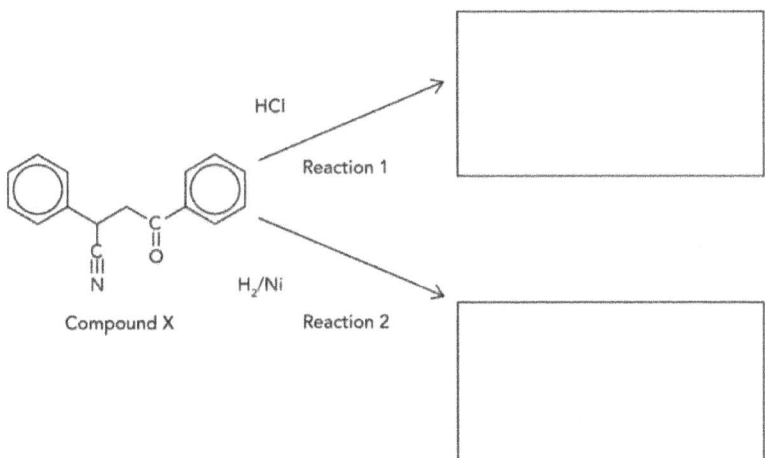

a. **Compound X** has several functional groups.

   i. State two of these functional groups. [2]

   ii. Describe a simple chemical test which can be used to identify one of the functional groups you have stated. [2]

b. The flowchart below shows some reactions of **Compound X**.

Complete the flowchart.

- Draw the structure of the product of **Reaction 1**.
- Draw the structure of the product of **Reaction 2**. [4]

c. The product from reaction 2 reacts with dilute hydrochloric acid to form **Compound Y**. Draw the structure of **Compound Y**. [1]

6 QUESTIONS — Organic chemistry and analysis

114. Ethylbenzene is a compound which occurs naturally in coal tar and petroleum. The flowchart below shows some reactions of ethylbenzene.

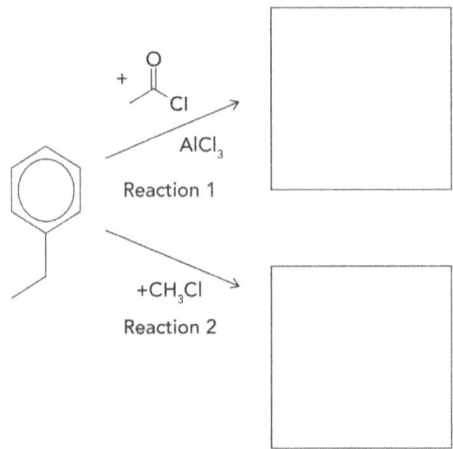

a. Complete the flowchart above.

   - Draw the structure of the major product of Reaction 1.
   - Draw the structure of the major product of Reaction 2. [2]

b. i. Give the molecular formula of the minor product formed in Reactions 1 and 2. [1]

   ii. Suggest the name of the mechanism involved in reaction 2. [1]

c. A student suggested that during Reaction 1, another **compound X**, of molecular mass 191 g mol^{-1} may also be also formed.

   Draw the structure of **compound X**. [1]

d. Suggest what you would observe if acyl chloride is reacted with 2,4-DNP. [1]

115. A chemist analysed a sample of a **compound P**. Following spectra and information were obtained through the analysis.

   Mass spectrum: Molecular ion peak at m/z = 73.0

   ^{1}H NMR spectra

   The number next to each signal represent the number of ^{1}H responsible for that signal.

   With no D$_2$O.

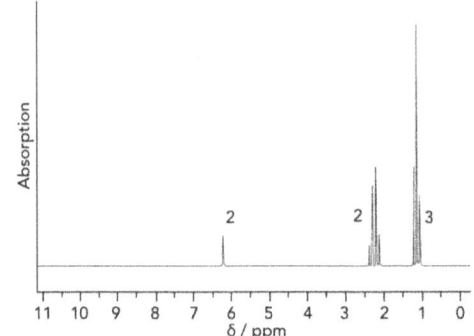

With D$_2$O added.

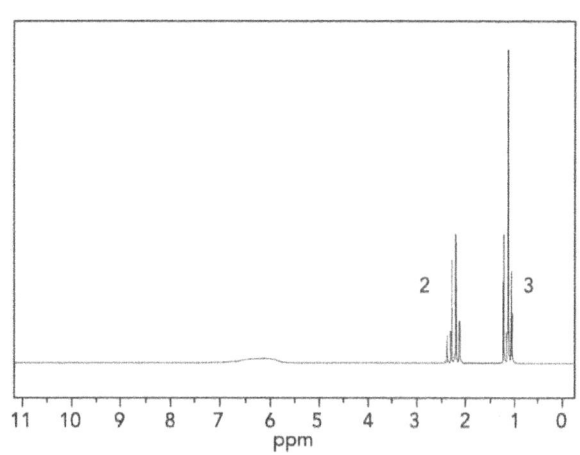

^{13}C NMR spectrum.

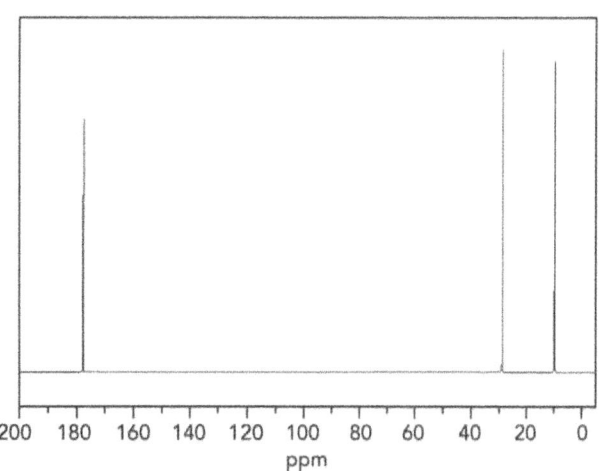

Using the information provided, suggest a structure for **compound P**. Give your reasoning. [5]

116. Give chemical explanations for the following statements.

   a. It is relatively easier to change water in liquid form to a gas than to change water into hydrogen and oxygen.[1]

   b. CH$_3$CHBrCHClCH$_3$ has four optical isomers. [1]

   c. Thin-layer chromatography allows identifications of components in a mixture. [1]

   d. D$_2$O removes any peaks due to NH and OH protons from a ^{1}H NMR spectrum. [1]

   e. 2.6 g of C$_6$H$_5$COCl contains 1.11 x 10^{22} molecules. [1]

   f. Benzene undergoes electrophilic substitution reactions but not electrophilic addition reactions. [1]

   g. 0.05 mol of both CO$_2$ (g) and NH$_3$ (g) occupy a volume of 1.2 dm^3. However, the amount of each gas in grams is 2.2 g and 0.85 g respectively. [1]

QUESTIONS
Organic chemistry and analysis

h. There are eight times the number of moles of gas X in 2 dm³ at 400 kPa and 25 K than in 2 dm³ at 200 kPa and 100 K. [1]

117. **Compound A** reacts with **compound B** as shown in the equation below.

Compound A + Compound B →

H—C(H)(H)—C(H)(CH₃)—C(OH)(H)—CN

Compound C

A silver-grey solid was observed when Tollens' reagent was added to **compound A**. Its ^{13}C NMR spectrum has 3 peaks and its mass spectrum has its m/z peak at 72.

a. Using the information given above, deduce the structure of **compound A**. [2]

b. State the molecular formula of **compound B**. [1]

c. Outline the mechanism for the above reaction, showing curly arrows and relevant dipoles. [3]

d. State two uses of infrared spectroscopy. [2]

e. A chemist reacted 4.32g of **compound A** and obtained 3.72g of **compound C**. A student incorrectly stated that the percentage yield of the reaction is 80%. Work out the mass in grams of **compound C** that needs to be formed for the percentage yield to be 80% provided 4.32g of **compound A** was used. [3]

118. You are given the following information about **compound Q**, with empirical formula C_3H_6O.

Mass spectrum:

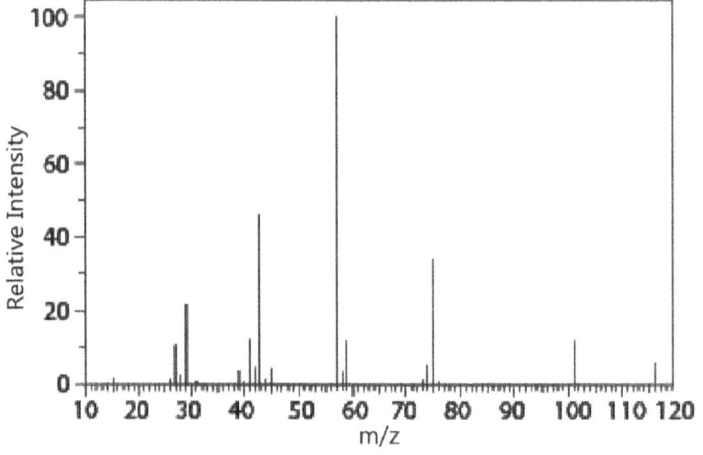

¹H NMR spectra The number next to each signal represent the number of ¹H responsible for that signal.

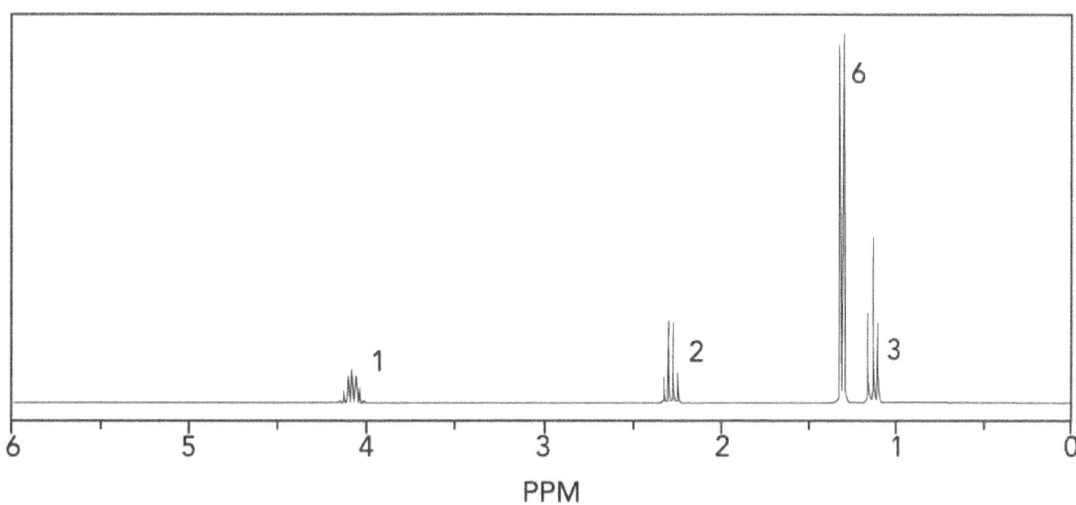

¹³C NMR spectrum

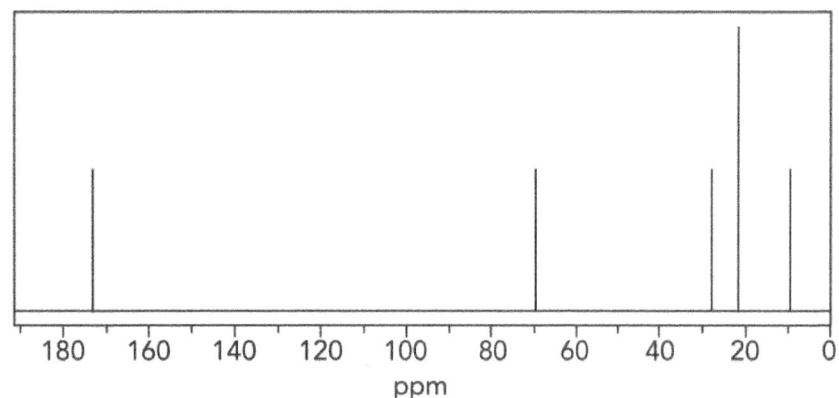

Use the information provided to suggest a structure for **compound Q**.

Give your reasoning. [6]

119. A student reacted 4.00 g of butanoic acid with propan-1-ol to form an organic **compound W** that is a liquid.

QUESTIONS
Organic chemistry and analysis

a. i. Draw the structure of **compound W**. [1]

ii. Outline the steps that could be carried out to obtain a pure sample of the organic product from the reaction mixture. [4]

b. i. Write a balanced equation for the reaction that takes place. [2]

ii. Calculate the minimum mass, in grams, of propanol required to completely react with the butanoic acid. [3]

iii. The student was then asked to prepare a 10.0 cm³ sample of a very dilute aqueous solution of compound W of concentration 5.00×10^{-2} mol dm^{-3}. Work out the mass in grams of compound W that the student would need to use. [2]

120. The following flow chart shows a two stage synthesis.

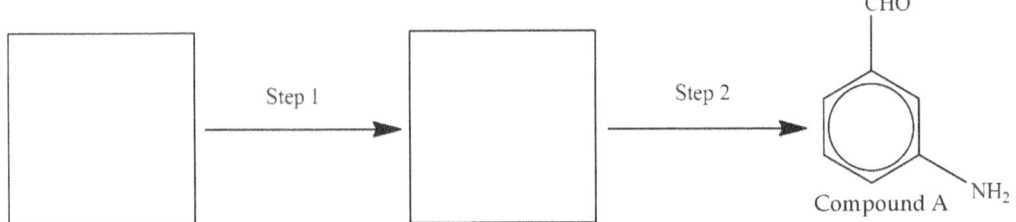

a. i. Complete the boxes above by drawing the major organic reactant / product. [2]

ii. State suitable reactant(s) and conditions, if any, for **step 1** and **step 2**. [2]

b. i. Outline the mechanism for **step 1**. Your mechanism should use curly arrows. [3]

ii. State the name of the mechanism involved in **step 1**. [1]

c. You are provided with two unlabelled samples of **Compound A** and phenylamine.

Describe a simple chemical test which would allow you to distinguish between the two samples. [2]

121. A chemist carried out several investigations on a sample of an ester to establish its identity. The results of his analysis are given below.

Elemental analysis by mass

C, 62.1%; H, 10.3%; O, 27.6%

Mass spectrum

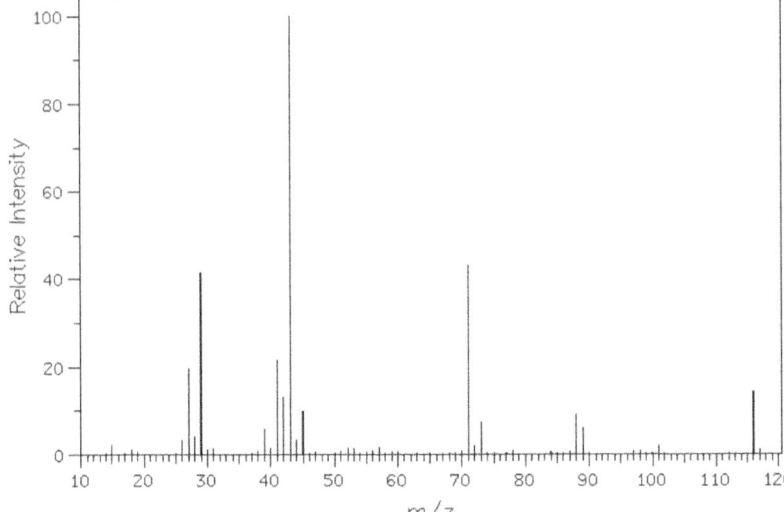

^{1}H NMR spectrum

The number next to each signal represent the number of ^{1}H responsible for that signal.

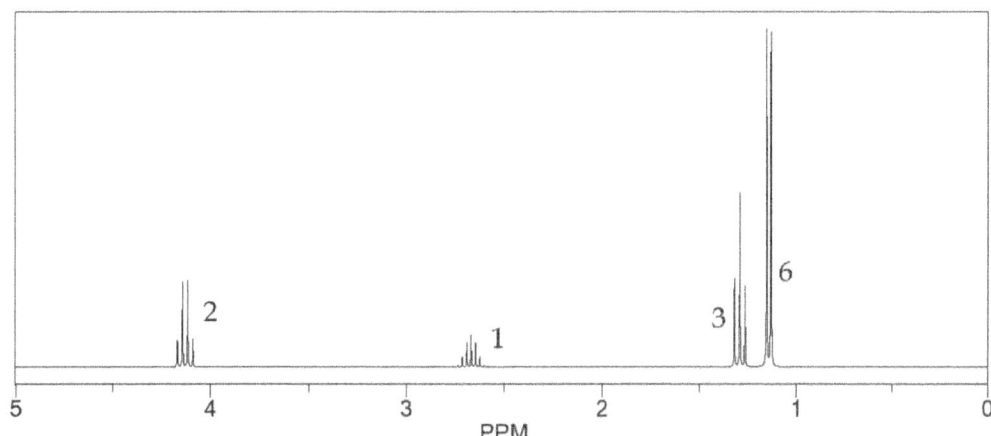

Use the results to identify the ester and draw its skeletal formula. Show **all** your reasoning. [6]

122.

a. State one advantage of the Friedel-Crafts reaction in synthesis. [1]

b. Benzaldehyde is often used as a precursor in the formation of other chemicals. It is also used to achieve an almond flavor. Some of the reactions of benzaldehyde are given in the chart shown below.

i. In the spaces provided, draw the skeletal formula of the major organic product of the following reactions. [4]

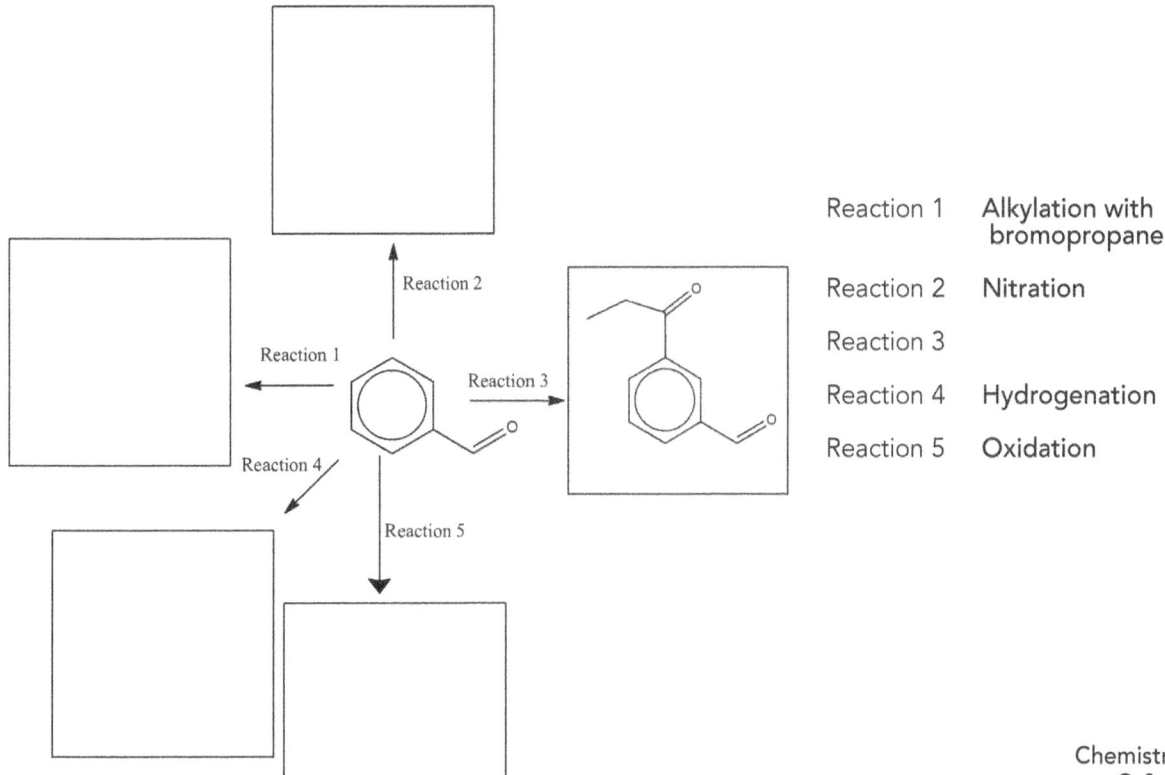

Reaction 1   Alkylation with bromopropane
Reaction 2   Nitration
Reaction 3
Reaction 4   Hydrogenation
Reaction 5   Oxidation

QUESTIONS — Organic chemistry and analysis

   ii. State the reagent(s) and conditions needed for **Reaction 3**. [3]

   iii. The product from **Reaction 5** was reacted with aqueous sodium hydroxide. Draw the skeletal formula of the major organic product and state the molecular formula of the other product(s), if any. [2]

   iv. The product from **Reaction 5** was then reacted with sulphur dichloride oxide. Write an equation using skeletal formula for the reaction that takes place. [2]

123. Propionitrile is used in the production of Flopropione, an antispasmodic agent.

    Propionitrile has the formula CH_3CH_2CN.

    a. State another chemical name for propionitrile. [1]

    A student was asked to prepare a sample of propionitrile. Two unlabelled containers containing bromoethane and potassium cyanide and a third container containing ethanol was provided.

    b. Describe a simple chemical test that can be used to distinguish between the two compounds. [3]

    c. i. Write an equation for a reaction that could be carried out to form CH_3CH_2CN from bromoethane. [2]

       ii. Draw the labelled diagram of a suitable reaction apparatus for the above reaction. [4]

       iii. Draw the reaction mechanism for the reaction you have described. [4]

       iv. State the type of reaction involved. [1]

    When some reactions are carried out, the reaction mixture may be heated by placing it in a water bath.

    d. Explain why this is necessary. [1]

124. $CH_3CHBrCHBrCH_2CH_3$ show optical isomerism.

    a. Draw the 3D structures of all the optical isomers of this compound. [4]

    b. State the number of peaks you would observe in the carbon-13 NMR spectrum of this compound. [1]

    A student used a sample of $CH_2BrCH_2CH_2CH_2CH_3$ to form compound **Q**. When compound **Q** was analysed using infrared spectroscopy, it had a broad absorption peak between 2400 and 3290 cm^{-1} and it **effervesced when reacted with sodium carbonate**.

    c. i. Devise a two-step synthesis of compound **Q** from 1-bromopentane. Give details of appropriate reagents and relevant conditions. Write an equation for each step, showing clearly all organic compounds. [4]

       ii. State the type of mechanism involved in the first step of the synthesis. [1]

       iii. Draw the mechanism for the first step. Use curly arrows and show relevant dipoles. [4]

       iv. Two peaks with m/z values 71 and 59 were observed in the mass spectrum of compound **Q**. State the molecular formula of the fragments responsible for these peaks. [2]

125. Serine and aspartic acid are examples of two naturally occurring amino acids.

Serine

Aspartic acid

a. State the general formula of an alpha amino acid. [1]

b. State the formulae of the R groups of serine and aspartic acid. [2]

c. Draw the skeletal formula of the product formed when aspartic acid reacts with nitric acid. [1]

d. Dipeptides are formed when the carboxyl group of one amino acid react with the amine group of another amino acid.

   i. Draw the structures of the dipeptides formed by serine and aspartic acid. [4]

   ii. Identify the chiral centres in one of the dipeptides you have drawn. [2]

   iii. State the number of optical isomers of the dipeptide. [1]

126. 5-hydroxypentanoyl chloride can be used to form an ester without the need for any other reactant.

   a. Write an equation using displayed formula for this reaction [2]

   b. 5-hydroxypentanoyl chloride can also be used to synthesise a polymer. Draw the displayed formula of one repeating unit of the polymer and name the type of the polymer and the polymerisation involved. [3]

   c. The following compound may be formed starting from 5-hydroxypentanoyl chloride.

   Devise a two-step synthesis of this compound from 5-hydroxypentanoyl chloride. Give details of appropriate reagents and relevant conditions. Write an equation for each step, showing clearly all organic compounds. [4]

127. A student analysed a sample of a **compound X**. Following spectra and information were obtained through the analysis.

   Elemental analysis by mass

   C, 40.7%; H, 5.08%; O, 54.2%

6 QUESTIONS — Organic chemistry and analysis

¹H NMR spectra

The number next to each signal represent the number of ¹H responsible for that signal.

With D_2O.

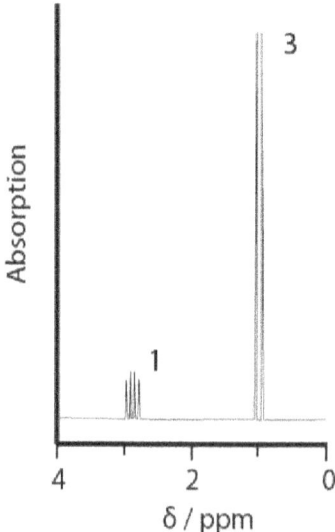

¹³C NMR spectrum.

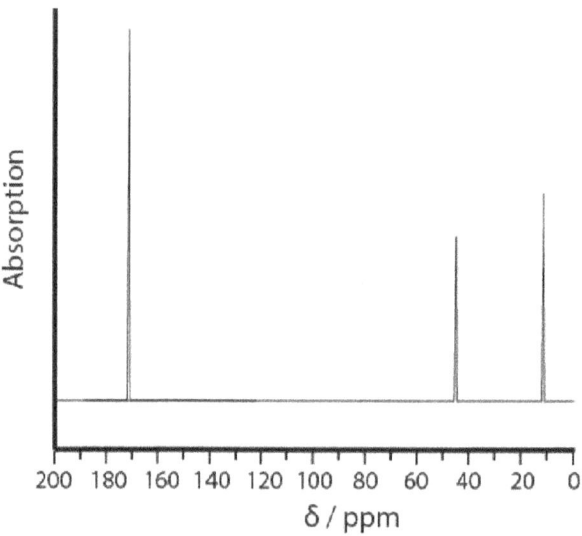

Mass spectrum.

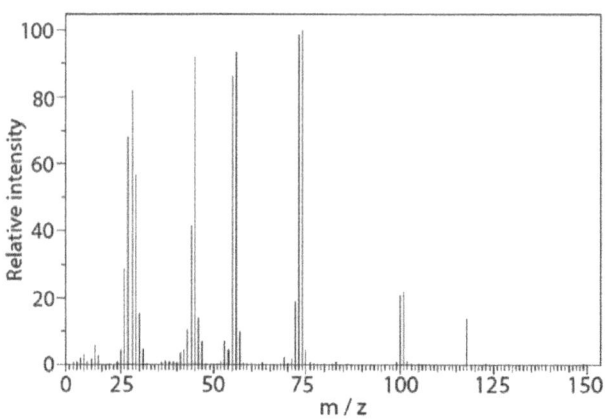

Using the information provided, suggest a structure for **compound X**. Give your reasoning. [6]

128. Thin-layer chromatography (TLC) and gas chromatography (GC) are two types of analytical techniques.

TLC was carried out on two mixtures of amino acids. Following the procedure, the two chromatograms shown below were produced.

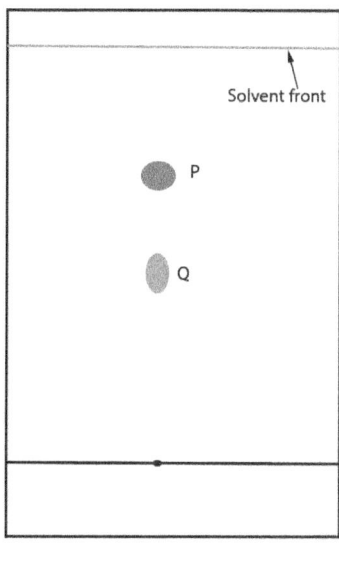

Mixture 1   Mixture 2

You are provided with the R_f values of some amino acids.

Amino acid	R_f
Alanine	0.38
Aspartic acid	0.24
Cysteine	0.40
Tyrosine	0.45
Methionine	0.55
Leucine	0.73
Serine	0.27

a. Determine the components of mixtures 1 and 2. [5]

b. Another chromatogram was produced for a different mixture. In this chromatogram, the solvent front had moved a distance of 7.23 cm. A student suspected that the mixture might contain aspartic acid. Calculate the distance aspartic acid would move if present in the mixture. [1]

c. Suggest one advantage of TLC over GC. [1]

d. State one type of information which GC provides but TLC does not. [1]

129. A student was given a mixture containing four different pesticides.

QUESTIONS
Organic chemistry and analysis

The student carried out gas chromatography on the sample. The resulting chromatogram is shown below. The number next to each peak gives the area of each peak. One of the components in the sample is Malathion.

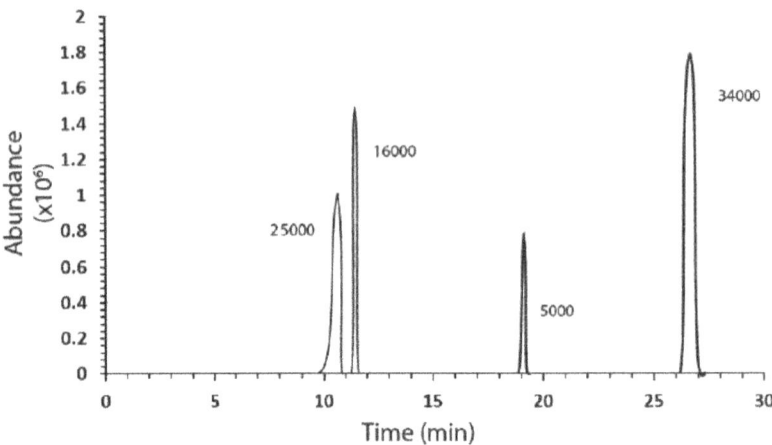

You are provided with the following data.

Pesticide	Retention time (min)
Chlorothalonil	8.89
Chlorpyrifos-methyl	9.78
Malathion	10.77
Dicofol	11.49
Endrin	14.85
Tetradifon	19.15
Azoxystrobin	26.83

Gas chromatography was carried out on different concentrations of Malathion. The peak areas under the curves were calculated for each concentration to obtain the table below.

Concentration ($\times 10^{-2}$ mol dm^{-3})	Area ($\times 10^3$)
3.0	12
4.2	20
3.8	35
5.3	40

a. Identify the components and determine the concentration of Malathion in the sample. [7]

b. Determine the area of the peak resulting from a Malathion sample of concentration 0.050 mol dm^{-3} to two significant figures. [2]

QUESTIONS 6

Organic chemistry and analysis

130. You are provided with a sample of a **compound Y**, of molecular mass 60 g mol^{-1}.

   a. Devise a two-step synthesis of the following compound from **compound Y**. Give details of appropriate reagents and relevant conditions. Write an equation for each step, showing clearly all organic compounds. Draw a labelled diagram of the apparatus you would use for step 1. [7]

$$H-\underset{\underset{H}{|}}{\overset{\overset{H}{|}}{C}}-\underset{\underset{H}{|}}{\overset{\overset{H}{|}}{C}}-\overset{\overset{O}{\|}}{C}-Cl$$

   b. **Step 2** also formed an organic product of molecular mass 156.6 g mol^{-1} along with HCl. Suggest a possible structure for this product. Draw its displayed formula. [2]

131. You are provided with an unlabelled mixture containing amino acids. One of the four tables below correctly outlines the mixture. The tables are incomplete.

Mixture 1		
	Retention time	Ratio of
X	80.2	27
Y	21.5	12
Z	47.0	

Mixture 2		
	Retention time	Ratio of
X	22.7	12
Y		39
Z	47.0	110

Mixture 3		
	Retention time	Ratio of
X	21.5	12
Y	38.4	
Z	47.0	110

Mixture 4		
	Retention time	Ratio of
X	22.7	
Y	38.4	27

Gas chromatography was carried out and the peak integration values of X:Y:Z were found to be 2000:4500:18300.

The three amino acids in the mixture are listed in the following table.

Amino acid	Molecular formula	Retention time (min)
Threonine	$C_4H_9NO_3$	21.5
Serine	$C_3H_7NO_3$	22.7
Proline	$C_5H_9NO_2$	30.0
Glycine	$C_2H_5NO_2$	38.4
Valine	$C_5H_{11}NO_2$	47.0
Histidine	$C_6H_9N_3O_2$	80.2

One of the amino acids in the mixture can react with a different amino acid that is **not** present in the mixture to form the dipeptide with the formula NH₂CH(CH₂OH)CONHCH(CH(CH₃)CH₃)COOH.

Identify the amino acids in the mixture. Give all your reasoning. [7]

132. A student wanted to prepare the following **compound X** for a school project.

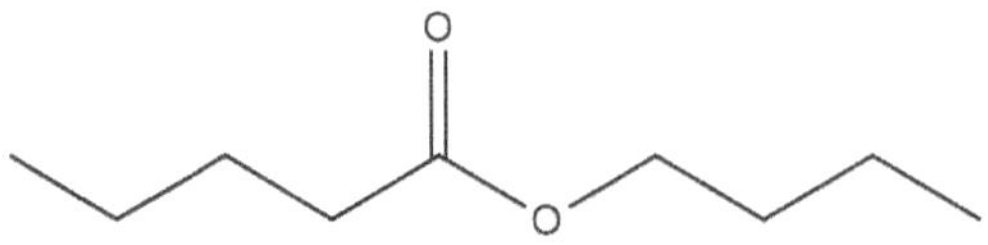

   a. i. Name the compound. [1]

   ii. Name its functional group. [1]

   b. **Compound X** may be formed via two synthetic routes as shown.

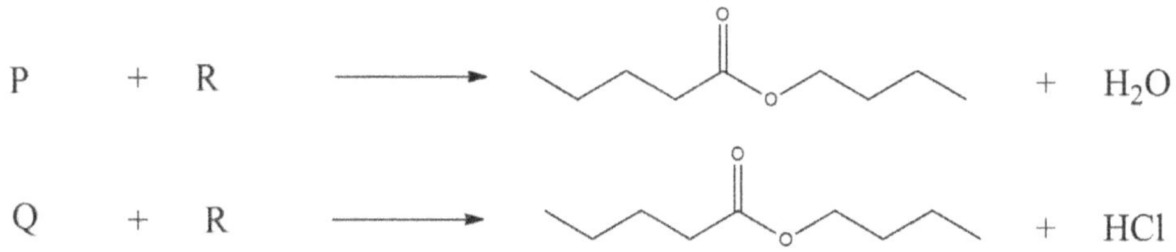

   P + R ⟶ [ester structure] + H₂O

   Q + R ⟶ [ester structure] + HCl

   Identify P, Q and R. Name them and draw their skeletal formulae. [4]

   c. **Compound X** was heated with aqueous potassium hydroxide.

   i. Write a balanced equation for the reaction. [2]

   ii. State the type of the reaction. [1]

133. The following compound, **D**, has isomers.

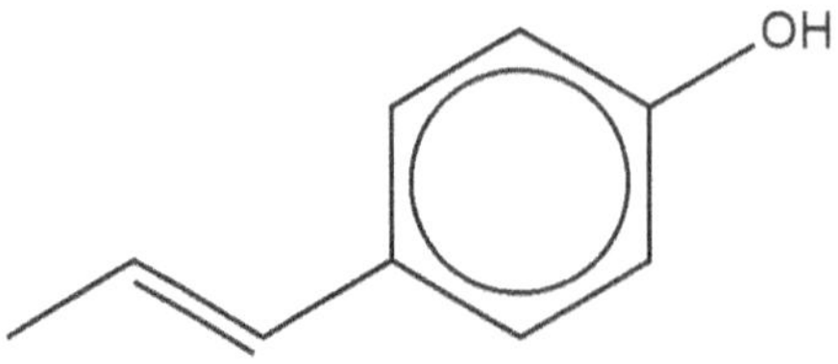

   a. Name two functional groups present in **compound D**. [2]

   b. Write equations using skeletal formulae for the reaction, if any, between **compound D** and the following reagents.

   i. NaOH [2]

ii. Na$_2$CO$_3$ [1]

c. i. Draw the displayed formula of an isomer of **compound D**. [1]

ii. Name the type of isomerism involved. [1]

iii. Name the type of isomer you have drawn. State your reasoning. [2]

134. NO$_2^+$ can attack a phenol ring to form a compound with molecular mass 139 g mol^{-1}. This compound has two structural isomers. This reaction occurs in the dark.

a. i. Draw the skeletal formula of the two structural isomers. [2]

ii. Explain the position of any groups in the structures you have drawn. [2]

b. State the type of reaction taking place. Explain. [2]

135. The products of a reaction may be determined by taking into account the nature of the groups involved. Some groups are electron-donating while some others are electron-withdrawing.

Electron-donating groups	Electron-withdrawing groups
CH$_3$	CF$_3$
OCH$_3$	COOH
OH	CN
NH$_2$	

a. Draw the product(s) of the following reactions: [4]

i.

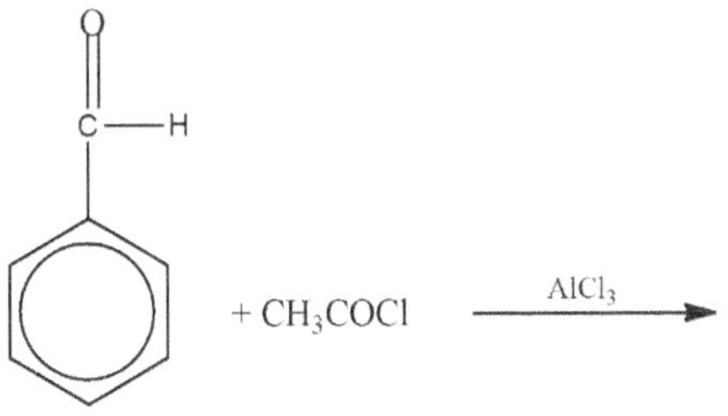

[benzaldehyde + CH$_3$COCl, AlCl$_3$]

ii.

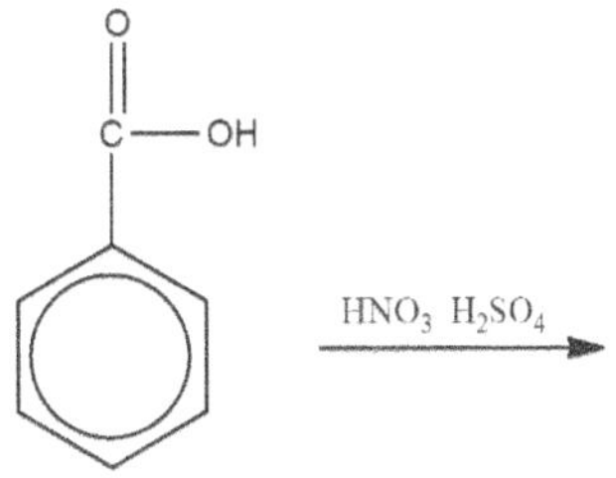

[benzoic acid, HNO$_3$ H$_2$SO$_4$]

iii.

[toluene, HNO$_3$ H$_2$SO$_4$]

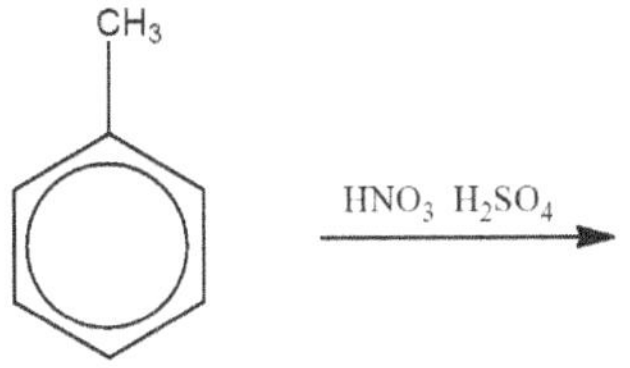

b. i. Describe the effect of electron-donating and electron-withdrawing groups on the formation of products. [2]

ii. State an instance when this might be important. [1]

c. State the formula of an electron-withdrawing group whose ion may attack carbonyl compounds in a nucleophilic addition reaction. [1]

136. Imipramine is a drug used to treat major depression. It was the first tricyclic antidepressant to be developed. Its structure is shown below.

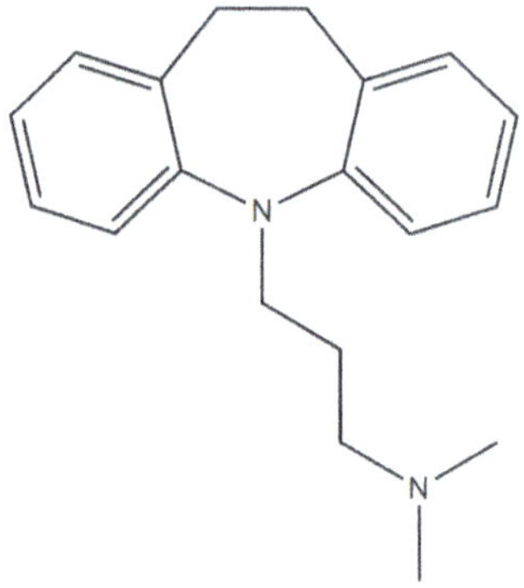

a. Name the type of formula used to show imipramine. [1]

b. Deduce the molecular formula of imipramine. [1]

c. State the number of peaks in its carbon-13 NMR spectrum. On the diagram shown above, identify the carbon atoms that are in the same environment. [2]

d. Explain why imipramine may react with acids. [2]

137. Glycerol is a colourless and odorless liquid that is widely used in the pharmaceutical industry. Its structure is shown below.

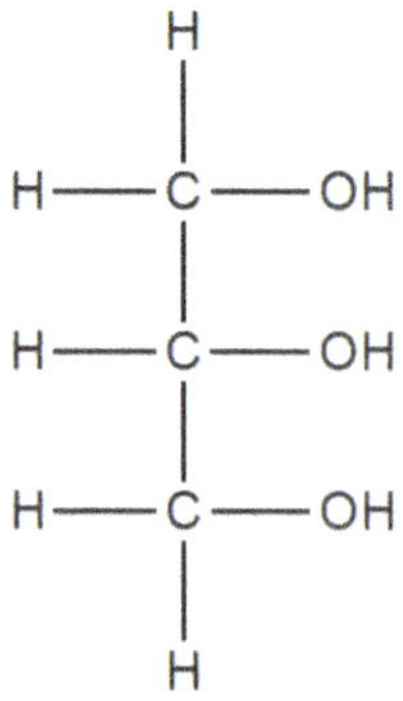

a. Glycerol reacts with the following fatty acid to form a major organic product and a minor product.

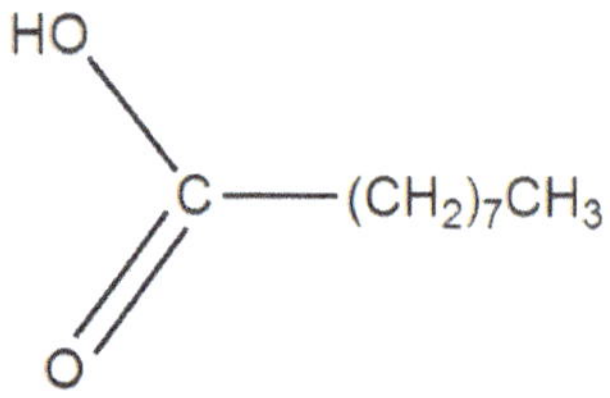

i. Write a balanced equation for the reaction between glycerol and fatty acid. [3]

ii. State the name of the type of reaction involved. [1]

iii. State the formula of another compound that could react with glycerol to form the same major organic product. [1]

b. i. The major organic product from the above reaction was heated with aqueous sodium hydroxide. Write a balanced equation for the reaction. [3]

ii. State the name of the type of reaction involved. [1]

138. Some amino acids are made in the human body. Glycine, is an example of a conditionally essential amino acid which means it cannot be made by the body under certain circumstances.

When alpha amino acids form salts with transition metals, four coordinate complexes are formed. Copper (II) forms copper diglycinate with glycine. Glycine is the simplest alpha amino acid whose R group is a hydrogen atom.

a. Draw the displayed formula of copper diglycinate, $C_4H_8CuN_2O_4$. [5]

b. i. Glycine was heated with butanol in the presence of hydrogen chloride. Write an equation, using displayed formula for the reaction that is taking place. [2]

ii. The major organic product from the reaction was reacted with hot aqueous sodium hydroxide. Draw the displayed formula(e) of the product(s) formed. [2]

139. You are asked to prepare a sample of the following **compound M**, in the laboratory.

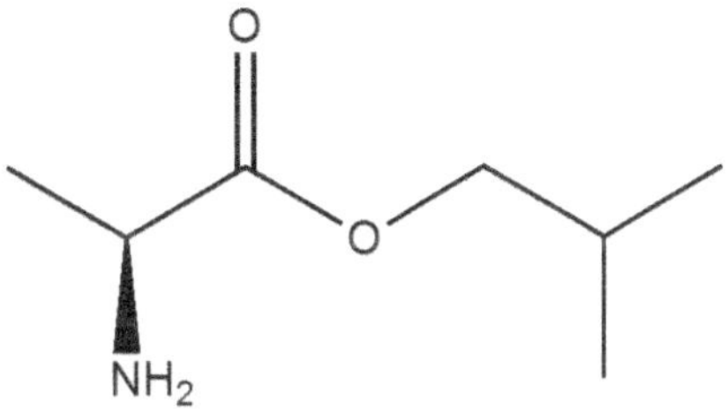

a. Draw the displayed formulae of the reactants you would use in the preparation. [2]

b. Draw the labelled setup of an apparatus you could use to separate **compound M** from the mixture of products formed. Explain why this apparatus is suitable. [5]

c. 2.56 g of **compound M** was formed during the reaction. Calculate the amount in moles of the compound formed. [2]

140. A student was supplied with a sample of **compound W** whose structure is shown below.

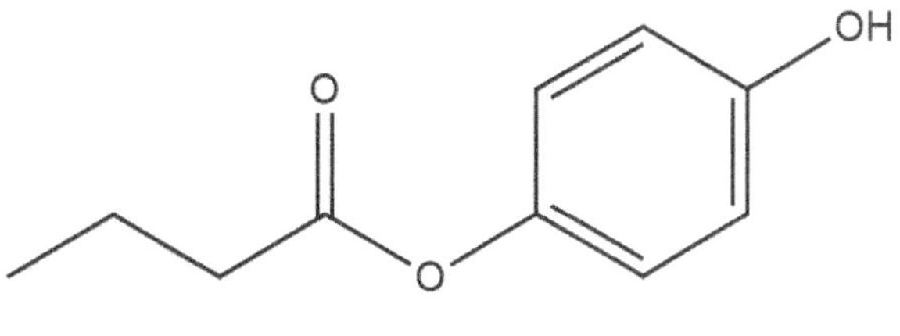

QUESTIONS
Organic chemistry and analysis

a. Name **compound W** and write its molecular formula. [2]

b. i. Draw the skeletal formulae of two reagents that can be used to prepare **compound W** in the laboratory and name them. [4]

ii. The student was given two unlabelled samples of the two reagents. Describe a simple chemical test that can be used to distinguish between the two reagents. [2]

iii. The student was asked to prepare a 4.20 g sample of **compound W**. Deduce the mass in grams of each reagent that would be required. Include an equation in your answer. [4]

141. You are asked to prepare a sample of nitrophenol in the laboratory.

   a. Name the reagents you would need. [2]

   b. Explain why 2-nitrophenol is (one of) the likely product(s) from the reaction. State the name of another isomer of 2-nitrophenol that may be formed. [3]

   c. i. Outline the mechanism for this reaction showing curly arrows. [3]

   ii. Name the type of mechanism. [1]

142. The displayed formula of (R)-2-hydroxy-2-methylbutanenitrile is given below.

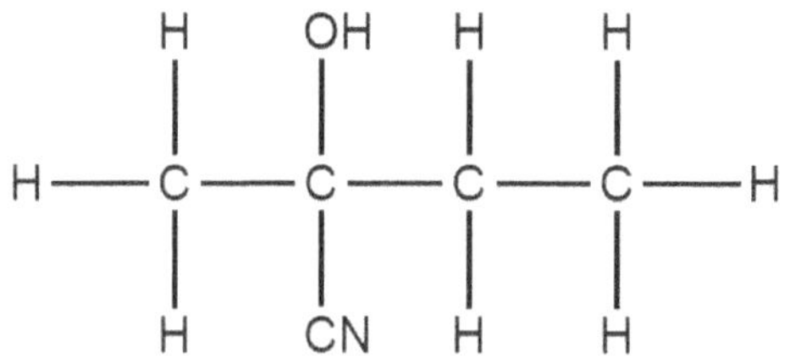

   a. Devise a one-step synthesis of (R)-2-hydroxy-2-methylbutanenitrile from a carbonyl compound. Give details of appropriate reagents. Write an equation for the reaction. [4]

   b. i. Outline the mechanism for the above reaction, showing curly arrows and relevant dipoles. [4]

   ii. Name the type of mechanism. [1]

   c. i. Draw the 3D structure of an isomer of (R)-2-hydroxy-2-methylbutanenitrile. [2]

   ii. Name the type of isomerism. [1]

143. Benzyl benzoate is used as a repellent for ticks and mosquitoes. It is classified as one of the most important medications in a basic health system. Its structure is shown below.

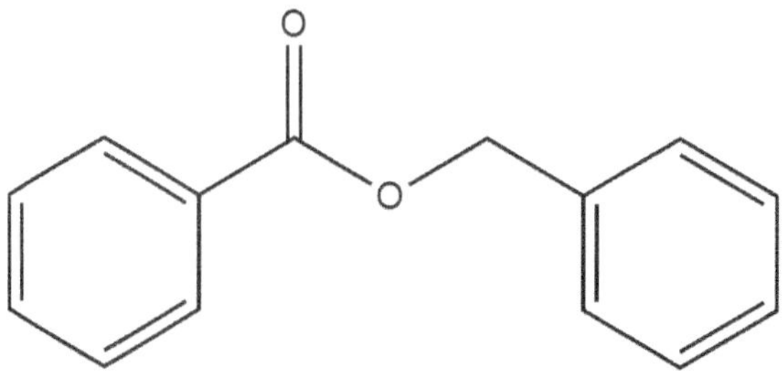

QUESTIONS 6

Organic chemistry and analysis

a. Benzyl benzoate may be synthesised using two reagents, one of which has a molecular mass of 140.5 g mol^{-1}. Draw the skeletal formulae of the two reagents. [2]

b. Benzyl benzoate was heated with aqueous nitric acid. Draw the skeletal formulae of any product(s) formed. [2]

c. The reagent with the molecular mass of 140.5 g mol^{-1} from part(a) may be synthesised using a molecule with a molecular mass of 108 g mol^{-1}. Devise a two-step synthesis. Give details of appropriate reagents. Write equations for the reactions. [5]

d. i. State the number of peaks in the carbon-13 NMR spectrum of benzyl benzoate. [1]

   ii. Identify the different carbon environments on a drawing of the formula of benzyl benzoate. [2]

144. A student was given a sample of 1-(3-nitrophenyl)ethanone. Its structure is as shown.

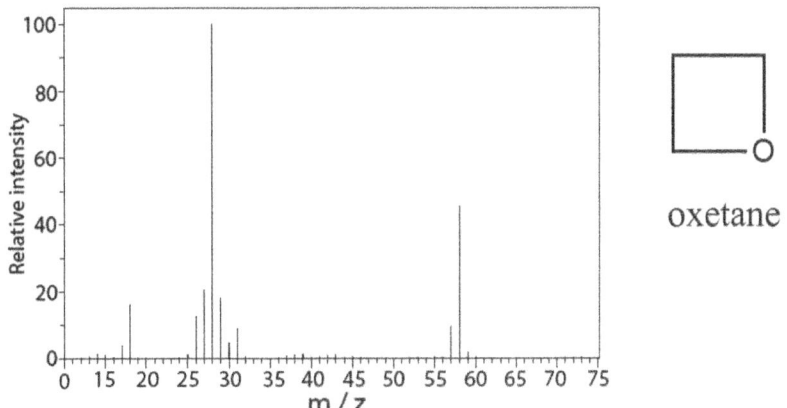

The sample was reacted with a reducing mixture.

a. Write a balanced equation for the reaction taking place. You may show the reducing agent as [H] in your equation. [3]

b. i. 0.20 g of the major product from the reaction in part (a) was reacted with ethanoyl chloride to form a major product **G** and a minor product **H**. Write a balanced equation for the reaction. [3]

   ii. The percentage yield from the reaction was 53.2%. Determine the mass of **G** in grams, formed. Give your answer to a suitable number of significant figures. [5]

145. A student carried out mass spectrometry on a sample of oxetane. The mass spectrum is shown below.

a. Draw the displayed formula of oxetane. [1]

b. Explain the peaks at m/z = 58, 28 and 30. Use equations to support your reasoning. [4]

QUESTIONS
Organic chemistry and analysis

c. Describe the splitting patterns you would observe on a proton NMR spectrum of oxetane. Label the different proton enviornments of oxetane on its displayed formula. [3]

d. State the number of peaks you would observe on a carbon-13 NMR spectrum of oxetane. [1]

146. Gas chromatography was carried out on an unknown mixture of alkanes. The resulting gas chromatogram is shown in **figure 1**

Table 1

Compound	Retention time (min)
Hexane	1.2
Octane	2.1
Decane	6.8
Dodecane	30.0
Tetradecane	142.8

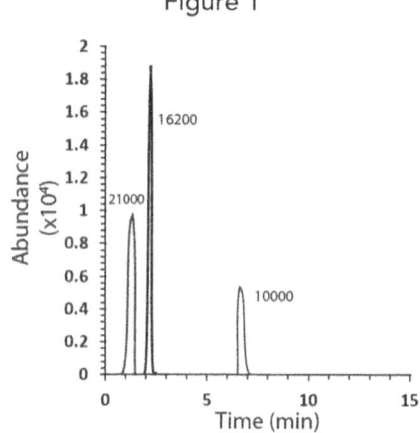

Figure 1

Concentration / mol dm^{-3}	Area		
	Hexane	Dodecane	Tetradecane
0.002	2300	5000	1600
0.03	47000	90000	26000
0.05	50000	185000	52000

Gas chromatography was also performed on different mixtures containing known components of known concentrations. The results are shown in **Table 2**.

a. Determine the identity of each component in the unknown mixture. [3]

b. Using the data provided, determine the concentration of one of the components in the mixture. [4]

147. Napthalene is used in the manufacture of mothballs. It has a very characteristic odour that is detectable at very low concentrations. Napthalene and a carboxylic acid may be used to form 1-(2-naphthyl)ethanone.

naphthalene   1-(naphthalen-2-yl)ethanone

a. Devise a two-step synthesis of 1-(2-naphthyl)ethanone from naphthalene. Give details of appropriate reagents and relevant conditions. Write an equation for each step, showing clearly all organic compounds. [6]

b. A student used 2.30 g of the carboxylic acid. Calculate the mass in grams of 1-(2-naphthyl)ethanone that would need to be produced in order to obtain a percentage yield of 67%. [4]

148. The mass spectrum of acrylic acid is shown in figure 1.

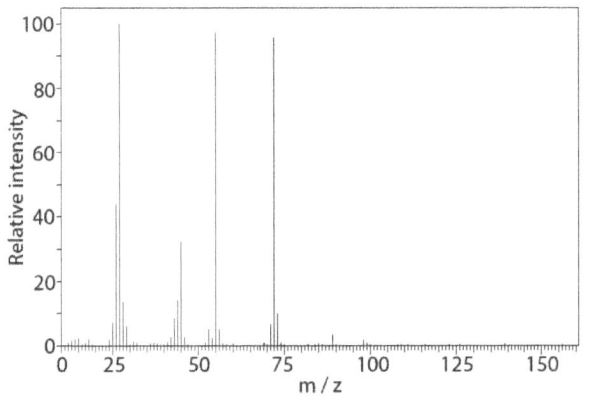
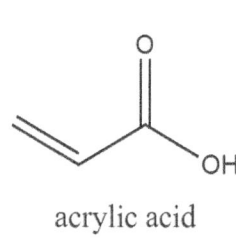

Figure 1 The mass spectrum of acrylic acid

a. Explain the three peaks with the highest relative intensity in the mass spectrum. Use equations to support your answer. [5]

b. i. Acrylic acid was reacted with pentan-1-ol. Draw the skeletal formula of the major organic product, **X**, of the reaction. [1]

ii. **Compound X** was then reacted with hot aqueous sodium hydroxide. Draw the product(s) formed. [2]

149. Write balanced chemical equations for the reactions, if any, between the following species.

a. Propanoic acid and sodium carbonate [2]

b. Ethanoic acid and magnesium [2]

c. Magnesium oxide and butanoic acid [2]

d. Calcium hydroxide and ethanoic acid [2]

150. 2-hydroxy-3-propionylbenzoic acid can be synthesised by completing **reactions 1** and **reaction 4** in the flowchart given in figure 1.

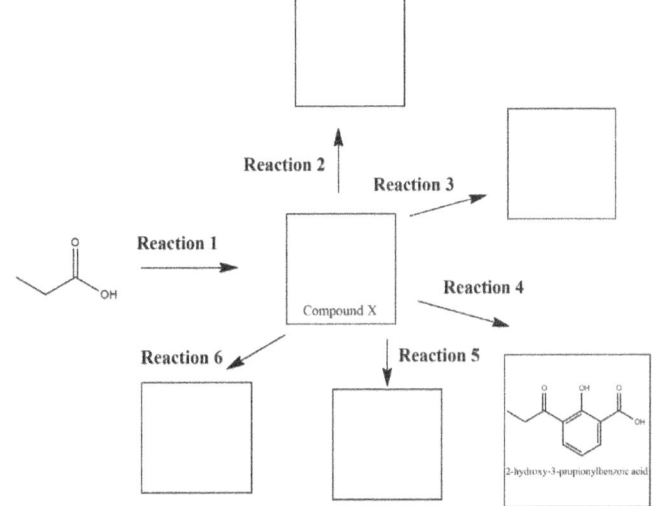

a. Write equations for reactions 1 and 4. Give details of relevant conditions. [5]

b. Using the table given below, draw the skeletal formulae of the major organic products of the other reactions. [4]

Reaction	Reagents
2	X + propylamine
3	X + ammonia
5	X + pentan-1-ol
6	X + water

c. State the name of the mechanism of reaction 6. [1]

151. The skeletal formula of 1-(2-hydroxyphenyl)ethanone is shown below.

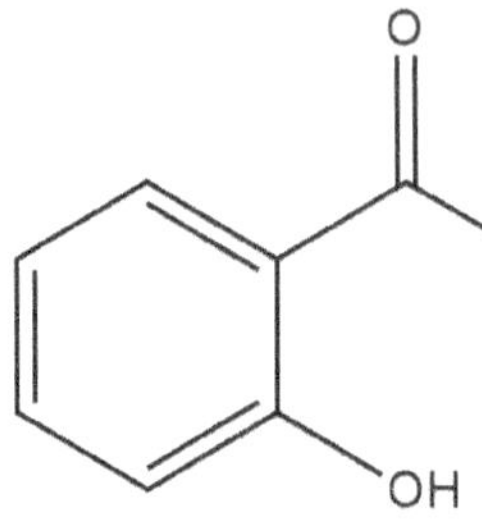

A student reacted 1-(2-hydroxyphenyl)ethanone with sodium tetrahydridoborate (III).

a. Write a balanced equation for the reaction. You may show sodium tetrahydridoborate (III) as [I I] in your equation. [2]

b. Outline the mechanism for the reaction, showing curly arrows and relevant dipoles. Name the type of mechanism. [5]

152. The structure of valine is shown below.

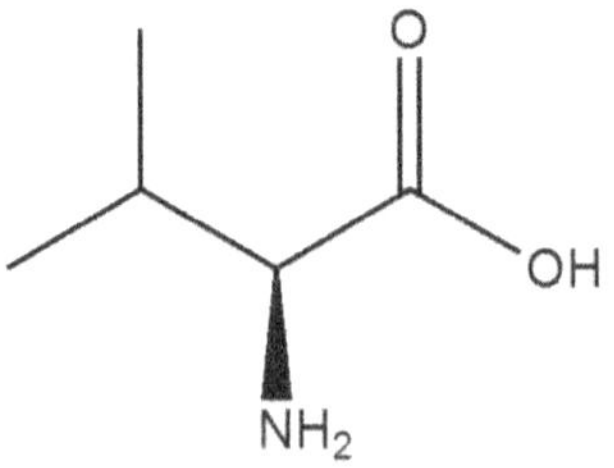

a. What type of a compound is Valine? [1]

b. Draw the displayed formulae of the organic product formed when Valine reacts with each of the following:

i. Sodium carbonate [1]

ii. Bromoethane in the molar ratio 1:1 [1]

iii. Propan-1-ol in the presence of a concentrated sulfuric acid catalyst [1]

c. i. Draw the skeletal formula of the organic product formed when Valine reacts with bromoethane in the molar ratio Valine : bromoethane = 1:2. [1]

ii. State the molecular formula of the minor product and the number of moles formed. [2]

d. Draw the 3D structures of the optical isomers of Valine. [2]

153. Benzene-1,4-diamine is used in the production of hair dyes and as a component in engineering polymers.

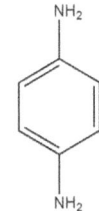
benzene-1,4-diamine

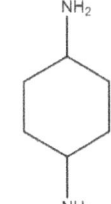
cyclohexane-1,4-diamine

a. Benzene-1,4-diamine may be synthesised starting from a compound with an empirical formula of $C_3H_2NO_2$. Give details of appropriate reagents and relevant conditions. Write an equation showing clearly all organic compounds. You may use [H] to show reducing agents and [O] to show oxidising agents, if required in your equation. Draw a labelled diagram of any apparatus required for synthesis. [7]

b. Suggest which of benzene-1,4-diamine and cyclohexane-1,4-diamine is the weaker base. Explain. [4]

154. The structure of 2-(2-amino-3-(4-hydroxyphenyl)propanamido)-3-phenylpropanoic acid is shown in the following figure.

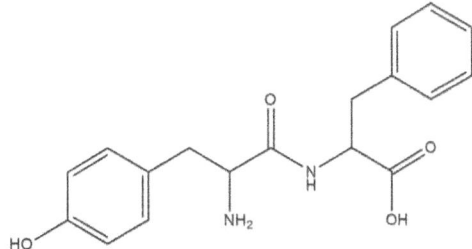

a. State the molecular formula of the compound. [1]

b. Determine the mass in grams of 1.0×10^{18} molecules of this compound. [3]

c. State the type of the two compounds that may be used in the synthesis of 2-(2-amino-3-(4-hydroxyphenyl)propanamido)-3-phenylpropanoic acid. [1]

155. A section of a polymer is shown in the following figure.

a. State the name of the type of polymer. [1]

b. The polymer was heated with dilute hydrochloric acid. Draw the skeletal formulae of the products formed. [2]

156. About 40 million tonnes of PET is produced worldwide. A section of the polymer is shown below.

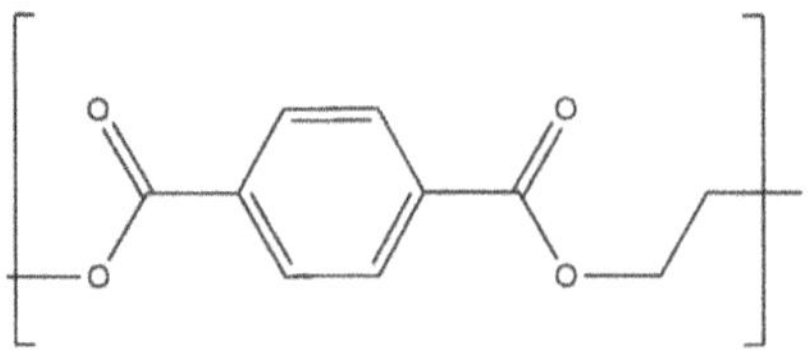

a. State the name of the type of polymer. [1]

b. The polymer was heated with aqueous sodium hydroxide. Draw the skeletal formulae of the products formed. [2]

157. A student proposed the following reaction mechanism for the reduction of 3-methylbutanal.

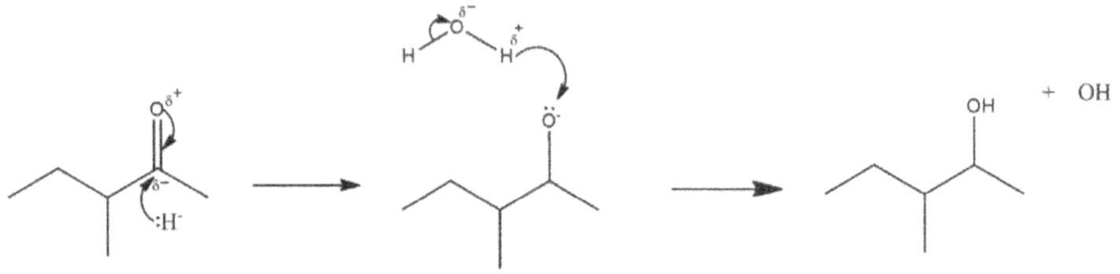

a. There are several mistakes in the suggested mechanism. Redraw the mechanism correcting any mistakes. [3]

b. State the name of the type of mechanism. [1]

158. The displayed formula of hexanenitrile is shown in the following figure.

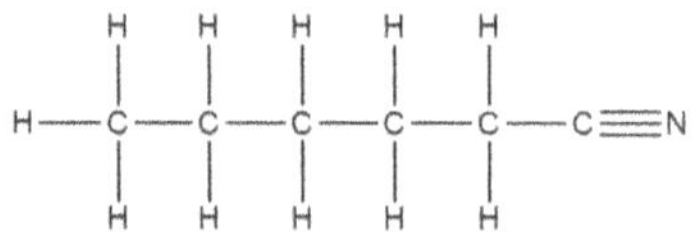

hexanenitrile

a. Devise a one-step synthesis of hexanenitrile from 1-chloropentane. Give details of appropriate reagents. [3]

b. Outline the mechanism for this reduction. Use curly arrows and show any dipoles. [3]

159. The displayed formula of 3-methylbutan-1-amine is shown in the following figure.

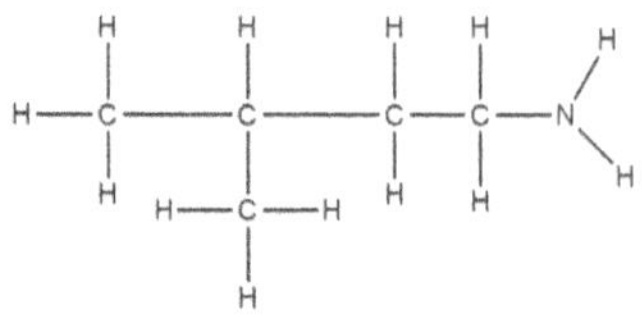

3-methylbutan-1-amine

a. Draw the skeletal formula of 3-methylbutan-1-amine. [1]

b. Devise a one-step synthesis of 3-methylbutan-1-amine from a nitrile. Give details of appropriate reagents. Write an equation for the reaction. [3]

160. Devise a one-step synthesis of hexanoyl chloride from a carboxylic acid. Give details of appropriate reagents. Write an equation for the reaction. [3]

161. The skeletal formula of phenylalanine is shown in the following figure.

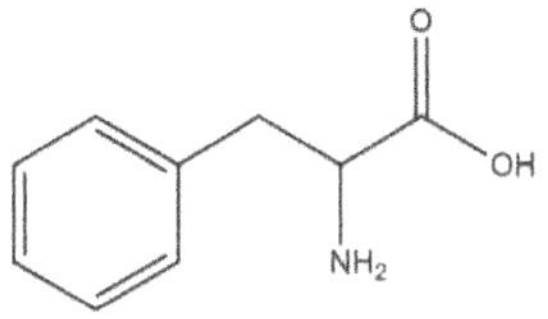

a. Draw the 3D structures of the two isomers of phenylalanine. [2]

b. Name the type of isomerism. [1]

162. The structure of naphthalene-2,7-diamine is shown below. This compound may be synthesised from **compound Q** which has a molecular mass of 218.

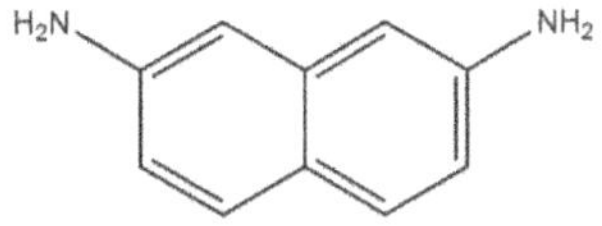

a. Devise a one-step synthesis of naphthalene-2,7-diamine from **compound Q**. Give details of appropriate reagents. Write an equation for the reaction. You may use [O] to represent oxidising agents and [H] to represent reducing agents if required in your equation. [3]

b. Outline the steps that could be carried out to obtain a pure sample of the organic product from the reaction mixture. [2]

c. Use electron repulsion theory to predict the shape of the bonds around the nitrogen atom of the NH_2 group. Give the bond angle. [2]

163. Propanenitrile can be synthesised by reacting KCN with bromoethane in an ethanol solution. The boiling point of propanenitrile is 98 °C and that of bromoethane is 38 °C.

Draw a labelled diagram to show how you would safely set up apparatus for the synthesis and describe a method to obtain a pure sample of the product from the reaction mixture. [6]

164. Salicylic acid is a naturally occurring pain-killer that can be extracted from willow bark. However, it is less effective than aspirin. The structure of salicylic acid is shown below.

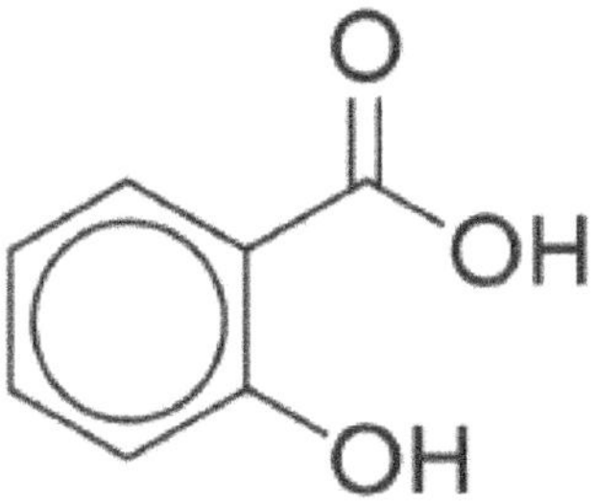

6 QUESTIONS — Organic chemistry and analysis

Aspirin has a melting point of 136 °C.

Molar mass of aspirin = 180 g mol^{-1}.

A student was asked to prepare a sample of aspirin by reacting salicylic acid with ethanoic anhydride.

a. Write an equation using displayed formula for the reaction that takes place. [2]

b. 2.00g of salicylic acid was used. Calculate the amount, in mol, of salicyclic acid present. [2]

Aspirin is prepared by heating ethanoic anhydride, salicylic acid and phosphoric acid (or sulfuric acid) under reflux.

c. Draw a labelled diagram of the apparatus that could be used to manufacture aspirin. [4]

d. Suggest the role of phosphoric acid in the reaction. [1]

e. Draw the setup of an apparatus which the student could use to isolate aspirin. [2]

f. Describe how the purity of the aspirin sample may be determined. In your answer you should include details of any apparatus that may be required. [5]

1.20g of aspirin was produced.

g. Calculate the percentage yield of aspirin. [3]

165. The structure of 2-(p-tolyl)ethanamine is shown in the following figure. 2-(p-tolyl)ethanamine may be synthesised using the following two routes.

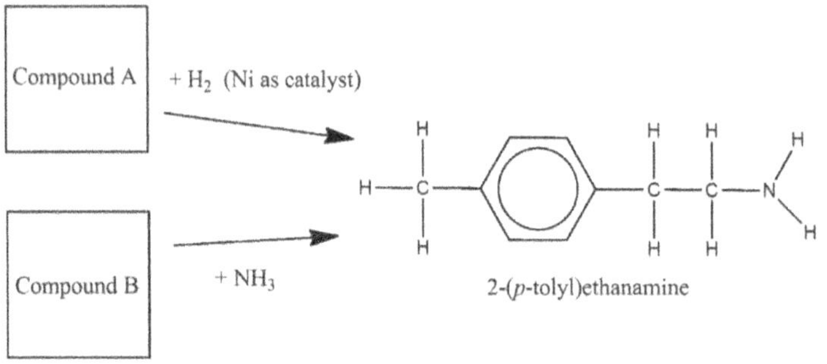

Compound A + H$_2$ (Ni as catalyst) → 2-(p-tolyl)ethanamine

Compound B + NH$_3$ → 2-(p-tolyl)ethanamine

a. Write equations for each of the reactions involving compound A and compound B. [4]

b. A student reacted compound A with aqueous hydrochloric acid. Write a balanced equation for the reaction taking place. [2]

166. 2-methylbutanoyl chloride can be synthesised from 2-methylbutanenitrile in two stages.

2-methylbutanenitrile → Reaction 1 → → Reaction 2 → 2-methylbutanoyl chloride

Devise a two-step synthesis. Give details of appropriate reagents. Write equations for the reactions. [4]

167. The structure of pivalaldehyde is shown below.

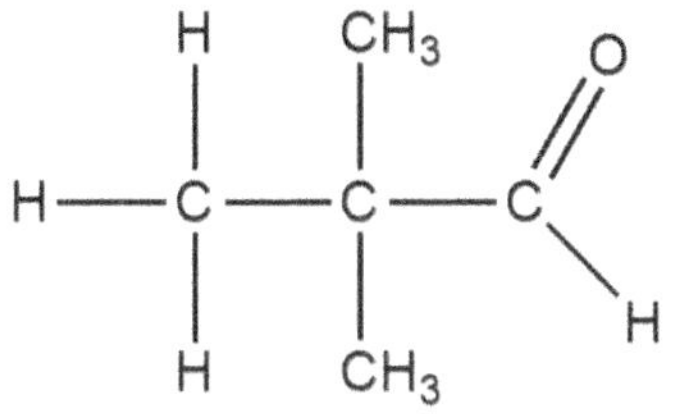

pivalaldehyde

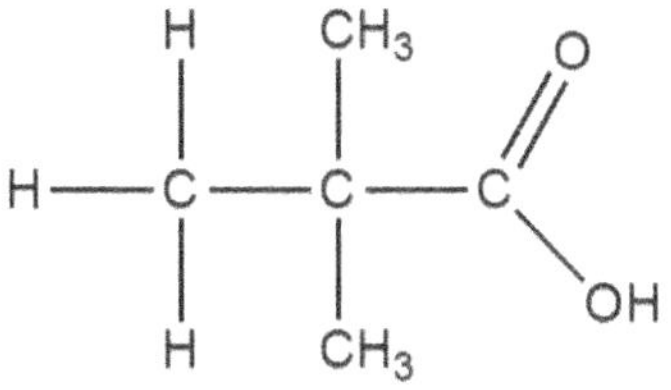

pivalic acid

Boiling points:
pivalaldehyde 74 °C
pivalic acid 163.7 °C

a. Devise a one-step synthesis of pivalaldehyde from an alcohol. Give details of appropriate reagents and conditions. Write an equation for the reaction. Draw a labelled diagram of an apparatus you could use for the synthesis. [7]

b. Outline a step that could be carried out to obtain a pure sample of the organic product from the reaction mixture. [1]

168. 1-aminobutan-2-ol can be synthesised from propanal. Devise a two-step synthesis of 1-aminobutan-2-ol from propanal. Give details of appropriate reagents and relevant conditions. Write an equation for each step, showing clearly all organic compounds. [4]

169. Name the type of mechanism involved in each of the following reactions.

a. Ethyl benzene with NO_2^+ [1]

b. Bromoethane with CN^- [1]

c. Formation of propan-1-ol from propanal [1]

d. Propyl benzene with Cl_2 in the presence of $AlCl_3$ [1]

170. Carboxylate anions of the following **compound R** act as important neurotransmitters.

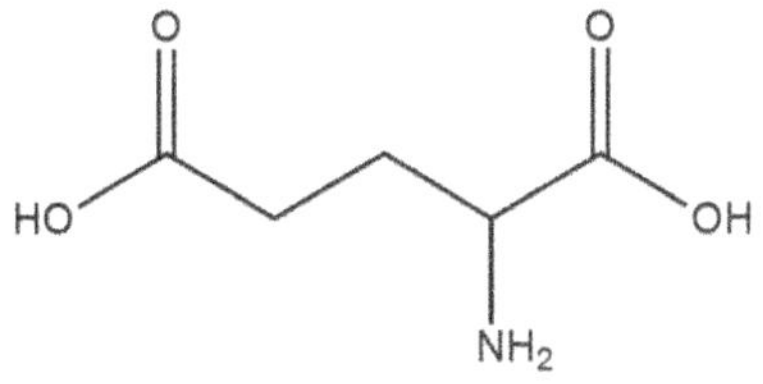

a. State the type of **compound R**. [1]

b. An excess of aqueous sodium hydroxide was added to a sample of solid **compound R**. The resulting reaction mixture was then purified to obtain a pure sample of the ionic organic product, monosodium glutamate.

i. Draw the structure of the ionic organic product. [1]

ii. Outline the steps that can be carried out to obtain a pure sample of the product from the reaction mixture. [2]

QUESTIONS
Organic chemistry and analysis

171. Devise a two step synthesis of ethyl 4-methylpentanoate from 4-methylpentanenitrile. Give details of appropriate reagents and conditions. Write an equation for each reaction. [4]

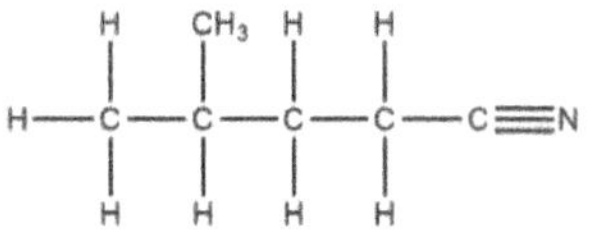

4-methylpentanenitrile 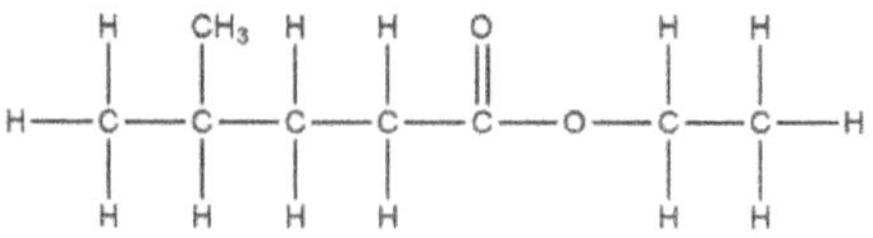 → *ethyl 4-methylpentanoate*

172. The structure of **compound K** is shown in the following figure. A student synthesised this compound from 4-methylpentan-1-ol.

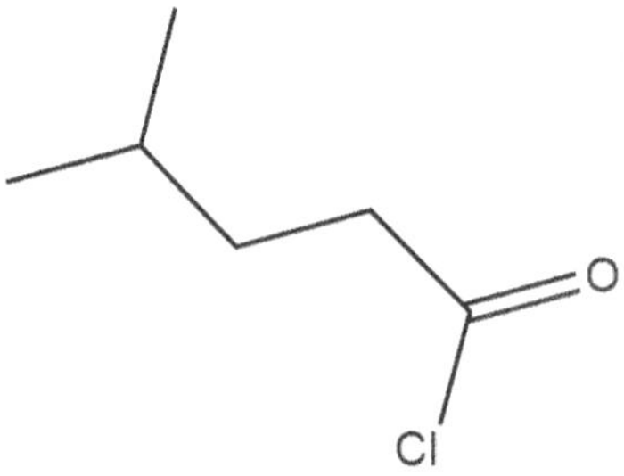

   a. Name **compound K**. [1]

   b. Devise a two step synthesis of **compound K** from 4-methylpentan-1-ol. Give details of appropriate reagents and conditions. Write an equation for each reaction. [4]

   c. A student found out that the percentage yield of the first step in the reaction was 53% and that of the second step was 73%. Determine the mass in grams of 4-methylpentan-1-ol you would need to use to synthesise 5.00 g of **compound K**. Give your answer to 2 significant figures. [5]

173. 200 cm³ of 5.00×10^{-2} mol dm⁻³ hexanoic acid reacts with Na_2CO_3 as shown in the following equation.

   $$2CH_3(CH_2)_4COOH + Na_2CO_3 \rightarrow 2CH_3(CH_2)_4COONa + H_2O + CO_2$$

   a. Determine the volume of CO_2 formed in cm³. [4]

   b. State the chemical name of an alcohol that may be used to synthesise hexanoic acid. [1]

   c. A student added Tollens' reagent to a solution containing hexanal. Describe in terms of oxidation and reduction, the changes that place. [2]

QUESTIONS 6

Organic chemistry and analysis

QUESTIONS
Organic chemistry and analysis

Multiple Choice Questions

1. Which statement about this compound is NOT true?

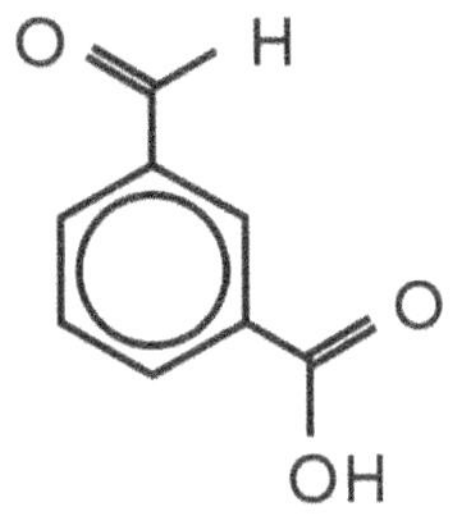

   A   The aldehyde group can be oxidised to a carboxylic acid group.

   B   It has a carboxylic acid functional group.

   C   The carboxylic acid group will react with bases.

   D   The bond angle around each carbon in the benzene ring is $270°$.

2. Which statement about NMR spectra is **not** true?

   A   Proton NMR spectra give more information than carbon-13 NMR spectra.

   B   Peak areas in a proton NMR spectrum does not provide any information about a compound.

   C   In proton NMR spectroscopy, the spin-spin coupling pattern of a peak give information about the number of protons on the adjacent carbon.

   D   There are four peaks in the carbon-13 NMR spectrum of phenol.

3. The displayed formula of a compound is shown below:

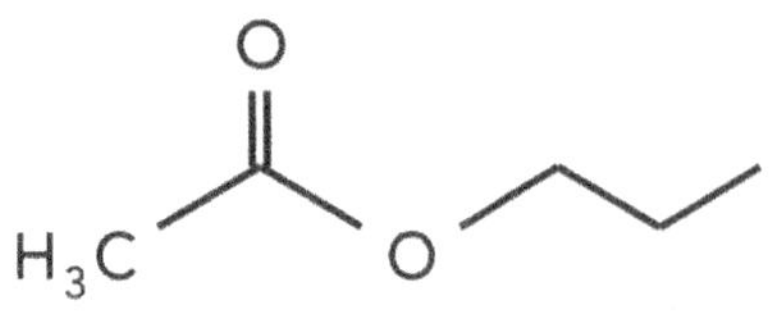

   Which statement about this compound is true?

   A   It is a polyamide.

   B   Its name is propyl ethanoate.

   C   There is an amide linkage in the compound.

   D   It has only one carbon atom.

4. Which statement about the following IR spectra is true?

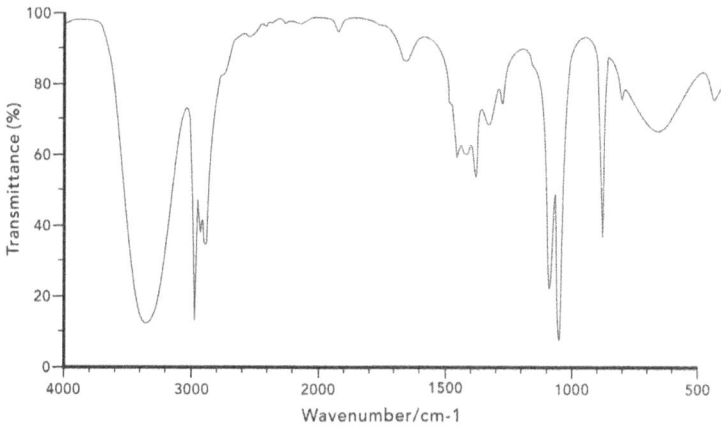

Figure 1

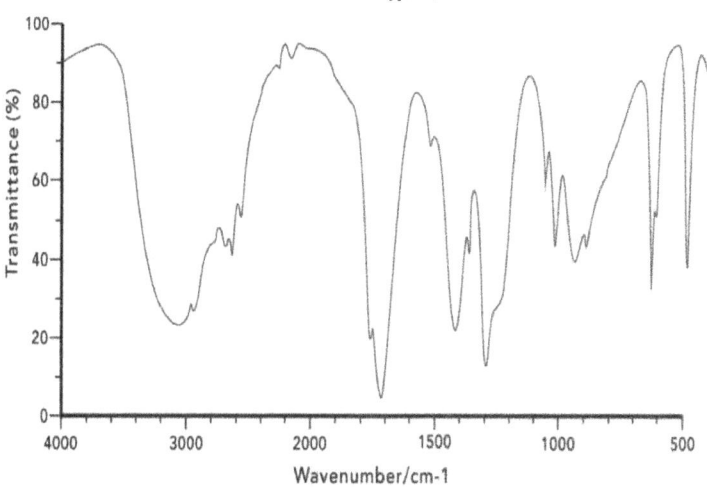

Figure 2

A   Both of them represent the same compound.

B   Figure 1 is that of a carboxylic acid.

C   Figure 2 is that of an alcohol.

D   Figure 2 is that of a carboxylic acid.

5. Which one of the following statements about benzene and phenol is true?

A   Phenol is more reactive than benzene because the hydrogen in phenol donates electrons to nearby species. This causes reactions to be more feasible.

B   Phenol is less reactive than benzene because it has a higher molecular weight and therefore reactions take time. The high electron density of the benzene's ring structure contributes to this high reactivity.

C   Phenol is more reactive than benzene because the lone pair of electrons on the oxygen atom is drawn to the benzene ring leading to a region of high electron density that is able to polarise other molecules.

QUESTIONS
Organic chemistry and analysis

    D   An aqueous solution of phenol will react with bromine water, forming a blue precipitate.

6. Which statement best describes the type of alcohol shown below:

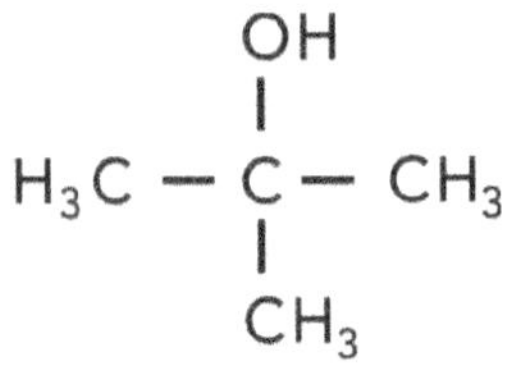

$$H_3C - \underset{\underset{CH_3}{|}}{\overset{\overset{OH}{|}}{C}} - CH_3$$

    A   Primary alcohol because there is only one OH group.

    B   Secondary alcohol because there are an even number of hydrogen atoms.

    C   Tertiary alcohol because most hydrogen atoms exist in groups of three.

    D   Tertiary alcohol because the carbon atom to which the OH group is attached is bonded to three other carbon atoms.

7. Name the compound shown:

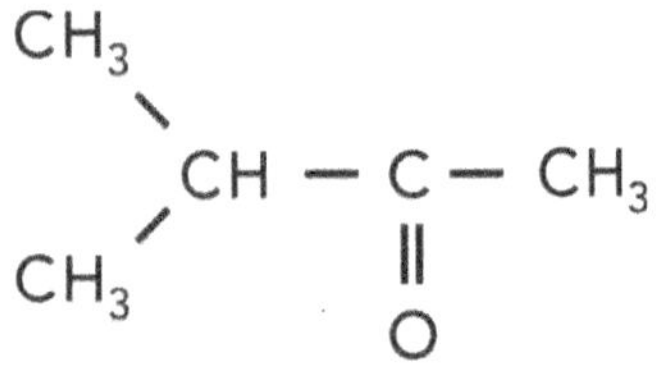

$$(CH_3)_2CH - \underset{\underset{O}{\|}}{C} - CH_3$$

    A   3-methylbutan-2-one

    B   methyl propanoate

    C   pentanoic acid

    D   2,4-dimethyl-2-butanone

8. Which of the following statements about benzene is/are NOT true?

    1:    Benzene reacts with bromine in the presence of oxygen.

    2:    Benzene can be nitrated using a nitryl cation.

    3:    Benzene only reacts under standard conditions.

    A   1 and 3

    B   1, 2 and 3

    C   Only 2

    D   2 and 3

Organic chemistry and analysis — QUESTIONS 6

9. What is the molecular formula of the compound shown?

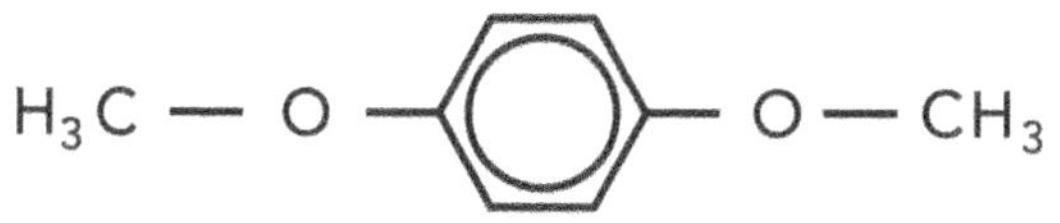

$$H_3C - O - \text{(benzene ring)} - O - CH_3$$

- A $C_8H_6O_2$
- B $C_8H_{10}O_2$
- C H_7COOCH_3
- D $C_8H_{14}O_2$

10. Which of the following compounds will have optical isomers?

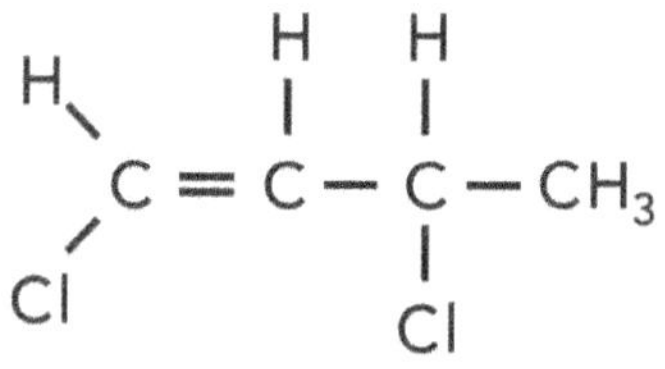

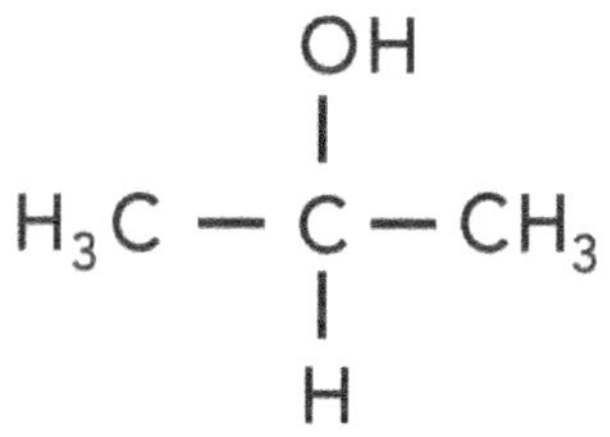

A.
$$\underset{Cl}{\overset{H}{\diagdown}}C = \underset{\underset{H}{|}}{C} - \underset{\underset{Cl}{|}}{\overset{H}{C}} - CH_3$$

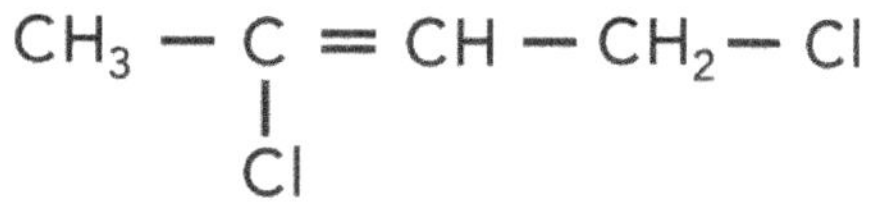

B.
$$CH_3 - \underset{\underset{Cl}{|}}{C} = CH - CH_2 - Cl$$

C.
$$H_3C - \underset{\underset{H}{|}}{\overset{\overset{OH}{|}}{C}} - CH_3$$

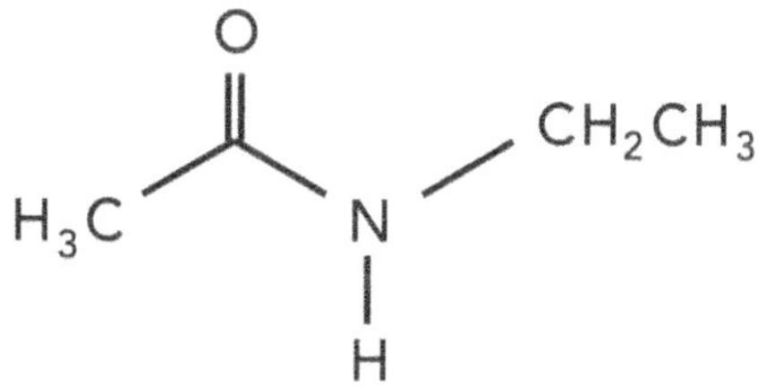

D.
$$H_3C - \overset{\overset{O}{\|}}{C} - \underset{\underset{H}{|}}{N} - CH_2CH_3$$

11. Which one of the following statements about carboxylic acids is NOT true?

- A   They react with bases to form salts.
- B   Hexanoic acid is an example of a carboxylic acid.
- C   When Brady's reagent is added to a carboxylic acid, a yellow or orange precipitate is formed.
- D   A carboxylic acid will react with an alcohol to form an ester.

12. When Brady's reagent is added to an unknown compound, an orange colour was observed. Ammoniacal silver nitrate was then added to the same compound which resulted in the formation of a "silver mirror" inside the test tube in which the reaction was carried. Which one of the following compounds could this be?

QUESTIONS
Organic chemistry and analysis

    A   Pentanal

    B   Methyl butanoate

    C   Pentanoic acid

    D   Hexanone

13. Esters can be hydrolysed using acids or alkalis. Butyl pentanoate can be hydrolysed using sodium hydroxide. What are the products formed?

    A   Butan-1-ol and pentanoic acid

    B   Butan-1-ol and sodium pentanoate

    C   Butan-1-ol and water

    D   Pentan-1-ol and sodium butanoate

14. Which of the following statements about optical isomers are true?

    1:    Optical isomers rotate plane-polarised light in the same direction.

    2:    A compound with a chiral carbon will show optical isomerism.

    3:    They are also known as *E/Z* isomers.

    A   1 and 2

    B   Only 3

    C   Only 2

    D   Only 1

15. The reaction between hexanal and $NaBH_4$ can be best described as:

    A   Nucleophilic addition

    B   Nucleophilic substitution

    C   Electrophilic addition

    D   Neutralisation

16. Which one of the following statements about carbon-13 NMR spectroscopy is true?

    A   The relative heights of the peaks is equal to the ratio of the number of carbon atoms in each environment.

    B   The number of peaks is equal to the number of different carbon environments.

    C   Carbon-13 NMR can only be used in conjunction with mass spectrometry and infrared spectroscopy.

    D   It can only be carried out with carbonyl compounds.

17. In chromatography, the solid stationary phase:

A   Always refers to the container in which the chromatography plate is held.

B   Separates by relative solubility.

C   Separates by adsorption.

D   Is only required in gas chromatography.

18. In chromatography, how can the R_f value be calculated? ("/" here means "divided by")

   A   Area of stationary phase / volume of liquid phase.

   B   Mass of the solvent / volume of the solvent.

   C   Distance moved by the solvent front / distance moved by the component.

   D   Distance moved by the component / distance moved by the solvent front.

19. Nitrobenzene can be formed by adding benzene to a nitrating mixture of concentrated nitric acid and concentrated sulphuric acid. While the benzene is added, at what temperature is the mixture kept and why?

   A   450 °C to help catalyse the reaction

   B   5-15 °C to prevent the risk of any explosion

   C   Below 50 °C so that the carbon dioxide formed can be removed slowly

   D   50 °C so that no more than one nitro group is substituted onto the benzene ring

20. During some reactions, the reacting mixture in a round-bottomed flask is usually heated in a water bath rather than using a bunsen burner placed at the bottom. Which one of the following statements best describes the reason for this?

   A   Some reactions need water as its solvent.

   B   It helps to heat the flask evenly all around without causing any hot spots.

   C   The round-bottomed flask is more balanced in a water bath.

   D   The water bath helps to retain any gases which may be formed.

21. Benzene normally takes part in substitution reactions instead of addition reactions, because

   A   It retains the delocalisation and stability of the ring structure.

   B   Benzene already has too many bonds.

   C   Any addition reaction would result in the formation of a highly reactive species.

   D   The length of C-C bonds and C=C bonds are different and any product formed through addition will be asymmetrical.

22. The molecular formula of chlorobenzene is

   A C_6H_5Cl

   B C_6H_6Cl

C $C_6H_{10}Cl$

D ClC_6H_8

23. The empirical formula of a compound with molecular weight 26 is CH. The molecular formula of this compound is:

   A $C_{26}H_{26}$

   B C_2H_2

   C C_3H_3

   D CH_{26}

24. Which of the following statements best describes why the Kekulé structure of benzene was incorrect?

   1:   All the carbon-carbon bonds in benzene are the same length, whereas C-C and C=C bonds have different lengths.

   2:   The delocalisation energy of benzene is less than expected (152 KJ mol^{-1} less than the delocalisation energy of cyclohexa-1,3,5-triene).

   3:   Benzene does not take part in electrophilic addition reactions expected from C=C bonds in alkenes.

   A   Only 1

   B   1 and 2

   C   Only 3

   D   1, 2 and 3

25. The following is a carbon-13 NMR spectrum of **compound X**. Which statement is most likely to be true?

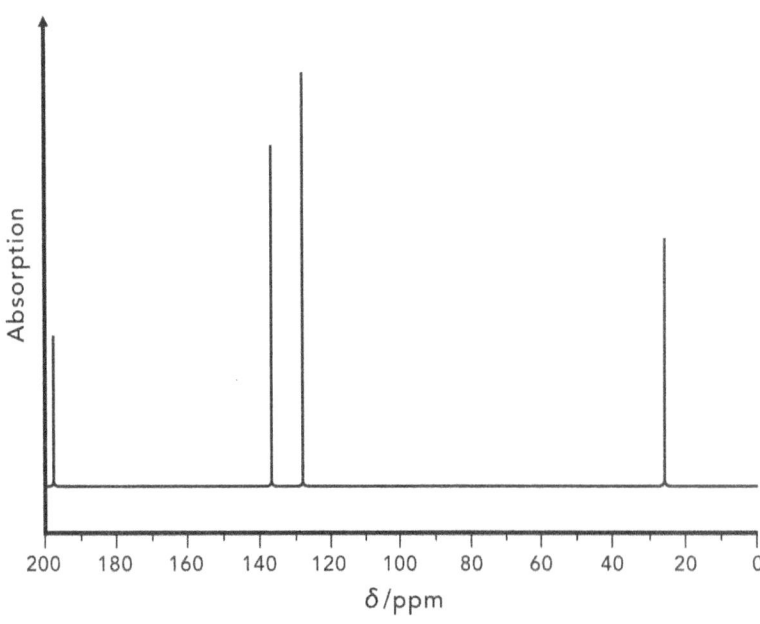

   A   The molecular mass of **compound X** could be 37.

B   **Compound X** consists of an atom of atomic mass 14.0.

C   The only functional group in **compound X** is benzene.

D   **Compound X** is but-3-en-2-one.

26. The molecular formula of water is H_2O. D_2O is a compound often used in proton NMR spectroscopy. Which one of the following statements is true about deuterium (D)?

　A   It is an isotope of hydrogen. Isotopes of the same element have the same number of protons but different numbers of neutrons.

　B   It is an isotope of hydrogen. Isotopes of the same element have the same number of neutrons but different numbers of protons.

　C   It is a stereoisomer of hydrogen with exactly the same number of protons and neutrons as hydrogen.

　D   It is an isotope of hydrogen. Isotopes of the same element have the same number of protons but different numbers of electrons.

27. A single peak was observed in the carbon-13 NMR spectrum of a compound. Which of the following statements could be true?

　1:   The compound could be cyclohexane or benzene.

　2:   The carbon-13 NMR spectra of both benzene and cyclohexane would have a peak at the same chemical shift.

　3:   The proton NMR spectra of both benzene and cyclohexane would have one peak.

　A   Only 1

　B   Only 3

　C   1 and 2

　D   1 and 3

28. In an infrared spectrum, around which wavenumber is a C=O absorption peak most likely to be found? (You may use the data sheet).

　A   3500 cm^{-1}

　B   2700 cm^{-1}

　C   1700 cm^{-1}

　D   1000 cm^{-1}

29. Gas chromatography can be used to identify chemicals based on their retention times. Which statement identifies one of the major limitations of gas chromatography?

　A   Gas chromatography can only be performed in a vacuum.

　B   Gas chromatography will only work when used in conjunction with mass spectrometry.

　C   Supplies of gas are limited and will soon run out.

　D   Many chemicals may have the same retention time making identification of compounds difficult.

QUESTIONS

Organic chemistry and analysis

30. Which of the following options describe the directing effects of the groups, NO_2, NH_2 and OH?

	NH_2	NO_2	OH
A	3-directing	2,4-directing	3-directing
B	2,4-directing	3-directing	2,4-directing
C	2,4-directing	2,4-directing	3-directing
D	2,4-directing	3-directing	3-directing

31. What is retention time in gas chromatography?

    A   Time taken to prepare a sample for chromatography.

    B   The time taken before a new chromatography can be performed after one has already been done.

    C   Time taken for a component in the sample to travel from the column inlet to the detector.

    D   Time taken for a component in the sample to evaporate back into the column inlet once the sample has been injected.

32. Optical isomerism is a type of stereoisomerism. Which of the following statement(s) about optical isomers is / are true?

    1: They rotate plane-polarised light in different directions.

    2: One optical isomer may be pharmacologically active whereas the other may present with unwanted side effects.

    3: Optical isomers do not exist in nature and can only be formed in the laboratory.

    A   1 and 2

    B   Only 1

    C   Only 2

    D   Only 3

33. What is the molecular formula of sodium tetrahydridoborate (III)?

    A $NaBH_4$

    B $NaBr_4$

    C NaH_4BO_3

    D HBr_4

34. Which one of the following statements about carbonyl compounds and $NaBH_4$ is true?

    A   Aldehydes cannot be reduced, but ketones can be reduced to primary alcohols using $NaBH_4$.

    B   Ketones are reduced to secondary alcohols using $NaBH_4$. $NaBH_4$ acts as an oxidising agent.

C Reduction of carbonyl compounds using NaBH$_4$ results in the formation of an alcohol and water.

D Ketones are reduced to secondary alcohols and aldehydes to primary alcohols using NaBH$_4$.

35. Hexan-3-one reacts with NaBH$_4$ to form which of the following

   A Hexan-3-ol and water

   B Hexan-3-ol ONLY

   C Hexanal ONLY

   D Hexan-2-ol ONLY

36. Name the following compound:

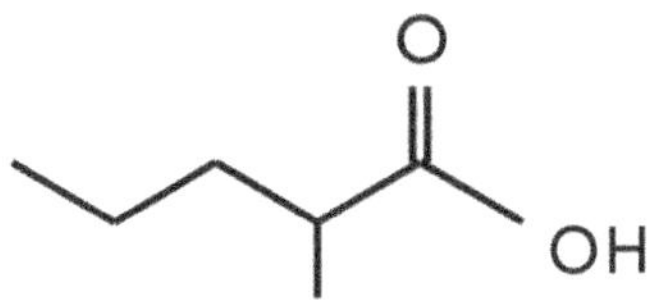

   A 2-methylpentanoic acid

   B 2-methylpentanal

   C 4-methyl pentanoic acid

   D Hexanoic acid

37. Which of the following statement is **not** true?

   A In a dative covalent bond, both electrons in the bond are provided by only one of bonding atoms.

   B Methylamine has a lone pair of electrons.

   C Amines react with acids to make salts.

   D Butylamine has the formula $CH_3CH_2CH_2CH_2NH_3$

38. Which of the following statements best describes amines?

   1: Amines are derivatives of ammonia.

   2: Amines are weak acids.

   3: Amines can be formed by gently warming halogenoalkanes with an excess of ammonia, using ethanol as the solvent.

   A Only 1

   B 1 and 3

   C Only 3

   D 1, 2 and 3

39. Which of the following statements about carbonyl compounds are true?

1: Aldehydes can be reduced (to primary alcohols) and oxidised (to carboxylic acids).

2: Ketones cannot be reduced but can be oxidised.

3: Ketones can only be reduced (to secondary alcohols) but not oxidised.

A   Only 1

B   1 and 2

C   Only 3

D   1 and 3

40. Which of the following best outlines a method for preparing aromatic amines?

   A   Nitrobenzene is reduced using a mixture of tin and concentrated hydrochloric acid, heated under reflux. The excess acid is then neutralised.

   B   Nitrobenzene is oxidised using ammonia heated to 50 ^{0}C in a water bath.

   C   Ammonia is reacted with a weak acid, resulting in the formation of a salt, which may be an amine.

   D   3-chlorobutane is reacted with an excess of ammonia to form an aromatic amine.

41. Which statement best describes the meaning of aliphatic and aromatic hydrocarbons?

   A   Aromatic hydrocarbons consist of a ring structure such as a benzene ring, whereas aliphatic hydrocarbons do not.

   B   Aliphatic hydrocarbons consist of a ring structure such as a benzene ring, whereas aromatic hydrocarbons do not.

   C   Aromatic hydrocarbons consist of at least one carbon to carbon double bond. Aliphatic hydrocarbons only contain single bonds.

   D   Aliphatic hydrocarbons consist of at least one carbon to carbon double bond. Aromatic hydrocarbons only contain single bonds.

42. OH and NH groups can be detected using proton NMR spectroscopy. Choose the statement that describe how this is done:

   A   Two spectra are run, the second one with D_2O added. Any peaks due to OH and NH groups will be absent in the second spectrum.

   B   Two spectra are run, the second one with tetramethylsilane (TMS) added. Any peaks due to OH and NH groups will be absent in the second spectrum.

   C $CDCl_3$ is used to remove any peaks due to OH and NH groups.

   D   OH and NH peaks are very easy to identify by looking at the chemical shifts of their peaks.

QUESTIONS 6

Organic chemistry and analysis

43. The mass spectrum of 2-methylbutane is shown below.

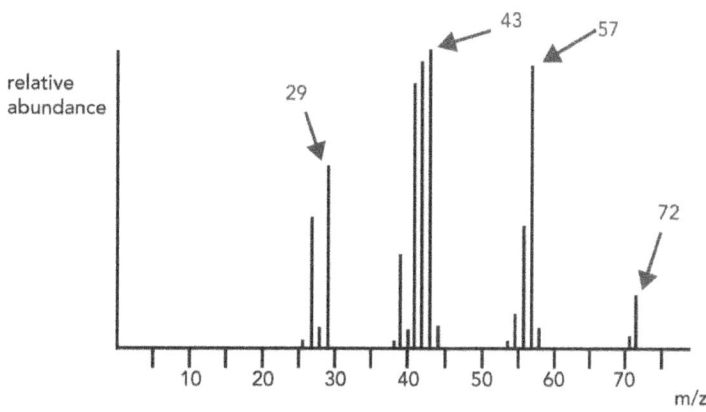

Which ion is responsible for the peak at 57?

A $CH_3CH_2CCH_3^+$

B $CH_3CH_2CHCH_2^+$

C $CH_3CHCH_2CH_3^+$

D $CH_2CCH_2CH_3^+$

44. Compound X reacts with hydrogen in the presence of a catalyst Y to form compound Z. What are the identities of X, Y and Z?

	X	Y	Z
A	Ethanenitrile	Nickel	Ethylamine
B	Ethanenitrile	Sulfuric acid	Ethylamine
C	Ethylamine	Nickel	Ethanenitrile
D	Ethylamine	Sulfuric acid	Ethanenitrile

45. What mechanisms are involved in the following reactions?

Reaction 1 $CH_3CH_2CH_2CH_2Cl + KCN \rightarrow CH_3CH_2CH_2CH_2CN + KCl$

Reaction 2 $(CH_3)_2CO + HCN \rightarrow CH_3C(OH)(CH_3)CN$

	Reaction 1	Reaction 2
A	Nucleophilic substitution	Electrophilic addition
B	Nucleophilic substitution	Nucleophilic addition

QUESTIONS
Organic chemistry and analysis

C	Nucleophilic addition	Nucleophilic substitution
D	Electrophilic substitution	Nucleophilic addition

46. Methylbenzene reacts with acyl chloride in the presence of a halogen carrier. Which of the following equations is/are correct?

1:

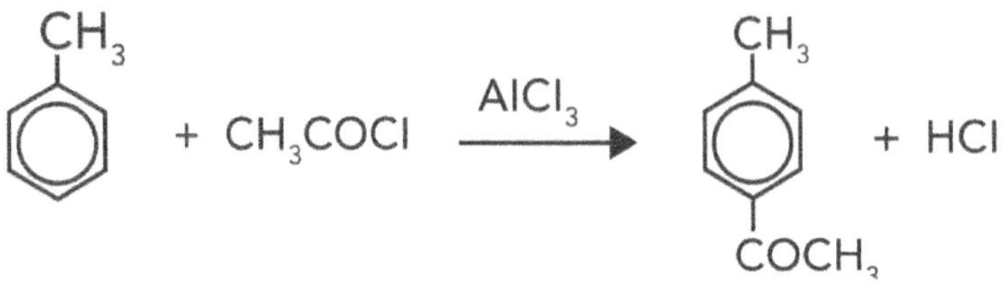

$$\text{C}_6\text{H}_5\text{CH}_3 + \text{CH}_3\text{COCl} \xrightarrow{\text{AlCl}_3} \text{4-CH}_3\text{C}_6\text{H}_4\text{COCH}_3 + \text{HCl}$$

2:

$$\text{C}_6\text{H}_5\text{CH}_3 + \text{CH}_3\text{COCl} \xrightarrow{\text{AlCl}_3} \text{3-CH}_3\text{C}_6\text{H}_4\text{COCH}_3 + \text{HCl}$$

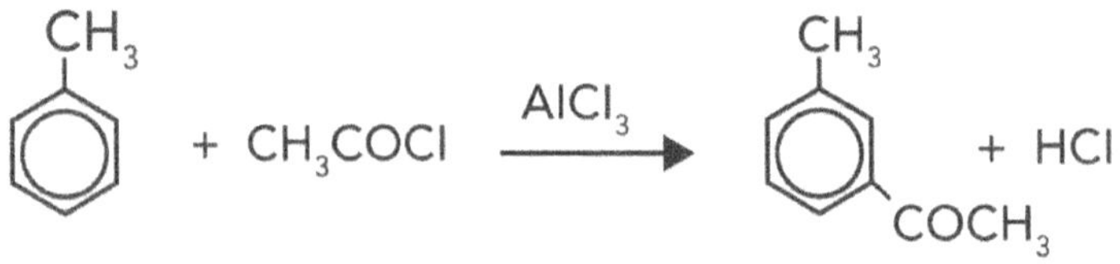

3:

$$\text{C}_6\text{H}_5\text{CH}_3 + \text{CH}_3\text{COCl} \xrightarrow{\text{AlCl}_3} \text{(4-CH}_3\text{, 1-COCH}_3\text{)C}_6\text{H}_4 + \text{HCl}$$

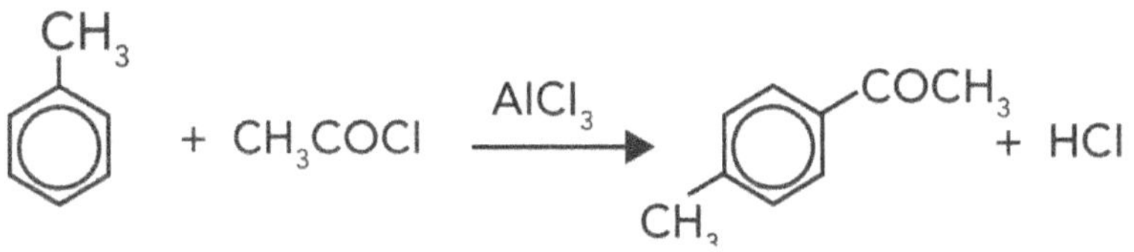

A   1, 2 and 3

B   Only 1 and 3

C   Only 1

D   Only 3

47. The following compound is 1,2-dimethylcyclopentene.

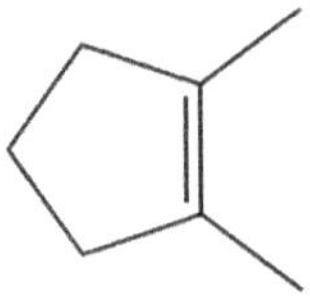

Which of the following statements is/are true?

1:    Its proton NMR spectrum consists of three peaks.

2: Its proton NMR spectrum consists of two peaks.

3: Its molar mass is 98.0 g mol⁻¹.

A   Only 1

B   1 and 3

C   Only 2

D   2 and 3

48. The following reaction was carried out in the laboratory, using concentrated sulfuric acid as a catalyst. The reaction mixture was heated to just above 50 °C.

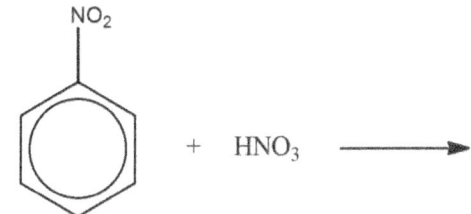

Which of the following statements about the carbon-13 NMR spectrum of the major organic product of the above reaction is true?

1: It will have three peaks.

2: All of its peaks will be within the range 0 to 50 ppm.

3: All of its peaks will be within the range 110-160 ppm.

A   Only 1

B   Only 2

C   Only 3

D   1 and 3

49. The following compound is a dioxane. The three statements that follow relate to the carbon-13 NMR spectrum of this compound.

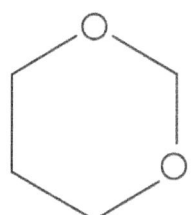

Which of the following statements is/are true?

6 QUESTIONS
Organic chemistry and analysis

1: It consists of three peaks.

2: It will have one peak within the range 0-50 ppm.

3: It will have two peaks within the range 50-90 ppm.

A   Only 1

B   Only 2

C   1 and 2

D   1, 2 and 3

50. The structure of 1,4-dioxane is shown below.

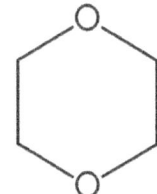

The following table relates to the proton and carbon-13 NMR spectra of the above compound.

Which of the following statements is true?

	Carbon-13 NMR	Proton NMR
A	Two peaks	One singlet
B	One peak	One singlet
C	Two peaks	Two singlets
D	One peak	Two singlets

51. You are asked to prepare the following compound in the laboratory.

$$CH_3-C\overset{\displaystyle O}{\underset{\displaystyle C_6H_5}{\|}}$$

Which of the following statements describe the method you would use in the preparation?

A Benzene is reacted with ethanoyl chloride using aluminium chloride as the catalyst. The mixture is heated to about 60 °C for around 3 minutes.

B Benzene is reacted with ethanal using aluminium chloride as the catalyst. The mixture is heated to about 60 °C for around 3 minutes.

C Benzaldehyde is reacted with ethanoyl chloride using aluminium chloride as the catalyst. The mixture is heated to about 60 °C for around 3 minutes.

D Benzaldehyde is reacted with ethanal using aluminium chloride as the catalyst. The mixture is heated to about 60 °C for around 3 minutes.

52. Thin-layer chromotography was carried out on a sample containing two components, A and B. The resulting chromotograph is shown in chromatograph 1. A different, unidentified sample known to contain component A was then analysed. The resulting chromatograph is shown in chromatograph 2.

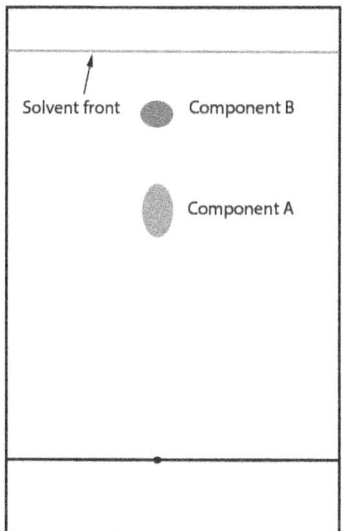
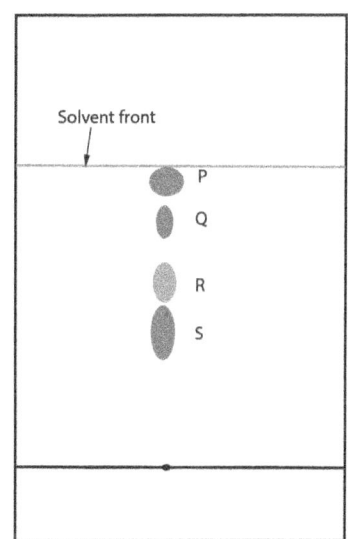

Chromatograph 1   Chromatograph 2

Determine which area contain component A in chromatograph 2.

A P

B Q

C R

D S

53. Which of the following statement(s) is/are **not** true?

1: Nitration of phenol requires concentrated HNO_3 and will not take place with dilute HNO_3.

2: Nitration of phenol requires a concentrated H_2SO_4 catalyst.

3: NO_2 is described as an electron-donating group.

A   Only 1

B   Only 2

C   Only 3

D   1, 2 and 3

54. A student reacted nitrobenzene with nitric acid using concentrated sulfuric acid as a catalyst. The reaction mixture was heated to just above 50 °C. The incomplete equation for the reaction is shown below.

+ HNO₃ ⟶

Which of the following major organic product is most likely to be formed during the reaction?

A

B

C

D

55. The following is a section of a proton NMR spectrum. The number next to the peak is the relative peak area of the peak. Which of the following statement(s) about the proton enviornment shown by the peak could be true?

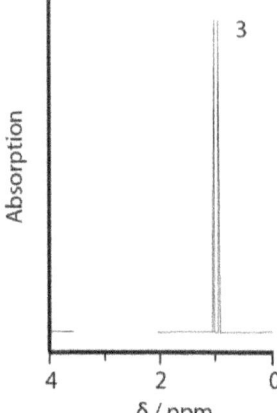

1: There are three protons in the proton environment.

2: There are two protons in the proton environment.

3: There is only one proton in the proton environment.

A   Only 1

B   Only 2

C   Only 3

D   1, 2 and 3.

56. Which of the following two reagents may react to form the compound shown when heated in the presence of a catalyst?

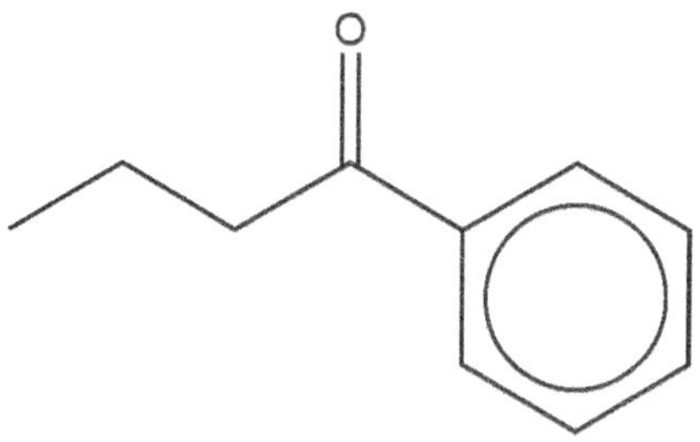

A   Benzene and butanoic acid

B   Benzene and butanoic anhydride

C   Benzene and butanoyl chloride

D   Benzanoyl chloride and butane

57. Which of the following is the most likely product formed when phenol reacts with dilute nitric acid?

A   2-nitrophenol

B   3-nitrophenol

C   2,4,6-trinitrophenol

D   2,3-dinitrophenol

58. A student reacted pentanoic acid with thionyl chloride. Which of the following statements could be true?

1: The mass spectrum of the major product has its molecular ion peak at m/z = 130.5.

2: One of the products reacted with ammonia to form ammonium chloride.

3: Sulfur monoxide was given off during the reaction.

A   Only 1

B   Only 2

C   1 and 2

D 2 and 3

59. Which of the following compound may react with a mixture of KCN and HCl to form the compound shown below?

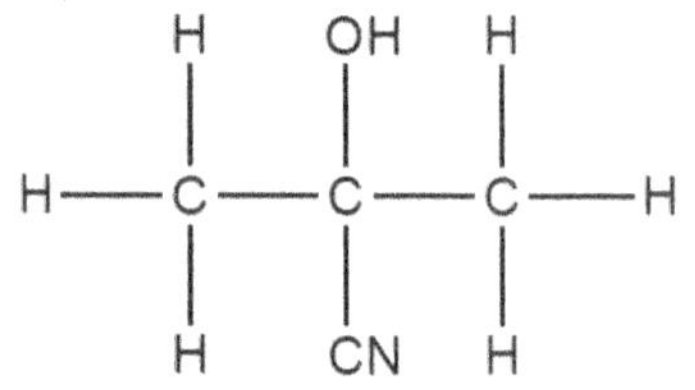

A   Propanone

B   Propan-2-ol

C   2-chloropropane

D   Propanal

60. Which of the following reagents may react together to form the compound shown below?

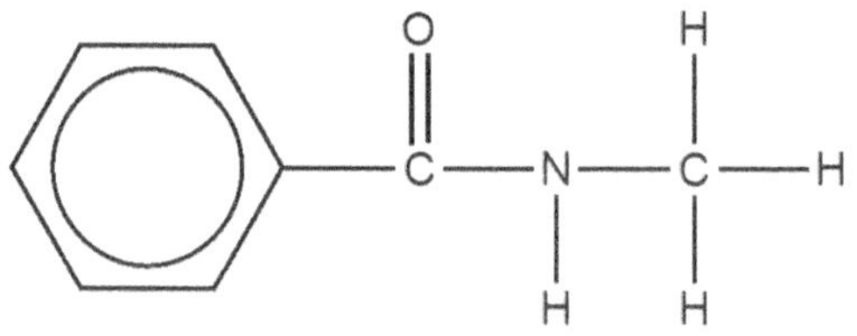

	Reagent 2	Reagent 2
A	Benzanoyl chloride	Methylamine
B	Benzanoyl chloride	Ammonia
C	Benzanoyl chloride	Methylammonium chloride
D	Benzoic acid	Methylamine

61. How many additional stereoisomers are there of $H_2C=CHCH=CH_2$?

A   0

B   2

C   4

D   6

62. A student was asked to prepare a sample of butyl butanoate and ethyl butanoate and was provided with X number of organic compounds and some potassium dichromate and sulfuric acid.

What is the minimum number X can take?

- A 1
- B 2
- C 3
- D 4

63. You are given three labelled samples of magnesium carbonate, sodium hydroxide and aqueous silver nitrate in ethanol and four unlaballed samples of phenol, propanoic acid, chloroethane and propanone.

    How many of the unlabeled samples can you identify?

    - A 1
    - B 2
    - C 3
    - D 4

64. A student completely combusted 30 cm³ a gaseous hydrocarbon. The reaction required 270 cm³ of oxygen and produced 180 cm³ of carbon dioxide. All the volumes are at room temperature and pressure.

    What is the formula of the hydrocarbon?

    - A C_6H_{14}
    - B C_6H_{12}
    - C C_8H_{16}
    - D C_8H_{18}

65. Reactions 1 and 2 can be used to prepare (E)-4-(phenyldiazenyl)phenol.

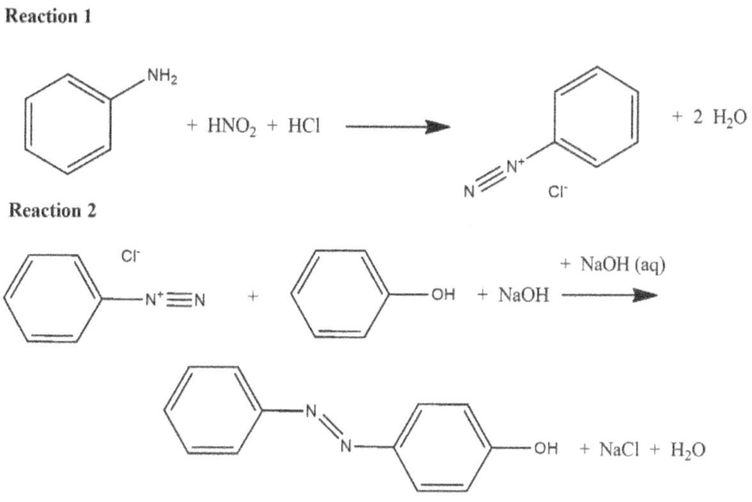

    Reaction 1

    C₆H₅NH₂ + HNO₂ + HCl ⟶ C₆H₅N₂⁺Cl⁻ + 2 H₂O

    Reaction 2

    C₆H₅N₂⁺Cl⁻ + C₆H₅OH + NaOH + NaOH (aq) ⟶ (E)-4-(phenyldiazenyl)phenol + NaCl + H₂O

    The percentage yield of (E)-4-(phenyldiazenyl)phenol is 57%. What mass of phenylamine is required to prepare 7.40 g of (E)-4-(phenyldiazenyl)phenol?

QUESTIONS
Organic chemistry and analysis

- A   6.1 g
- B   3.5 g
- C   2.0 g
- D   7.0 g

66. Phenol reacts with dilute nitric acid.

    Which of the following correctly shows the skeletal formula of the major intermediate formed in the mechanism?

    A, B, 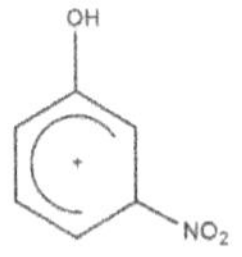 C, 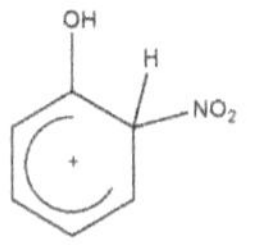 D [skeletal formulae of nitrated phenol intermediates]

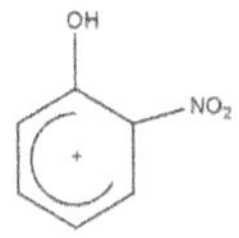

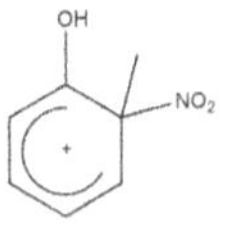

67. The displayed formula of Serine is shown in the following figure.

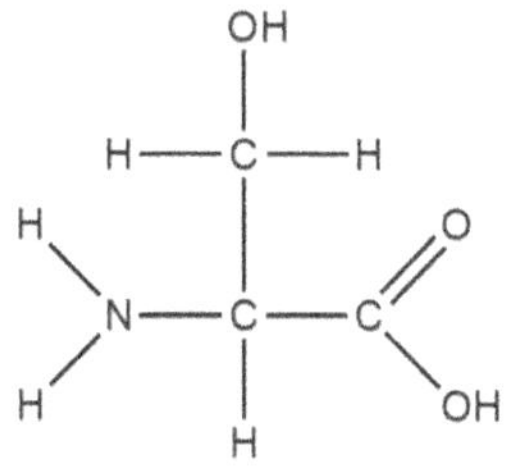

    [displayed formula of Serine]

    Which of the following correctly shows the skeletal formula of the dipeptide formed by joining two Serine molecules?

    A, B, 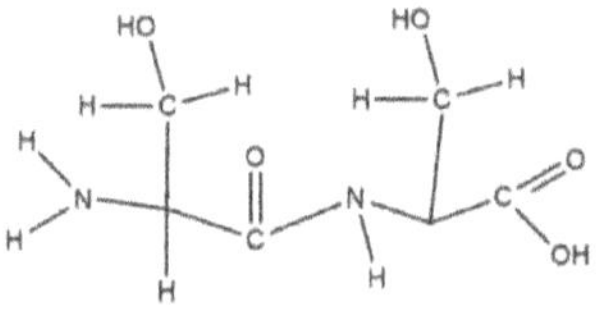 C, 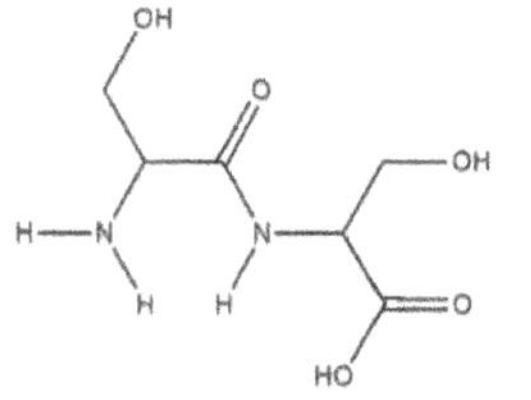 D [skeletal formulae of dipeptides]

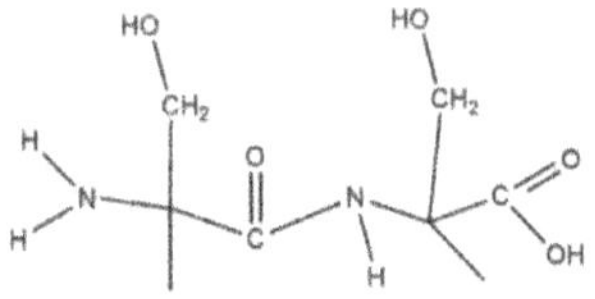

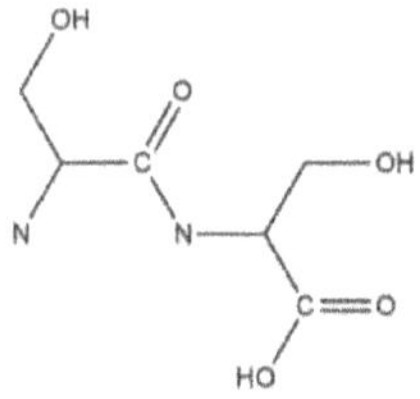

68. A student was given 2.50 g of an alcohol of molar mass 116 and 2.10 g of a carboxylic acid of molar mass 88. The two compounds were reacted together to obtain 1.30 g of the major organic product.

    What is the percentage yield of the reaction?

A   29.6%

B   32.5%

C   29.3%

D   61.9%

69. A student wanted to form 4.30 g of an ester of molar mass 178. An excess of an alcohol of molar mass 94.0 was reacted with another compound. 1 mole of alcohol forms 1 mole of HCl as a minor product of the reaction. The percentage yield of the reaction is 67.3 %.

What is the mass in grams of the other compound that should react?

A   1.96 g

B   3.02 g

C   4.33 g

D   2.91 g

70. Which of the following options is correct?

	$C_6H_5CONH_2$	$HCONH(CH_3)$
A	Primary amide	Secondary amide
B	Secondary amide	Primary amide
C	Primary amide	Primary amide
D	Secondary amide	Secondary amide

71. A **compound S**, of molar mass 130 reacted with an alcohol to form an ester of molar mass 102. What could be the formula of the alcohol?

A C_2H_5OH

B C_3H_7OH

C C_4H_9OH

D $C_5H_{11}OH$

72. Gas chromatography was carried out on a mixture containing unknown compounds. The resulting chromatogram is shown below.

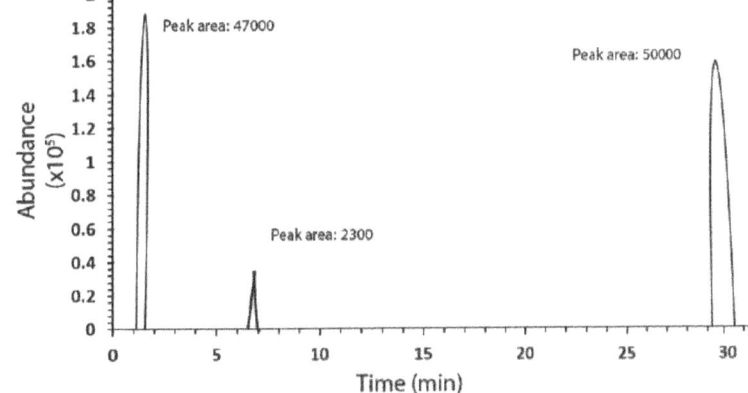

QUESTIONS

Organic chemistry and analysis

Which of the following statements is true?

1: The concentration of one of the compounds is 0.47 mol dm^{-3}.

2: The amount of compound with the retention time of 6.8 minutes is very little compared to others.

3: The abundance of one of the compounds is 1.88.

A   Only 1

B   Only 2

C   Only 3

D   2 and 3

QUESTIONS 6

Organic chemistry and analysis

Answers

All the answers for the questions on Module 5 and Module 6 are given in this spread. Please pay attention to how marks are awarded. There are explanations for the answers to the Multiple Choice Questions where appropriate.

When checking your answers, it will help if you mark them just like an examiner would. Keep in mind that only certain points will score you the marks for any given question.

ANSWERS

Physical Chemistry and Transition Elements

MODULE 5

1.
   a. 3 minutes and 27 seconds = 207 seconds.

   $207 / 23 = 9$ half-lives✓. Therefore divide 0.500 by 2 nine times. $0.500 / 2^9 = 9.77 \times 10^{-4}$ mol dm^{-3}.✓

   b. Order with respect to P = 1✓ ; order with respect to Q = 0✓.

   c. rate = $k[P]$✓✓

   (Note that Q is not included as its order is 0.)

2. Concentration has no effect on rate in this instance.✓

3.
   a. If the reaction produces a gas, the change in volume or pressure can be measured✓. Alternatively, loss of mass of the reactants can be measured.✓ For reactions that involve acids or bases, pH changes may be worked out using a pH meter✓. Reaction rates may also be measured by taking into account colour changes (colorimetry) or the formation of precipitates. When a precipitate is being formed, you time how long it takes for the mixture to become opaque (exampe: thiosulfate / acid reaction). Change in concentration of a reactant or product per unit time gives the reaction rate.✓ (Rate of reaction = change in concentration / time).

   "Change" here means difference. For example, if there were 5 mol dm^{-3} of a substance at the start of the reaction and it dropped down to 1 mol dm^{-3} within 2 seconds, the *average reaction rate* would be 2 mol dm^{-3} s^{-1}. (5 minus 1, divided by 2).

   (Note that, similar questions will not be asked in an exam. The purpose of this question is to test your understanding.)

   b. The gradient of the graph is calculated by drawing a tangent at the given time. ✓ This gives the rate of reaction.

   c. The gradient of the tangent is calculated at the point where t=0 ✓(the starting point of the graph).

   d. Moles = concentration x volume (in dm^3) = $0.0400 \times 0.200 = 0.00800$ moles of butanol. ✓

   Molar mass of butanol = $(12.0 \times 4) + (1.0 \times 10) + 16.0 = 74.0$✓

   Mass(g) = moles x molar mass = $0.00800 \times 74.0 = 0.592$ g of butanol (in 200 cm^3).

   Concentration $0.592 \times 5 = 2.96$ g dm^{-3}.✓

   (Molar masses should be given to one decimal place in your workings. Note that 0.592g is in 200 cm^3 and therefore it needs to be multiplied by 5 to get the mass in 1000 cm^3)

   e. OH groups form hydrogen bonds with water molecules.✓

ANSWERS 5

Physical Chemistry and Transition Elements

4.
   a. $[Cu(Cl)_4]^{2-}$ ✓
   b. Tetrahedral. ✓ 109.5°. ✓

5.
   a. 6 ✓
   b. Octahedral. ✓
   c. $1s^2\ 2s^2\ 2p^6\ 3s^2\ 3p^6\ 3d^7$ ✓ (Explanation: The electrons which were present on the 4s sub-shell were lost when the ion formed).
   d. 107° ✓
   There are three bonded pairs and one lone pair around O of H_2O in $[Co(H_2O)_6]^{2+}$ ✓
   Lone pairs repel more than bonded pairs. ✓

6.
   a. Rate of reaction is the change in concentration of a reactant or product per unit time. ✓
   b. It is a constant that links rate of reaction with concentrations of reactants raised to the powers of their orders in the rate equation. ✓
   c. Value of the rate constant, k, goes up with temperature. ✓ Rate of reaction also increases likewise. At higher temperatures, particles collide more often with more energy leading to more successful collisions. ✓

7. If the order is 0 with respect to a given reactant, rate is unaffected by changing its concentration. ✓ If it is 1, rate is proportional to the concentration (if the concentration doubles, rate doubles. If it increases by threefold, rate increases in the same way); If it is 2, rate is proportional to the concentration to the power 2. ✓ For example, if the concentration increases by 2, rate increases by 4 (2^2). If it increases by 3, rate increases by 9 (3^2).

8. The concentration of the reactant or product can be measured at regular time intervals as a reaction proceeds. A graph of concentration against time can then be plotted. The rate at any time is the gradient of the curve at that time. ✓ (To measure gradient of a curve, a tangent is drawn at that point and its gradient is found). Units of rate of reaction: $mol\ dm^{-3}\ s^{-1}$. ✓ (Note that, when speaking about rate equations and orders of reaction, we use the units, $mol\ dm^{-3}\ s^{-1}$, for the rate of reaction. However reaction rates can be measured using alternative units as well.)

9. $1\ mol\ dm^{-3}\ s^{-1}$ ✓

10.
   a. i. The time taken for the concentration to reduce to half is its half-life. ✓

      ii. Half-life is 7 seconds. ✓

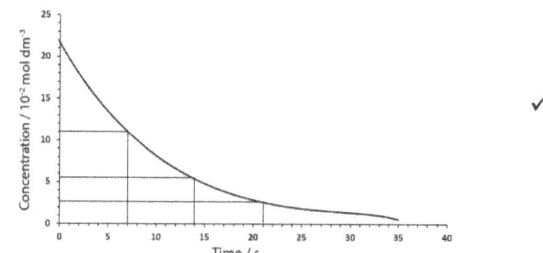

iii. First order ✓

iv. No change. Half-life of a first order reactant is constant, regardless of its concentration. ✓

b. As this is a decomposition reaction, there will be no other reactants. As the only reactant is first order, this will be a first order reaction. Therefore, $k = (\ln 2) / \text{half-life}$ ✓ $= (\ln 2) / 7 = 0.0990$ ✓. At 8 seconds, the concentration of X is 10×10^{-2} (note the y-axis's label on the graph) $= 0.1$ mol dm^{-3} ✓. Rate equation will be rate $= k[X] = 0.0990 \times 0.1 = 9.90 \times 10^{-3}$ ✓ mol dm^{-3} s^{-1} ✓.

11.
   a. $1s^2\ 2s^2\ 2p^6\ 3s^2\ 3p^6\ 4s^1\ 3d^{10}$ ✓

   ✓ equation
   ✓ balancing

   b. $[Cu(H_2O)_6]^{2+}$ (aq) $+ 4NH_3$ (aq) $\rightleftharpoons [Cu(NH_3)_4(H_2O)_2]^{2+}$ (aq) $+ 4H_2O$ (l)

   c. Ligand substitution ✓

   d. Pale blue solution turns into pale blue precipitate with the addition of some ammonia. ✓ When an excess is added, the precipitate dissolves forming a deep blue solution. ✓

   e. $[Cu(NH_3)_4(H_2O)_2]^{2+}$ is an octahedral complex ion. ✓

12. Zero order: concentration decreases at a constant rate. ✓ First order: concentration halves at equal time intervals. ✓ Second order: initially, concentration decreases rapidly and the rate of decrease then slows down. ✓

13.
   a. From the rate equation, P is a second order reactant.

   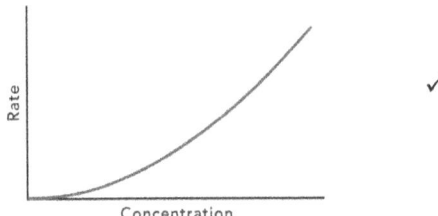
   ✓

   b. R is a first order reactant.

   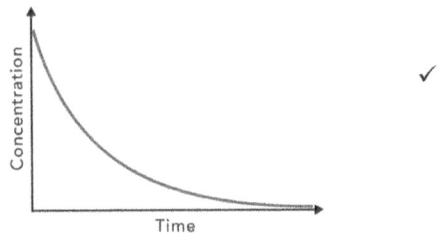
   ✓

   c. As Q does not appear in the rate equation, it should be a zero order reactant.

Physical Chemistry and Transition Elements — ANSWERS 5

14. In a zero order reactant, the half-life decreases with time.✓ In a first order reactant, the half-life is constant.✓ In a second order reactant, the half-life increases with time.✓

15.

   a. $HN_3 \rightleftharpoons H^+ + N_3^-$ (equilibrium sign✓ rest of the equation✓)

   b. Hydrazoic acid used in the preparation of the buffer solution is diluted / aqueous.✓

   c. HN_3 reacts with KOH forming N_3^-/KN_3 OR $HN_3 + KOH \rightarrow KN_3 + H_2O$✓

   Some HN_3 remains / HN_3 is in excess✓.

   d. $n(HN_3) = 0.0605 \times 0.1 = 0.00605$ mol✓

   $n(KOH) = 0.0200 \times 0.250 = 0.005$ mol ✓   (The molar ratio of KOH to HN_3 in the equation is 1:1. Therefore 1 mole of KOH reacts with 1 mole of KN_3.)

   $0.00605 - 0.005 = 0.00105$ moles✓ of HN_3 is in excess. 0.005 mol of KN_3 will be formed.

   $[H^+(aq)] = K_a \times [HA(aq)] / [A^-(aq)] = 1.9 \times 10^{-5} \times (0.00105/0.005) = 3.99 \times 10^{-6}$✓

   (Note: in this calculation, substituting the molar values will give the same answer as substituting the concentrations. This is because the volume is the same for both HN_3 and KN_3 and therefore will cancel out in the above expression).

   $pH = -\log[H^+] = -\log 3.99 \times 10^{-6} = 5.4$ (answer✓ 2 d.p.✓)

16.

   a. It is the power to which the concentration of a reactant is raised in the rate equation.✓ (The order tells us to what extent the concentration of that particular reactant affects the rate of reaction).

   b. Because its concentration has no effect on the reaction rate.✓

17.

   a. rate $= k[F][G]^2$ ✓ (The rate determining step coefficients of reactants are directly related to the order of those reactants)

   b.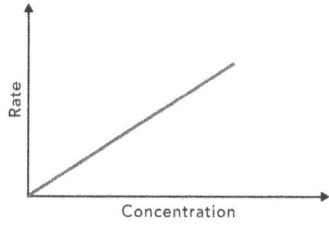
   ✓

   c.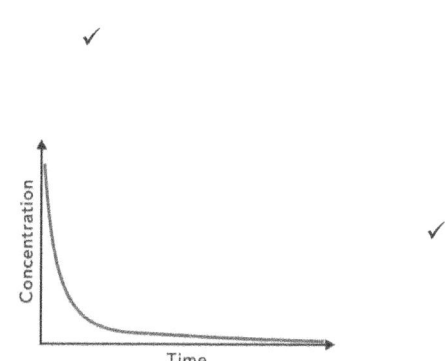
   ✓

5 ANSWERS Physical Chemistry and Transition Elements

d.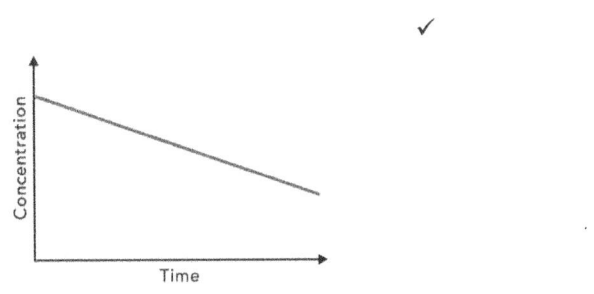

18.

   a. Comparing experiments 1 and 2, when the concentration of O_2 doubles (while the other concentrations remain the same), the reaction rate doubles.✓ So the order with respect to O_2 is 1.✓
Comparing experiments 1 and 3, when the concentration of NO doubles (while the other concentrations remain the same), the reaction rate increases by 4.✓ So the order with respect to NO is 2.✓

   b. rate=$k[NO]^2[O_2]$✓

   c. k = rate / $[NO]^2[O_2]$ = mol dm^{-3} s^{-1} / (mol dm^{-3})^2(mol dm^{-3}) = dm^6 mol^{-2} s^{-1} ✓

   d. rate = $k[NO]^2[O_2]$
k = rate / $[NO]^2[O_2]$ = 0.36 / 0.36^2 × 0.15 = 18.5✓   dm^6 mol^{-2} s^{-1}

   e. rate=18.5(2.02×10^{-3})^2(1.05×10^{-3})= 7.93×10^{-8} mol dm^{-3} s^{-1} ✓

19. Rate ∝ [reactant]^2.✓ That is if the concentration doubles, rate increases by 4 (2^2). If it increases by 3, rate increases by 9 (3^2).

20.

   a. If the forward reaction is exothermic: when temperature increases, K_c decreases, position of equilibrium shifts towards reactants, when temperature decreases, K_c increases, position of equilibrium shifts towards products.✓ If the forward reaction is endothermic: when temperature increases, K_c increases, position of equilibrium shifts towards products, when temperature decreases, K_c decreases, position of equilibrium shifts towards reactants.✓

   b. The expression for the equilibrium constant can be worked out just by using a balanced equation. Experimental data are not required.✓

   c. If the concentration of reactants are doubled, the position of equilibrium will move towards the products, according to Le Chatelier's principle. Therefore, in the expression of K_c, if the concentration of reactants double, the bottom of the expression will get bigger (as the denominator represents the reactants). In order to maintain the same value for K_c, the top will get bigger as a result (the top getting bigger means more product is being produced: result of equilibrium position moving to the product's side). If the pressure of an equilibrium system is doubled, the position of equilibrium will move so as to reduce the pressure. In a gaseous reaction, it will move in the direction that contains fewer gas particles. If there are more gas particles in the reactant's side, equilibrium will shift towards the products. Changes in pressure cause changes in concentration in gases. As the pressure increases, so does the concentration. Value of K_c does not change with changes in concentration and pressure or when a catalyst is introduced. ✓ (Only last sentence needed for the mark)

21.

   a. It is the slowest step in the reaction mechanism of a reaction involving many stages.✓

b. Any reactant in the rate equation appears in the rate-determining step.✓ Note that, zero order reactants do not have to be included in the rate equation as they have no effect on the reaction rate.

Consider a reaction, with the rate equation: rate = $k[CO]^2[H_2O]$

The rate-determining step of this reaction will read:

$2CO + H_2O \rightarrow$ *

(*To work out the rest of the rate-determining step, more information is required).

If the order with respect to a reactant is 2, the coefficient of that reactant in the rate-determining step will be 2. If it is 1, it will be 1 and if it is 0, it will not appear in the rate-determining step.

22.

a. Comparing experiments 1 and 2, when the concentration of A doubles (while the other concentrations remain the same), reaction rate is unaffected.✓ So the order with respect to A = 0.✓

Comparing experiments 1 and 3, when the concentration of B increases by 4 times (while the other concentrations remain the same), the reaction rate increases by 16 times.✓ So the order with respect to B = 2.✓

b. rate= $k[B]^2$✓

c. rate = $k[B]^2$

k = rate / $[B]^2$ = 0.0320 / 0.0150 to the power 2 = 142✓ dm^3 mol^{-1} s^{-1}

k= rate/$[B]^2$ = mol dm^{-3} s^{-1} / (mol $dm^{-3})^2$ = dm^3 mol^{-1} s^{-1}✓

23. A base is a proton acceptor.✓

24.

a. An acid and an alkali react to form a salt and water.

   $HCl(aq) + NaOH(aq) \rightarrow NaCl(aq) + H_2O(l)$✓

   $H^+(aq) + OH^-(aq) \rightarrow H_2O(l)$✓

b. An acid reacts with a metal to form a salt and hydrogen gas.

   $2Na(s) + 2HNO_3(aq) \rightarrow 2NaNO_3(aq) + H_2(g)$✓

   $2Na(s) + 2H^+(aq) \rightarrow 2Na^+(aq) + H_2(g)$✓

c. An acid reacts with a base to form a salt and water

   $H_2SO_4(aq) + CaO(s) \rightarrow CaSO_4(aq) + H_2O(l)$ ✓

   $2H^+(aq) + CaO(s) \rightarrow Ca^{2+}(aq) + H_2O(l)$ ✓

d. An acid reacts with a carbonate to form a salt, water and carbon dioxide.

H_2SO_4 (aq) + $CuCO_3$ (s) → $CuSO_4$ (aq) + CO_2 (g) + H_2O (l) ✓

$2H^+$(aq) + $CuCO_3$(s) → Cu^{2+}(aq) + CO_2(g) + H_2O(l) ✓

25. In a dynamic equilibrium, the rate of forward reaction is equal to the rate of reverse reaction. ✓ It exists in a closed system. ✓ (A closed system is one where reactants or products cannot enter or leave the system).

26.

   a. $K_c = [C]^c[D]^d / [A]^a[B]^b$ ✓ (Remember the expression for K_c can be worked out just by using a balanced equation).

   b. The units of each K_c expression are different from one another. Therefore the K_c of one equilibrium will be different from another. ✓ To work out the units, each concentration term in the expression is replaced by its unit. The unit can then be simplified by "cancelling out" the units that are the same in the top and bottom of the fraction.

   c. If the value of K_c is 1, neither reactants nor products are favoured. ✓ If it is greater than 1, the position of equilibrium lies towards the products (product-favoured). ✓ If it is less than 1, the position of equilibrium lies towards the reactants (reactant-favoured). ✓

27. Moles = mass (g) / molar mass = 2.56 / 40.0 = 0.0640 moles ✓. Concentration (mol dm^{-3}) = amount (moles) / volume (dm^3) = 0.064 / 0.023 = 2.8 mol dm^{-3}. (answer ✓ 2 s.f. ✓) Note that, as the volume is given in the question to 2 s.f., your answer cannot be any more accurate than 2 s.f.

28.

   a. It is the series of steps, which as a whole forms the overall reaction. ✓

   b. It is a species formed in one of the steps of a multi-step reaction ✓, which is then used up in a later step ✓. An intermediate is not seen as a reactant or a product in the overall equation.

   c. Any reactant in the rate equation also appears in the rate-determining step. ✓ The order with respect to any reactant in the rate equation tells you how many moles of that reactant is present in the rate-determining step. ✓ For example, if the order with respect to a reactant C is 2 in the rate equation (Ie: C^2), C would appear as 2C in the rate-determining step. *(In some cases, an intermediate may be present in the rate-determining step that does not appear in the rate equation. However, this will not be the case with most examples you will come across).*

29.

   a. i. Cs(g) + Cl(g)   [Step C forms] ✓

      Cs^+(g) + Cl^-(g)   [Step E forms] ✓

      ii. Standard enthalpy change of formation.

      First ionisation energy of caesium.

      First electron affinity of chlorine.

      3 correct = ✓✓
      2 correct = ✓

   b. -433 = 79 + 121 + 376 + (-346) + LE ✓

      LE = -433 - 79 - 121 - 376 + 346

Physical Chemistry and Transition Elements — ANSWERS 5

LE = -663 kJ mol^{-1} ✓

c. Mg^{2+} has a greater charge density / smaller ionic radius than Cs^+ ✓

O^{2-} has a greater charge density than Cl^- ✓

Greater attraction between Mg^{2+} and O^{2-} ions ✓

30.

a. k = rate / ($[ClO_2]^2[OH^-]$)

1.75 x 10^{-3} / (3.20 x 10^{-2} x 2.90 x 10^{-3}) ✓ = 18.9 ✓

Units = mol dm^{-3}s^{-1} / ((mol dm^{-3})^2 x (mol dm^{-3})) ✓

mol dm^{-3}s^{-1} / (mol^2 dm^{-6} x mol dm^{-3}) = mol dm^{-3} s^{-1} / mol^3 dm^{-9} = dm^6 mol^{-2} s^{-1} ✓

b. Rate constants can only be calculated using experimental data. ✓ (Rate constant measures the "speed" or rate of a reaction).

31.

a. An acid is a proton donor. ✓

b. i. An acid dissociates when added to water, releasing H$^+$ ions. ✓ The extent of dissociation depends on how strong the acid is. A strong acid such as HCl dissociates completely, whereas a weak acid like CH_3COOH dissociates only partially.

ii. HCl(g) + aq → H$^+$(aq) + Cl$^-$(aq)   (H$^+$ ion shown ✓. Rest of the equation ✓.)

c. 2HCl(aq) + Mg(s) → $MgCl_2$(aq) + H_2(g) (Formulae ✓. Balancing ✓.)

d. Ethanoic acid. CH_3COOH. ✓   (Any other named weak acid scores the mark).

32.

a. N_2(g) + $3H_2$(g) ⇌ $2NH_3$(g).   (equation ✓ balancing ✓)

b. In this reaction the forward reaction is exothermic and the reverse reaction is endothermic. ✓ At a lower temperature, K_c will be large as the reaction will move in the exothermic direction and k will be small as the reaction rate is less at a lower temperature. ✓ At a higher temperature, K_c will be small as the reaction will move in the endothermic direction and k will be large as the reaction rate is high at a high temperature. ✓ Therefore, compromise conditions are used so that a reasonable yield of ammonia is achieved at a reasonable rate.

(For correctly mentioning the effect of temperature on K_c or k, one mark is awarded, at any point in the answer.)

c. Any factor other than temperature. Example: concentration or pressure ✓.

d. NH_4^+ ✓; $(NH_4)_2CO_3$ ✓

(The charge of ammonium is 1+ and carbonate is 2-. Therefore you need two ammonium ions to balance the formula.)

e. The reaction will oppose the change by cooling. Therefore it will move towards the endothermic direction✓. Position of equilibrium will shift to the left.✓

f. +92 kJmol^{-1} ✓

33. K_c is not affected by changes in concentration, pressure or the presence of a catalyst. It only changes with temperature.✓

34.

a. $CO + NO_2 \rightarrow CO_2 + NO$ ✓✓

b. 0. ✓ (Note that CO does not appear in the rate equation, which means its order is 0: the concentration of CO would have no effect on the rate of reaction.)

c. $2NO_2 \rightarrow NO_3 + NO$ ✓ (Rate-determining step)

$NO_3 + CO \rightarrow NO_2 + CO_2$ ✓

(Alternative answer: $2NO_2 \rightarrow N_2O_4$; $N_2O_4 + CO \rightarrow NO_2 + CO_2 + NO$)

Note that **any reactant present in the rate equation will appear on the left side of the rate-determining step**. This is always the case. The power to which these reactants are raised in the rate equation will dictate how many moles of these reactants are present in the rate-determining step. In this case NO_2 is raised to the power of <u>two</u> in the rate equation and therefore there are <u>two</u> moles of NO_2 in the rate-determining step.

In the reaction mechanism a new species, NO_3, is present. However, this is just an intermediete which means it is not a reactant or a product. It is formed in the first stage and used up in the next. So it just disappears and is not present in the overall equation.

To get the overall equation from these two steps, you would have to cross out common molecules from either side of the equation. For example, NO_3 is present on the right side of one equation and the left of another - so you cross out both NO_3's. You also get <u>two</u> moles of NO_2 on the left of one equation but <u>one</u> mole of NO_2 on the right of the other - so you cross out the extra NO_2 from the right of the second equation. Now, if you add the remaining reactants and products from both steps, you will get the overall equation.

d. rate = $k[NO_2]^2$

$7.54 \times 10^{-9} = k(0.004)^2$ ✓

$k = 7.54 \times 10^{-9} / 1.6 \times 10^{-5} = 4.71 \times 10^{-4}$ ✓

Units = mol dm^{-3}s^{-1} / (mol dm^{-3})^2 ✓

Units = mol dm^{-3}s^{-1} / mol^{2}dm^{-6}

Units = mol^{-1}dm^{3}s^{-1} ✓

ANSWERS 5

Physical Chemistry and Transition Elements

35.

a. pH = -log [H^+(aq)]. ✓

b. i. K_w = [H^+(aq)][OH^-(aq)]   [H^+] = 10^{-pH} = $10^{-6.63}$ = 2.34 × 10^{-7} mol dm^{-3} ✓

5.48 × 10^{-14} = 2.34 × 10^{-7} × [OH^-(aq)]   [OH^-(aq)] = 2.34 × 10^{-7} ✓

ii. At a pH of 7, [H^+(aq)] = 10^{-pH} = 10^{-7} mol dm^{-3}.

For pure water, at ionisation [H^+(aq)] = [OH^-(aq)]. K_w = [H^+(aq)][OH^-(aq)] = 10^{-7} × 10^{-7} = 10^{-14} mol^2 dm^{-6}. [H^+(aq)] = 10^{-7} mol dm^{-3}. [OH^-(aq)] = 10^{-7} mol dm^{-3} ✓

c. An increase in the ionic product of water indicate an increase in the concentration of H^+ and OH^- ions. ✓ Therefore, increasing the temperature moves the position of equilibrium to the right. ✓ The forward reaction must be endothermic. ✓

36.

a. The concentration of a strong base will be the same as the concentration of OH^- ions, as a strong base would dissociate completely in solution. ✓

Using the ionic product of water K_w = [H^+(aq)][OH^-(aq)], the concentration of H^+ can then be calculated. The ionic product of water is equal to 1.00 × 10^{-14} mol^2 dm^{-6} at 25°C. ✓

Using the H^+ concentration, pH can be calculated: pH = -log[H^+(aq)]. ✓

b. As NaOH is a strong base, there will be complete dissociation in solution.

Therefore: [OH^-(aq)] = [NaOH(aq)] = 0.035 mol dm^{-3}.

Using the ionic product of water: K_w = [H^+(aq)][OH^-(aq)] = 1.00 × 10^{-14}.

[H^+(aq)] = K_w / [OH^-(aq)] = 1.00 × 10^{-14} / 0.035 = 2.86 × 10^{-13} mol dm^{-3}. ✓

pH = -log[H^+(aq)] = -log(2.86 × 10^{-13}) = 12.54 ✓

37.

a. +5 ✓

The charge of the NO_3^- ion is 1-. Oxidation number of oxygen is -2. Since there are three oxygen atoms present, the total would be -6. To get the overall ionic charge as 1-, nitrogen would have to be +5.

b. HNO_3 is a strong acid and therefore completely dissociates.

This means [H^+(aq)] = [HNO_3(aq)] = 2.42 × 10^{-3} mol dm^{-3}. ✓

pH = -log[H^+(aq)] = -log (2.42 × 10^{-3}) = 2.62 ✓

38.

a. A buffer solution can be made from a weak acid ✓ and a salt of the weak acid ✓. For example, ethanoic acid and potassium ethanoate (CH_3COOH and CH_3COOK). OR from an excess of a weak acid and a strong alkali such as CH_3COOH and NaOH (in this case, the acid and the alkali react to form the salt of the weak acid).

ANSWERS — Physical Chemistry and Transition Elements

b. A strong acid dissociates completely in solution (example: H_2SO_4). ✓ A weak acid dissociates partially in solution (example: CH_3COOH). ✓

c. The following equation can be used to calculate the pH of a buffer solution:

$[H^+(aq)] = K_a \times [HA(aq)] / [A^-(aq)]$ ✓

$pH = -\log [H^+(aq)]$ ✓

H^+ (aq) ion concentration is calculated using the first equation. Second equation will give you the pH. ✓

39.

a. +5 ✓ (The phosphate ion has a charge of 3-. Since there are four oxygen atoms each with -2, leading to a total of -8, phosphate should be +5 to make the total charge of the ion 3-).

b. Amount of $Co_3(PO_4)_2(s)$ = 1.54 / 366.7 = 4.20 × 10^{-3} ✓

Amount of hydrous salt = 4.2 × 10^{-3} × 3 = 1.26 × 10^{-2} ✓

Molar mass of hydrated salt = 3.00 / 1.26 × 10^{-2} = 238 ✓

Molar mass of anhydrous salt = 1.64 / 1.26 × 10^{-2} = 130 ✓

Mass of water = 3.00 - 1.64 = 1.36g ✓

Moles of water = 1.36 / 18.0 = 7.56 × 10^{-2} mol ✓

Water of crystallisation = 7.56 × 10^{-2} / 1.26 × 10^{-2} = 6 ✓

Molar mass of anion = 238 - (58.9 + 6 × 18) = 71.1 ✓

(Cl has a molar mass of 35.5. So the anions should be 2Cl$^-$)

Formula of salt is $CoCl_2.6H_2O$ ✓

40.

a. Weak acid - weak base:

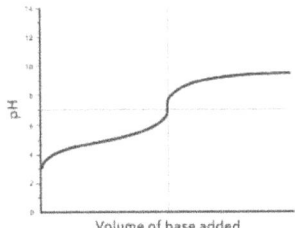

The initial pH is about 3 as the acid in the conical flask is weak ✓. As the alkali is added from the burette, the H^+ ions of the acid react with the alkali. As the concentration of H^+ ions decrease, the pH increases (becomes more alkaline). At the final pH of about 10, almost all of the H^+ ions of the acid have reacted ✓.

b. Strong acid - weak base:.

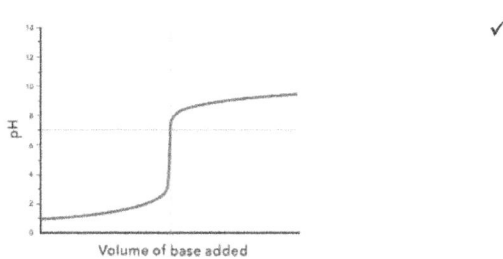

The initial pH is about 1 as the acid in the conical flask is strong✓. As the weak alkali is added from the burette, the H⁺ ions of the acid react with the alkali. As the concentration of H⁺ ions decrease, the pH increases (becomes more alkaline). At the final pH of about 10, almost all of the H⁺ ions of the acid have reacted✓ (Strong acids have a pH of around 1 and strong alkalis have a pH of around 13).

c.

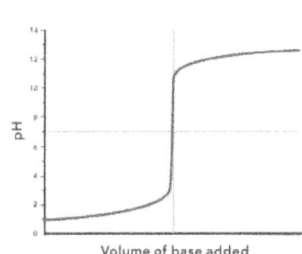

✓

d.

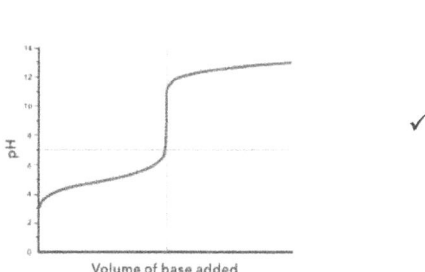

✓

41. A high K_a value indicates a higher level of dissociation: stronger acid. A low K_a value indicates a lower level of dissociation: weaker acid.✓

42.

a. Sulfuric acid is a strong acid AND boric acid is a weak acid.✓

H_2SO_4:

pH = -log 1.7 x 10⁻³ = 2.77.✓

(Explanation: Because sulfuric acid is a strong acid, the concentration of H⁺ ions is the same as the concentration of the acid; complete dissociation. Note that sulfuric acid is a dibasic acid - able to donate two protons. Protons from partial ionisation of HSO_4^- has been ignored in this calculation).

H_3BO_3:

$K_a = 10^{-9.23} = 5.89 \times 10^{-10}$ mol dm⁻³ ✓

$[H^+] = \sqrt{(K_a \times [H_3BO_3])}$

$[H^+] = \sqrt{(5.89 \times 10^{-10} \times 1.7 \times 10^{-3})}$ = 1.00 x 10⁻⁶ ✓

pH = -log (1.00 x 10⁻⁶) = 6.00 ✓

b. i. A buffer solution minimises changes in pH ✓
when small amounts of acid or base are added;

ii. $CH_3COOH \rightleftharpoons H^+ + CH_3COO^-$

ANSWERS

Physical Chemistry and Transition Elements

On addition of alkali
CH$_3$COOH reacts with any added alkali/base/OH$^-$
Equilibrium shifts to the right, forming H$^+$ ✓

On addition of acid
CH$_3$COO$^-$ reacts with added acid / H$^+$;
Equilibrium shifts to the left, forming CH$_3$COOH ✓

43.

   a. Acid dissociation constant measures the extent of dissociation of an acid in solution. K_a = [H$^+$(aq)] [A$^-$(aq)] / [HA(aq)] ✓

   b. K_a = [H$^+$(aq)][A$^-$(aq)] / [HA(aq)] ✓. Unit of K_a is always mol dm^{-3} ✓

   c. i. pK_a = $-\log_{10} K_a$ ✓

   K_a = 10^{-pK_a} ✓

   ii. pK_a = $-\log_{10} K_a$ = $-\log 1.75 \times 10^{-5}$ = 4.76 ✓

44. K_w = [H$^+$(aq)][OH$^-$(aq)] ✓

45.

   a. [H$^+$(aq)] = 10^{-pH} ✓

   b. i. Two species that transform into each other by gaining or losing a proton. ✓

   ii. Acid 1: CH$_3$COOH Base 1: CH$_3$COO$^-$ ✓   Acid 2: H$_3$O$^+$ Base 2: H$_2$O ✓

   c. False ✓. Because CH$_3$COOH is a weak acid ✓ and therefore there will only be very little dissociation ✓. K_c will be very small ✓.

46. a. pH = $-\log$ [H$^+$(aq)] = $-\log$ 0.00000453 = 5.34 ✓

   b. 2.57 ✓ c. 9.45 ✓ d. 4.22 ✓

47. The following equilibrium exists in a buffer solution:

   HA (aq) ⇌ H$^+$ (aq) + A$^-$(aq) ✓

   When an acid is added, concentration of H$^+$(aq) increases, resulting in the conjugate base A$^-$(aq) reacting with H$^+$(aq) ions. ✓ This causes equilibrium to shift to the left. This removes most of the extra H$^+$(aq) ions. When an alkali is added, the concentration of OH$^-$(aq) increases. H$^+$(aq) ions react with OH$^-$(aq) ions, ✓ shifting the equilibrium to the right, replacing most of H$^+$ (aq) ions that reacted.

48.

a. Carbonic acid - hydrogencarbonate ion system acts as a buffer in blood. The carbonic acid acts as the weak acid (H_2CO_3). Hydrogencarbonate acts as the conjugate base (HCO_3^-). ✓

b. $H_2CO_3 \rightleftharpoons H^+(aq) + HCO_3^-(aq)$. ✓

If the concentration of $H^+(aq)$ increases, it is removed by the conjugate base, shifting the equilibrium to the left. ✓ If $OH^-(aq)$ ions are added, it reacts with the small concentration of $H^+(aq)$ ions present ✓, as follows:

$H^+(aq) + OH^-(aq) \rightarrow H_2O(l)$. To restore the $H^+(aq)$ ions that have reacted, the equilibrium moves to the right. ✓

49.

a. The following equation may be used to calculate the half-life although it could also be worked out by trial and error. Note that the question gives concentrations and NOT the **amount** in moles. However, as the volume is constant in this case, using the concentrations will give you the same answer as if you were to use the amounts in moles. You are NOT required to use this equation in the exam although it is given here.

$$\text{Elapsed time} = \text{half-life} \times \frac{\log\left(\frac{\text{Initial amount}}{\text{Current amount}}\right)}{\log 2}$$

Four half-lives ✓ has elapsed and therefore 120 x 4 = 480s ✓. (Using the above equation would give you 480s directly. If you then wanted to find out how many half-lives that is, you would divide your answer by the half-life (120) to get 4.

b. rate = $k[C_6H_5CHO]^2[KOH]$ ✓✓

c. Yes, if a higher temperature is used more collissions would exceed the activation energy and be successful and therefore the rate constant will be high. ✓

d. Start by using the equation $Q = mc\Delta T$

Total volume of solution = 25 + 35 = 60.0 cm³. 60.0 cm³ has a mass of 60.0 g ✓ (refer to the density of solution given in the question - each 1 cm³ is equal to 1g).

Heat gained by the surrounding: $Q = mc\Delta T$ = 60 x 4.18 x 5.2 = 1304.16 J

Therefore the heat lost by the reaction = -1304.16 J ✓ (heat given out : exothermic)

Now calculate the no. of moles of KOH and HCl that reacted:

moles of KOH = 0.60 x 0.035 = 0.021 ; moles of HCl = 0.84 x 0.025 = 0.021 ✓

Write the balanced chemical equation for the reaction:

HCl + KOH → KCl + H_2O ✓

1 mole of HCl reacts with 1 mole of KOH to form 1 mole of H_2O.

ANSWERS
Physical Chemistry and Transition Elements

Find the energy released per mole: -1304.16 / 0.021 = -62103 J mol^{-1}.

Therefore ΔH_{neut} = -62.1 kJ mol^{-1} ✓

50. Using the formula [H$^+$] = 10^{-pH} (If the pH is known, always use this formula to find [H$^+$]). ✓

51.

a.

	Ba	I
Mass	31.6	58.4
Number of moles	31.6/137.3	58.4/126.9 (both columns correct ✓)
	0.230	0.460
Ratio of moles	1	2 (both columns correct ✓)

Empirical formula: BaI$_2$

b. (31.6 / 90) x 100 = 35.1% ✓

(90 in the working is obtained by adding 31.6 and 58.4 as that is the total weight of the sample.)

52.

a. CaCO$_3$ → CaO + CO$_2$ ✓

b. The relative formula masses are as follows: CaCO$_3$: 100.1, CaO: 56.1, CO$_2$: 44.

Atom economy = (Formula mass of the desired product / Total formula masses of the products) x 100

= (56.1 / 100.1 ✓) x 100 = 56.0% ✓

c. **First step: Find the moles of CaCO$_3$ present in 20.0 g.**

Moles = mass (g)/molar mass = 20 / 100.1 = 0.200 moles ✓

Second step: Find the theoretical yield of CaO using the equation.

The balanced equation for this reaction shows us that one mole of CaCO$_3$ forms one mole of CaO. Therefore, in a perfect world (theoretical), we would yield 0.200 moles of CaO ✓.

Third step: Find the number of moles in 8.20 g of CaO.

Moles = 8.20 / 56.1 = 0.146 moles of CaO ✓ (actual yield).

Fourth step: Find the percentage yield.

Percentage yield = (actual amount, in mol, of product / theoretical yield, in mol, of product) x 100.

= (0.146 / 0.200) x 100 = 73% ✓

d. Bubble the gas through calcium hydroxide solution.✓ Carbon dioxide would cause a white precipitate.✓ The white precipitate is due to the formation of calcium carbonate.✓

53.

a. $Na^+(g) + I^-(g) \rightarrow NaI(s)$ ✓

b. Value of lattice enthalpy, tells you how strong an ionic lattice is and is a measure of its ionic bond strength. (How much energy is released in making the ionic bonds).✓

c. i.

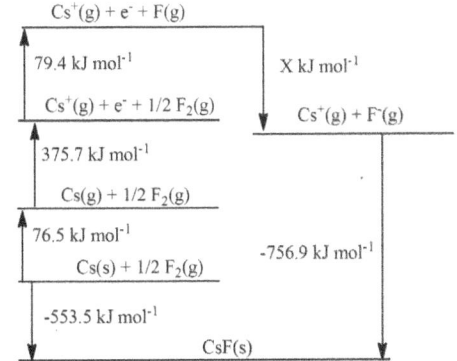

✓✓✓✓

Lose marks for missing:
- State symbols
- Electrons
- ΔH labels
- Arrow wrong way

ii. Enthalpy change of formation = Sum of all other enthalpy changes

$-553.5 = 76.5 + 79.4 + (-756.9) + X + 375.7$

$-553.5 = 76.5 + 79.4 - 756.9 + X + 375.7$ ✓

$X = -553.5 - 76.5 - 79.4 + 756.9 - 375.7 = -328.2$ kJ mol^{-1} (answer.✓ - sign.✓)

(Note that the atomisation of F and ionisation of Cs can be in either order in the cycle above)

54. The enthalpy change that takes place when a mole of a compound is completely dissolved✓ in water under standard conditions.✓

55.

a. The enthalpy change that takes place when a mole of isolated ions in their gaseous state✓ is dissolved in water to form a mole of aqueous ions✓ under standard conditions.

b. Ionic radius✓ and ionic charge✓. Smaller ionic radii results in greater attraction between the ions and water molecules✓, resulting in more exothermic enthalpy change of hydration values. Greater ionic charges result in greater attraction between the ions and water molecules resulting in more exothermic enthalpy change of hydration values.✓

c. $Mg^{2+}(g) + aq \rightarrow Mg^{2+}(aq)$ (equation✓ state symbols✓)

d. Al^{3+}, Mg^{2+}, Sr^{2+}, I^- ✓✓

(As you go down a group in the periodic table, the ionic radii increases.)

e. No, standard enthalpy change of hydration is always exothermic because you are always forming bonds which is an exothermic process.✓

5 ANSWERS — Physical Chemistry and Transition Elements

56.

    a. 2, 2, 2 ✓

    b. i.

$K_c = [N_2][H_2O]^2 / [NO]^2[H_2]^2$ ✓

ii.

$(mol\ dm^{-3})(mol\ dm^{-3})^2 / (mol\ dm^{-3})^2 (mol\ dm^{-3})^2$

$= (mol\ dm^{-3}) / (mol\ dm^{-3})^2$

$= 1 / mol\ dm^{-3}$

$= mol^{-1}\ dm^3$ ✓

iii.

At equilibrium:

Moles of H_2 = 0.12 mol

Moles of NO = 0.32 - 0.13 = 0.19 mol ✓

(Explanation: Because 0.13 mol of NO reacted, we subtract 0.13 from 0.32 to find out how much NO is remaining in the equilibrium mixture. We get 0.13 from 0.25-0.12. As the question says, 0.12 mol of H_2 remains; 0.13 of H_2 must have reacted since we started with 0.25 mol of H_2. Looking at the molar ratios in the equation, 0.13 mol of H_2 will react with 0.13 mol of NO, since it is a 1:1 ratio. Therefore 0.19 mol of NO must remain in the mixture).

Moles of H_2O = 0.13 mol ✓

(Explanation: If 0.13 mol of H_2 reacted, it will form 0.13 mol of H_2O)

Moles of N_2 = 0.065 mol ✓

(Explanation: 0.13/2 as the molar ratio of H_2O to N_2 is 2:1)

$K_c = 0.065 \times 0.13^2 / 0.19^2 \times 0.12^2 = 2.11\ mol^{-1}dm^3$ ✓

57. Equivalence point is the point in a titration at which volume of one solution reacts exactly with the volume of the other solution. ✓ End point is the point in a titration where there are equal concentrations of the weak acid and the conjugate base of the indicator. ✓

58. The energy change that takes place when an aqueous acid is neutralised by an aqueous base, forming one mole of H_2O ✓ (l) under standard conditions. ✓

59.

    a. Lattice enthalpy is the change in enthalpy when one mole of an ionic compound is formed from its gaseous ions ✓ under standard conditions ✓.

    b. Smaller ionic radii result in greater attraction between the ions resulting in a more exothermic lattice enthalpy ✓ and vice versa. Ions with higher charges bond together with a greater attraction resulting in a more exothermic lattice enthalpy and vice versa. Small ions with higher ionic charges have the most exothermic lattice enthalpies. ✓

    c. Because it is almost impossible to react together gaseous ions to form one mole of an ionic lattice. ✓

    d. Because there are no ions in covalent compounds. ✓ (Covalent compounds are formed when atoms share their electrons).

60.

a. $Al^+(g) \rightarrow Al^{2+}(g) + e^-$; $Sr^+(g) \rightarrow Sr^{2+}(g) + e^-$ (both equations correct✓✓ one correct✓)

Note that, in equations involving enthalpy changes, state symbols should always be included unless otherwise stated.

b. Endothermic✓

61.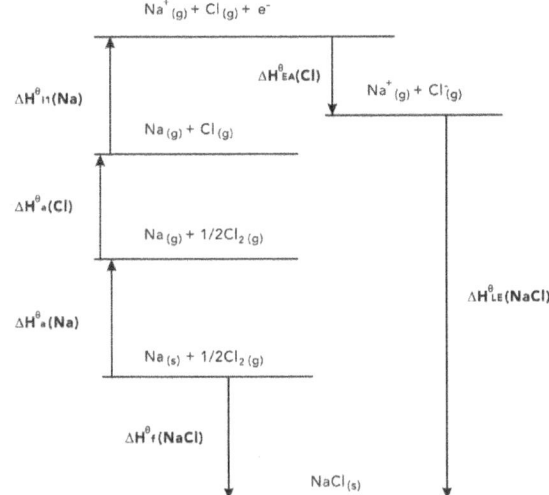

✓✓✓✓

Lose marks for missing:

-State symbols
-Electrons
-ΔH labels
-Arrow wrong way

62.

a. Lattice enthalpy✓

b.

i. $S(g) + e^- \rightarrow S^-(g)$ ✓✓

ii. Due to the positive charge of the nucleus, the electron that is being added is attracted to the outershell of the atom by the electrostatic force.✓ This releases energy.

63. Entropy measures the amount of disorder in a system, in quantitative terms.✓

64.

a. H_2✓

b. $\frac{1}{2}O_2(g) \rightarrow O(g)$ $\frac{1}{2}F_2(g) \rightarrow F(g)$ $Li(s) \rightarrow Li(g)$ (two correct✓✓ one correct✓)

(Note that the standard states of both oxygen and fluorine is gas and for lithium its solid.)

c. Endothermic.✓ Energy is required to break the bonds between the atoms.✓

ANSWERS
Physical Chemistry and Transition Elements

65.

a. Transition elements are elements that form at least one ion with an incomplete or partially filled d sub-shell ✓

Electronic configuration of Mn^{2+} is: $1s^2\ 2s^2\ 2p^6\ 3s^2\ 3p^6\ 4s^0\ 3d^5$ ✓ (Explanation: The 4s sub-shell here does not have any electrons - they were lost as the ion formed. Should you wish, you can avoid including the 4s sub-shell in the electronic configuration of Mn^{2+})

b. $[Cr(H_2O)_6]^{3+}$ reacts with an excess of ammonia as shown below:

$[Cr(H_2O)_6]^{3+}(aq) + 6NH_3(aq) \rightleftharpoons [Cr(NH_3)_6]^{3+}(aq) + 6H_2O(l)$ ✓✓

Blue solution turns into a purple solution. ✓

c. Shape: Octahedral; coordination number: 6. ✓

66.

a.

i. The change in enthalpy that takes place when one electron is removed from each atom in one mole ✓ of gaseous atoms to form one mole of gaseous 1+ ions ✓.

ii. $Na(g) \rightarrow Na^+(g) + e^-$ (Correct state symbols. ✓ Rest of the equation. ✓)

iii. As the ionisation energy number increases, the enthalpy change gets more endothermic. There is a big jump from the 1st to the 2nd ionisation energy. ✓ This is because the second electron is in a shell more closer to the nucleus than the first. ✓ This means that there is a stronger electrostatic attraction between the nucleus and the 2nd electron resulting in a higher amount of energy being needed to remove it from the atom. ✓ As the electrons are removed, the electron repulsion decreases. ✓ Due to the decreasing electron repulsion, it becomes increasingly difficult to remove the electrons.

(Note that when a question asks you to describe, you have to state what is present or what you can see from the information given. When a question asks you to explain, you have to state the reasons as to why something is the way it is.)

b.

i. Moles = mass(g)/molar mass = 26/ 22 = 1.1818 moles ✓

1.1818 x (6.02 x 10^{23}) = 7.1144 x 10^{23} ✓

ii. In 9.1 years, there are 3.5 half lives (9.1/2.6 = 3.5 half-lives). ✓

(Therefore divide 7.1144 x 10^{23} by two, three and a half times.)

7.1144 x 10^{23} / $2^{3.5}$ = 6.29 x 10^{22} ✓

67.

a. $Al^{2+}(g) \rightarrow Al^{3+}(g) + e^-$ ✓

b. Al_2O_3 ✓ only as none of the other compounds contain a 2- ion ✓. (In NaCl, chlorine is Cl^- and in $CaCl_2$, there are two Cl^- ions, which is not the same as Cl^{2-}.)

c. Breaking bonds: endothermic ✓ (you need energy to break something.) Making bonds: exothermic. ✓

Physical Chemistry and Transition Elements — ANSWERS 5

68. 25 °C temperature, 100 kPa pressure and 1 mol dm^{-3} concentration. ✓

69. When choosing a suitable indicator, the pH value of the end point should be as close as possible to the pH value of titration's equivalence point. ✓

70.

   a. **First step: calculate the amount, in mol, of thiosulfate that reacted.**

   Moles = concentration × volume(in dm^3) = 0.150 × 0.03425 = 5.1375 × 10^{-3} ✓

   Second step: determine the amount, in mol, of iodine that was used in the titration.

   It is evident from the second equation that two moles of thiosulfate are required to react with one mole of iodine. So the ratio of $I_2 : S_2O_3^{2-}$ is 1:2.

   Since 5.1375 × 10^{-3} moles of thiosulfate reacted, there should have been (5.1375 × 10^{-3}) /2 = 2.5688 × 10^{-3} moles of iodine. ✓

   Third step: determine the amount, in mol, of iodate ions that was required to produce this much of iodine.

   It is evident from the first equation that one mole of iodate ions are required to form three moles of iodine. So the ratio of $IO_3^- : I_2$ is 1:3.

   Since 2.5688 × 10^{-3} moles of iodine was produced, it should have required (2.5688 × 10^{-3}) / 3 = 8.5627 × 10^{-4} moles of iodate ions. ✓

   Final step: calculate the concentration, in mol dm^{-3}, of iodate used.

   Concentration = moles / volume (in dm^3) = 8.5627 × 10^{-4} / 0.025 = 0.0343 mol dm^{-3} ✓.

   (Use more significant figures in your working than is needed for the answer. For example, if your answer needs to be given to 3 s.f., use 5 s.f. in your working to avoid rounding errors. Answers should be given to an appropriate number of significant figures. Most answers may be given to 3 s.f.)

   b. In the first equation, I in IO_3^- has been reduced ✓ from +5 to 0 ✓ and I in I$^-$ has been oxidised from -1 to 0. In the second equation, I in I_2 has been reduced ✓ from 0 to -1 ✓.

71. Hess's Law states that the enthalpy change for any chemical change is the same, independent of the path taken as long as the initial and final reaction conditions are the same. ✓

72.

   a.

   K$^+$(g) + F$^-$(g)
   ↓ -320
   K$^+$(aq) + F$^-$(g)
   -813 ↓
   ↓ -524
   KF(s)
   X ↓
   K$^+$(aq) + F$^-$(aq)

   ✓✓✓

   Lose marks for missing:
   state symbols
   labels
   arrow wrong way

b. Sum of clockwise enthalpy changes = sum of anti-clockwise enthalpy changes

(If you find it hard to work out clockwise and anticlockwise enthalpy changes, you may use the following equation.)

Enthalpy changes of hydration = lattice enthalpy + enthalpy change of solution

-320 + (-524) = -813 + X ✓

-320 - 524 = -813 + X

X = -320 - 524 + 813 = -31 kJ mol^{-1} ✓

c. K^+ has more electron shells than F^-. ✓

73. The change in enthalpy when one mole of a compound is formed ✓ from its constituent elements in their standard states ✓ under standard conditions.

74. A reaction where both reduction and oxidation take place. ✓

75. The temperature ✓, entropy change of the system and the enthalpy change of reaction ✓. All these factors are combined together to give free energy change, ΔG.

76. Enthalpy is not the only factor determining whether a reaction will take place. An endothermic reaction will take place spontaneously if ΔS is positive ✓ and if the temperature is such that $T \Delta S > \Delta H$ ✓.

77.

a. Entropy of a solid is less than that of a liquid. ✓

b. Change of entropy = Sum of entropy of products - Sum of entropy of reactants ✓

c. Entropy measures the degree of disorder in a system. Everything in the universe is disordered to an extent so entropy can never be negative. ✓ However, change in entropy measures the difference between two entropy values, which could take a negative value, if the entropy of reactants are greater than that of products (for example when a solid forms from an aqueous solution in a precipitation reaction). ✓

78.

a. Gradient = $-E_a/R$, Y-intercept = $\ln A$ ✓

b. Gradient = $-E_a / R$

-7490 = $-E_a/8.314$ ✓

$-E_a$ = -7490 × 8.314 = -62300 J mol^{-1}
E_a = 62300 J mol^{-1} = 62.3 kJ mol^{-1} ✓
Y-intercept = $\ln A$ = 3.2
A = 24.5 ✓

79. The entropy will increase as liquids have a higher entropy value compared to solids. Therefore, the change of entropy (ΔS) will be positive. ✓

Physical Chemistry and Transition Elements — ANSWERS 5

80. Free energy change is what links enthalpy, entropy and temperature for a given process, where $\Delta G = \Delta H - T\Delta S$. ✓

81.

   a. $\Delta S = \sum S^\ominus_{(products)} - \sum S^\ominus_{(reactants)}$
   $\Delta S = (39.8 + 213.7) - 92.9$ ✓
   $\Delta S = 160.6$ J K^{-1} mol^{-1} ✓

   b. $\Delta G = \Delta H - T\Delta S$. ✓ G : kJ mol^{-1}. H: kJ mol^{-1}. S: kJ K^{-1} mol^{-1}. T: K. ✓

   (Note that units of entropy (s) is usually given in J K^{-1} mol^{-1}. However, as enthalpy (H) is given in kJ mol^{-1}, entropy has to be converted to kJ K^{-1} mol^{-1}, by dividing by 1000. When you do this, your calculated value for G will have the units kJ mol^{-1}.)

   c. $\Delta G = 178.2 - (15 + 273) \times (160.6/1000)$ ✓ $= 178.2 - 288 \times 0.1606 = 131.9$ kJ mol^{-1} ✓

   No, because ΔG is greater than 0. ✓

   (273 is added to the temperature to covert it into kelvin. As you can see here, entropy is divided by 1000 to covert the units into kJ K^{-1} mol^{-1} to match with the units of enthalpy.)

   d. $\Delta G = \Delta H - T\Delta S$
   $0 = 178.2 - T \times (160.6/1000)$ ✓
   $0 = 178.2 - 0.1606T$
   $0.1606T = 178.2$
   $T = 1109.6$ K ✓
   $1109.6 - 273 = 836.5$ °C ✓

   The minum temperature required can also be calculated using this formula.

   $$T = \frac{\Delta H}{\Delta S}$$

82. The change in entropy will be negative as the reactants are more disordered (high entropy) than the products. ✓ (Change of entropy = sum of entropy of products - sum of entropy of reactants. Reactants will have a higher entropy if they contain more gas particles than products).

83. To convert °C to K, add 273. To convert K to °C minus 273.

   i. 278 K ✓   ii. -255 °C ✓   iii. -256 °C ✓   iv. 290 K ✓   v. 0 K ✓   vi. -272 °C ✓   vii. 298 K ✓

84.

   a. Oxidation number decreases when a species is reduced. ✓

   b. Oxidation number increases when a species is oxidised. ✓

   c. Copper has been oxidised ✓ from 0 in Cu to +2 in $Cu(NO_3)_2$ ✓. Nitrogen has been reduced ✓ from +5 in NO_3 to +2 in NO ✓.

   (Note that NO_3 ion has a charge of 1-. The oxidation numer of combined oxygen is -2. Since there are three oxygen atoms present in NO_3, it adds up to -6. In order to get -1 as the overall charge for the ion, nitrogen must be +5. This way +5 - 6 = -1.)

85. Liquids have a higher entropy than a solid✓ (because the particles in a liquid can move freely). High level of disorder means higher entropy.

86. An agent that reduces another species. The reducing agent will add electrons to that species. (Reduction occurs when a species gains electrons).✓

87.
   a. Starting from bottom left, going in a clockwise direction:
   A, F, E, B, D, G, C, H. ✓✓✓✓ (1 mark for evey two correct)

   b. $-155 = 339 + 249 + x + 1960 - 141 + 790 - 4097$
   $x = -155 - 339 - 249 - 1960 + 141 - 790 + 4097$ ✓
   $x = +745$ kJ mol^{-1} (answer with correct sign✓)

88. A half-cell consists of an element in two oxidation states.✓

89. The change in entropy will be positive as the products are more disordered than the reactants.✓

90. A reaction is feasible if the change in free energy is less than or equal to zero ($\Delta G < 0$).✓

91.
   a. -2✓
   b. The charge of that ion.✓
   c. $+1$✓
   d. -1✓
   e. 0✓

92. An agent that oxidises another species. The oxidising agent will take electrons away from that species. ✓(Oxidation occurs when a species loses electrons).

93. A half-cell with a more negative standard electrode potential is more likely to lose electrons (become oxidised).✓

94. Note that for a given reaction to be feasible, the reduction half-cell has to have the greater electrode potential (than the oxidation half-cell). The reduction half-cell is the half-cell which will gain electrons and the oxidation half-cell is the one which loses electrons. If solid magnesium were to react with dilute sulfuric acid, solid magnesium (Mg(s)) will be converted to magnesium sulfate (which consists of Mg^{2+} ions). Therefore, the magnesium half-cell would move to the left, losing electrons - this makes it the oxidation half-cell. The hydrogen half-cell would move from left to right as H$^+$ ions of sulfuric acid is being reacted with magnesium to form H$_2$ gas. As it gains electrons, it is the reduction half-cell✓, and as 0 V is a greater value than -2.37 V, the reaction will be feasible✓. (The electrons which are being lost by Mg(s) are being gained by H$^+$(aq)).

   Note that your answer in the exam does not have to be this detailed. The explanation above is intended to provide you with a better understanding.

Physical Chemistry and Transition Elements — ANSWERS

95.

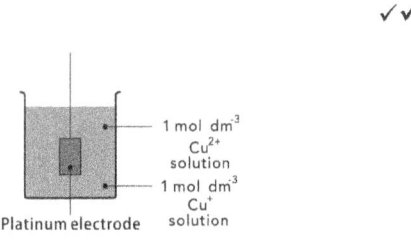

$Fe^{3+}(aq) + e^- \rightleftharpoons Fe^{2+}(aq)$ ✓   (Reduction equation is always used for half-cells)

96. $1s^2\ 2s^2\ 2p^6\ 3s^2\ 3p^6\ 4s^2\ 3d^2$ ✓

97.
   a.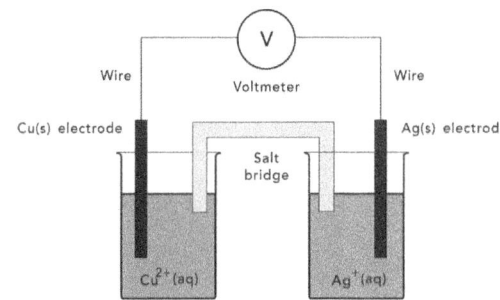

   Voltmeter and salt bridge ✓
   Correct Cu electrode ✓
   Correct Ag electrode ✓

   b. Standard cell potential = Standard electrode potential of positive terminal - standard electrode potential of negative terminal = 0.80 - 0.34 = 0.46 V ✓

   c. $Cu(s) + 2Ag^+(aq) \rightarrow Cu^{2+}(aq) + 2Ag(s)$ ✓

   d. Electrons carried in wire; ions carried in salt bridge. ✓

98.

   a. When two half-cells are combined, the one with the most negative electrode potential will be the negative terminal which will provide the electrons. Therefore the electrons will flow from the half-cell with the more negative electrode potential to the one with the less negative / more positive electrode potential. ✓

   b. The standard electrode potential of a half-cell can be measured by connecting it to a standard hydrogen half-cell. A voltmeter connected between the two half-cells will give the standard electrode potential of the half-cell. ✓

   c. Standard cell potential is calculated by subtracting the standard electrode potential of the negative terminal (the half-cell with the least positive electrode potential) from the standard electrode potential of the positive terminal. $E^\ominus_{cell} = E^\ominus$ (positive terminal) - $E^\ominus$(negative terminal). ✓

99.
   a.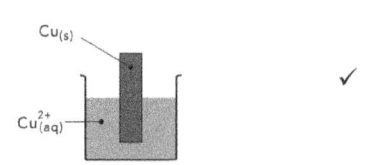

   ✓

   $Cu^{2+}(aq) + 2e^- \rightleftharpoons Cu(s)$ ✓

Physical Chemistry and Transition Elements

b.

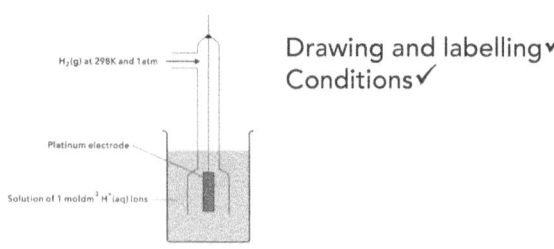

Drawing and labelling ✓
Conditions ✓

$2H^+(aq) + 2e^- \rightleftharpoons H_2(g)$ ✓

c. Two half-cells each with different electrode potentials can be connected together to make an electrochemical cell. One half-cell will release electrons and the other will gain electrons. ✓

100. In practice, the conditions used for a reaction are usually different from the standard conditions which were used to measure the standard electrode potential values. ✓ Standard electrode potentials relate to aqueous equilibria but in reality a lot of reactions will take place under other conditions and not in aqueous solution. ✓

101.

a. Vanadium forms an ion with an incomplete d sub-shell. ✓

b. Zn: $1s^2\ 2s^2\ 2p^6\ 3s^2\ 3p^6\ 4s^2\ 3d^{10}$ Cr: $1s^2\ 2s^2\ 2p^6\ 3s^2\ 3p^6\ 4s^1\ 3d^5$ Cu: $1s^2\ 2s^2\ 2p^6\ 3s^2\ 3p^6\ 4s^1\ 3d^{10}$
(Three correct ✓✓. Two correct ✓)

c. V_2O_5 ✓   - Allow multiples such as V_4O_{10}.   (Combined oxygen's oxidation number is -2. Note that vanadium's oxidation number is given in roman numberals in its name: (V) = five. With two vanadium atoms and five oxygen atoms, the numbers are balanced: +10 and -10.)

d.

i. $2SO_2(g) + O_2(g) \rightleftharpoons 2SO_3(g)$ (Formulae ✓. Balancing ✓.)

ii. Forward reaction is exothermic. ✓ If the temperature is increased, the system will oppose this change by decreasing the temperature. This will shift the position of equilibrium in the reverse / endothermic direction. ✓ Value K_p would decrease. ✓

iii. There are more gas particles on the left side of the equation. ✓ To achieve a higher yield, the pressure of the system would have to be increased, so the position of equilibrium moves towards the side with the least number of gas particles ✓ (products side).

102.

a. Transition metals form coloured compounds. ✓

b. Reactants side has more gas particles than the products side. ✓ The system would oppose a decrease in pressure by increasing the pressure. This would shift the position of equilibrium to the left. ✓

c. A catalyst does not affect K_p. ✓ It would only increase the rate of both the forward and backward reaction. This would mean that equilibrium is reached faster with a catalyst present. ✓

d. Moles = mass(g) / molar mass

Moles of N_2 = 49.0 / 28.0 = 1.75 moles. ✓ The molar ratio of N_2 to NH_3 in the equation is 1:2.

ANSWERS

Physical Chemistry and Transition Elements

Therefore, moles of NH_3 = 1.75 x 2 = 3.50 moles ✓

Mass of NH_3 = moles x molar mass = 3.50 x 17.0 = 59.5 g of NH_3 ✓.

e. Moles (of a gas) = Volume (in dm^3) / 24 ✓

Moles = 400 / 24 = 16.7 moles. For both gases, the number of moles formed is 16.7. ✓

(One mole of *any* gas occupies the same volume.)

103.

a. The coordination number is the total number of coordinate bonds formed between a metal ion and its ligands. ✓

b. Through a coordinate bond, also known as a dative covalent bond. ✓

104.

a. A ligand which donates only one pair of electrons to a metal ion to form one coordinate bond. ✓

b. A catalyst is a substance that increases the rate of a chemical reaction ✓ by providing an alternate reaction pathway ✓. It is not used up by the reaction.

c. A ligand is a molecule or ion that donate a pair of electrons to a transition metal ion to form a coordinate or dative covalent bond. ✓ (Coordinate bond is another term for a dative covalent bond)

d. A complex ion is a transition metal ion bonded to one or more ligands via coordinate bonds. ✓

e. A ligand which donates two or more lone pairs of electrons ✓ to the metal ion to form two or more coordinate bonds ✓.

f. A ligand which donates six lone pairs of electrons ✓ to the metal ion to form six coordinate bonds ✓.

105. Fe^{2+} (aq) + $2OH^-$(aq) → $Fe(OH)_2$ (s) ✓ (Pale green solution turns into a green precipitate. ✓ This precipitate turns rusty brown at the surface when exposed to air/oxygen).

106. There will be *cis-trans* isomers if there are 4 of one monodentate ligand and 2 of another, or 2 bidentate ligands and 2 monodentate ligands. This way, there will be two different ligands in the whole complex ion. ✓

107.

a. Species that have the same structural formula ✓ but have a different arrangement of the atoms in space ✓.

b. *Cis-trans* isomerism and optical isomerism. ✓

108. Cu^{2+} (aq) + $2OH^-$(aq) → $Cu(OH)_2$ (s) ✓. Pale blue solution turns into a pale blue precipitate. ✓

109.

a. Fe^{3+} (aq) + $3OH^-$(aq) → $Fe(OH)_3$ (s) ✓✓

b. Pale yellow solution ✓ turns into a rusty-brown precipitate. ✓

c. **Step 1: Determine the concentration of OH⁻ ions.**

As NaOH is a strong base, there will be complete dissociation, which means

[OH⁻(aq)] = [NaOH(aq)] = 0.035 mol dm⁻³. ✓

Step 2: Use K_w to find the concentration of H⁺ ions.

K_w = [H⁺(aq)][OH⁻(aq)] = 1.00 × 10⁻¹⁴ mol² dm⁻⁶.

[H⁺(aq)] = K_w / [OH⁻(aq)] = 1.00 × 10⁻¹⁴ / 0.035 = 2.857 × 10⁻¹³. ✓

Step 3: Calculate the pH using [H⁺(aq)].

pH = -log [H⁺(aq)] = -log (2.857 × 10⁻¹³) = 12.5 ✓   (Correct answer scores full marks)

(Note that the value for K_w is given in the data sheet in the exam.)

110. Reduced energy usage / costs ✓

111.

a. Moles = mass(g) / molar mass = 2.55/55.8 = 4.570 × 10⁻² moles ✓

b. MnO_4^-(aq) + 8H⁺(aq) + 5Fe²⁺(aq) → Mn²⁺(aq) + 5Fe³⁺(aq) + 4H₂O(l)

(Equation ✓ balancing ✓)

c. 4.570 × 10⁻³ moles ✓ of Fe²⁺ ions present in 10.0 cm³. (4.570 × 10⁻² moles / 10)

Using molar ratios from the equation: moles of MnO_4^- = (4.570 × 10⁻³) / 5 = 9.14 × 10⁻⁴ moles ✓

Concentration of MnO_4^-(aq) ions = (9.14 × 10⁻⁴) / 0.0183 = 0.0500 mol dm⁻³ ✓

d. Acidified potassium manganate(VII) ✓; accepts electrons. ✓

e. Transition elements form ions with two or more oxidation states. ✓

112.

a. [Co(H₂O)₆]²⁺ ✓.

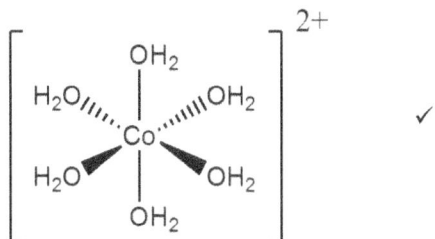

✓

b. i. Six ✓

ii. 90° ✓

Physical Chemistry and Transition Elements — ANSWERS 5

113. It is the oxidation number of the transition metal in the compound. ✓

114. A reaction where soluble ions are mixed together to produce an insoluble compound (precipitate). ✓

115. There will be *cis-trans* isomers if there are two of one monodentate ligand and two of another. This way, there will be two different ligands. ✓ (The complex needs to be a square-planar complex).

116.

   a. A ligand which can donate two lone pairs of electrons to the metal ion to form two coordinate bonds. ✓

   b. Ethane-1,2-diamine. ✓

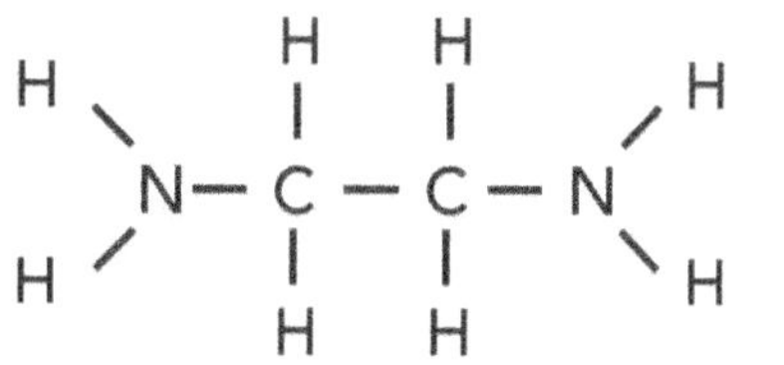

   c.

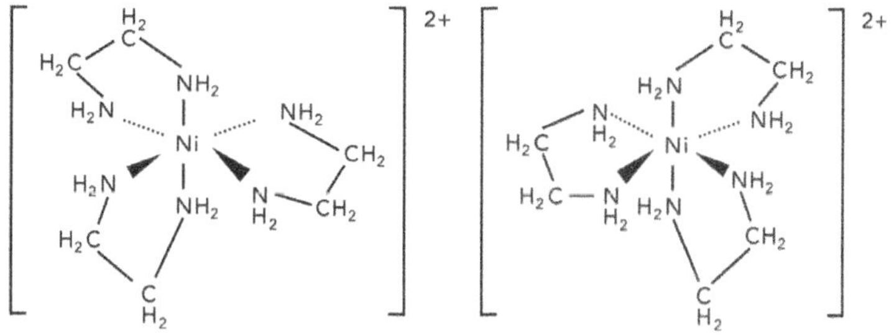

117.

   a. Enantiomers ✓

   b. Cis-platin. ✓

   c. Optical isomers are a type of stereoisomer that are non-superimposable mirror images of each other. ✓

118.

   a. i. Positive; Three moles of gas formed from one mole of liquid. ✓
       ii. Negative; Three moles of gas form two moles of a liquid. ✓
       (Explanation: Entropy of gases > liquids > solids).

   b. i. $\Delta S = \Sigma S^\theta_{products} - \Sigma S^\theta_{reactants}$ ✓
       Change in entropy = (252 + 214) - (49.9 + 5.7 + 223 x2)
       Change in entropy = 466 - 501.6 = -35.6 JK^{-1}mol^{-1} ✓
       ii. $\Delta G = \Delta H - T\Delta S$ ✓

119. Ethane-1,2-diamine ✓, $NH_2CH_2CH_2NH_2$ ✓.

5 ANSWERS  Physical Chemistry and Transition Elements

120.

  a. i. $MnO_4^-(aq) + 8H^+(aq) + 5e^- \rightarrow Mn^{2+}(aq) + 4H_2O(l)$ (species✓ balancing✓)

  ii. Purple to almost colourless.✓

  b. i. $KMnO_4$ moles = 0.1 / 158 = 6.329 x 10⁻⁴ mol✓. Ratio of MnO_4^- : H^+ is 1:8 from the equation. Therefore moles of acid required = 6.329 x 10⁻⁴ x 8 = 5.063 x 10⁻³ moles✓.

  Concentration = mol / volume = 5.063 x 10⁻³ / 0.030 = 0.169 mol dm⁻³ of HNO_3✓.

  pH = - log [H^+] = - log (0.169) = 0.8 ✓

  ii. Sulfuric acid is a dibasic acid AND nitric acid is a monobasic acid.✓

121. Square planar / Tetrahedral ✓

122. $2Cu^{2+}(aq) + 4I^-(aq) \rightarrow 2CuI(s) + I_2(aq)$ (species✓ balancing✓)

123. Carbon monoxide reduces the ability of haemoglobin to carry oxygen. Carbon monoxide binds more strongly to haemoglobin than oxygen and the binding is irreversible✓. This is an example of a ligand substitution reaction✓.

124.

  $[Cu(H_2O)_6]^{2+}(aq) + 4Cl^-(aq) \rightleftharpoons [CuCl_4]^{2-}(aq) + 6H_2O(l)$. (correct species✓ balancing✓) Pale blue solution to yellow solution. ✓

125. $[Cr(H_2O)_6]^{3+}(aq) + 6NH_3(aq) \rightleftharpoons [Cr(NH_3)_6]^{3+}(aq) + 6H_2O(l)$ ✓

  The blue solution changes to a purple solution.✓

126.

  a. Haemoglobin is responsible for carrying oxygen in the red blood cells.✓

  b. The Fe^{2+} ion in each haem group form four coordinate bonds with four nitrogen atoms, one coordinate bond with the protein globin and finally a coordinate bond forms with an oxygen molecule.✓ Therefore oxygen is transported by Fe^{2+} by the formation of a coordinate bond. However, carbon monoxide can replace this oxygen, binding more strongly to the Fe^{2+} than oxygen. This is an example of a ligand substitution reaction.

  c. A buffer system cannot "prevent" pH changes. It will only minimise the changes so the change in pH is not significant.✓

  d. When an acid is added, the concentration of H^+(aq) ions will increase. These H^+ ions will react with the HCO_3^-(aq) ions✓ shifting the position of equilibrium to the left✓.

  e. Moves towards the right.✓

f.

i. Carbonic acid is a weak acid.

First find the [H⁺(aq)] using the pH: $[H^+(aq)] = 10^{-3.27} = 5.370 \times 10^{-4}$ mol dm⁻³ ✓

For a weak acid, $K_a = [H^+(aq)]^2 / [H_2CO_3(aq)] = (5.370 \times 10^{-4})^2 / 0.035 = 8.24 \times 10^{-6}$ mol dm⁻³ ✓

(Note that we have ignored any further dissociation of HCO_3^-)

ii. $pK_a = -\log_{10} K_a = -\log_{10}(8.24 \times 10^{-6}) = 5.08$ ✓

127.

a. $[Cu(H_2O)_6]^{2+}$ ✓

b. i. $[Cu(H_2O)_6]^{2+}$ (aq) + $4NH_3$ (aq) ⇌ $[Cu(NH_3)_4(H_2O)_2]^{2+}$ (aq) + $4H_2O$ (l) (reactants✓ products✓)

ii. The pale blue solution changes to a pale blue precipitate on addition of a small amount of ammonia ✓ and then to a deep blue solution after an excess of ammonia is added. ✓

The pale blue precipitate is due to the formation of $Cu(OH)_2$.

iii. Precipitation reaction ✓ and ligand substitution ✓

c. i. $[Cu(H_2O)_6]^{2+}$ (aq) + $4Cl^-$ (aq) ⇌ $[CuCl_4]^{2-}$ (aq) + $6H_2O$ (l). (reactants✓ products✓)

ii. The pale blue solution changes to a yellow solution. ✓

iii. Ligand substitution ✓

iv. Tetrahedral ✓. Due to the large size of the chloride ligands and because of its strong electrostatic repulsions, only four chloride ligands can fit around a copper ion. ✓ In the case of water ligands, six can fit around the copper ion.

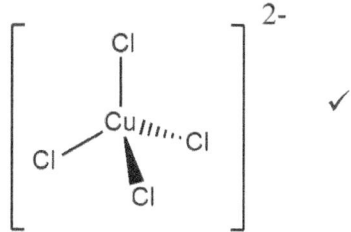

128. MnO_4^- (aq) + $8H^+$(aq) + $5Fe^{2+}$ (aq) → Mn^{2+} (aq) + $5Fe^{3+}$ (aq) + $4H_2O$ (l). ✓✓

129. Cis-platin is used in the treatment of cancer ✓. It prevents the cancer cells from reproducing by binding to their DNA ✓.

130. MnO_4^- is an oxidising agent. ✓

131.

a. $1s^2\ 2s^2\ 2p^6\ 3s^2\ 3p^6\ 4s^0\ 3d^6$ ✓ (The 4s orbital need not be written as it does not contain any electrons. It is shown here for the purpose of explanation.)

Fe^{2+} has a partially filled d sub-shell.✓

b. Haemoglobin.✓

132. a. Cu(s) → Cu^{2+} (aq) + 2e$^-$

   2Cu^{2+}(aq) + 4I$^-$(aq) → 2CuI(s) + I$_2$(aq)

   2S$_2$O$_3^{2-}$ (aq) + I$_2$ (aq) → 2I$^-$(aq) + S$_4$O$_6^{2-}$(aq)

   All three correct ✓✓
   Two correct ✓

b. The amount of thiosulfate that reacted can be calculated using

   moles = concentration x volume (in dm^3)

   Moles = 0.040 x 0.0155 = 6.20 x 10^{-4} mol of thiosulfate ✓

   (Refer to the equations given above)

   If 1 mol Cu forms 1 mol Cu^{2+}, it means 2 mol Cu forms 2 mol Cu^{2+}

   2 mol Cu^{2+} forms 1 mol I$_2$ (Using the second equation of part a)

   1 mol I$_2$ reacted with 2 mol S$_2$O$_3^{2-}$ (Using the third equation of part a)

   So 1 mol Cu = 1 mol S$_2$O$_3^{2-}$ ✓

   Therefore moles of thiosulfate = moles of copper = 6.20 x 10^{-4} mol

   Mass of copper in the 25.0 cm^3 solution = n x M = 6.20 x 10^{-4} x 63.5 = 0.03937g ✓

   Mass of copper in the 250 cm^3 solution = 0.03937g x 10 = 0.3937g ✓

   Percentage of copper = (0.3937 / 0.870) x 100 = 45.3 % ✓

c. Starch is used to identify the end-point in the titration. When starch is first added, a blue / black colour will be observed due to the presence of iodine. As iodine changes to iodide ions, the blue / black colour will disappear marking the end-point. ✓

133. Ligand substitution is a reaction where one ligand in a complex ion is replaced by another ligand. ✓

134.

a. i.

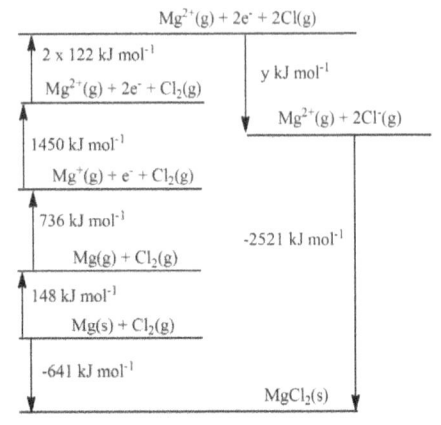

✓✓✓✓

Lose marks for missing:

-State symbols
-Electrons
-ΔH labels
-Arrow wrong way

ii. $\Delta H^\ominus_f (MgCl_2) = \Delta H^\ominus_a (Mg) + \Delta H^\ominus_a (Cl) \times 2 + \Delta H^\ominus_{I1} (Mg) + \Delta H^\ominus_{I2} (Mg) + \Delta H^\ominus_{EA1} (Cl) \times 2 + \Delta H^\ominus_{LE} (MgCl_2)$ ✓

Let $\Delta H^\ominus_{EA1} (Cl) = y$

$-641 = 148 + 122 \times 2 + 736 + 1450 + 2y - 2521$ ✓

(Explanation: y has been multiplied by 2 as there are 2 chlorine atoms).

$-641 - 148 - 244 - 736 - 1450 + 2521 = 2y$

$-698 = 2y$

$-349 = y$

Therefore first electron affinity of chlorine is -349 kJ mol^{-1}. ✓

b. i. Enthalpy change involved in the formation of a mole of an ionic compound ✓ from its gaseous ions under standard conditions ✓.

ii. Ionic charge and ionic radii. ✓

135.

a. $\Delta S = \sum S^\ominus_{(products)} - \sum S^\ominus_{(reactants)}$

$\Delta S = (198 + 131) - 238$

$\Delta S = 91$ J K^{-1} mol^{-1} ✓

b. $\Delta G = \Delta H - T \Delta S$

$0 = 90.7 - T \times (91/1000)$ ✓

$0 = 90.7 - 0.091T$

$0.091T = 90.7$

$T = 996.7$ K ✓

996.7 K $- 273 = 723.7$ °C

Minimum temperature at which the reaction is feasible is 723.7 °C. ✓

The minum temperature can also be calculated using this formula:

$$T = \frac{\Delta H}{\Delta S}$$

136.

a. i. $2S_2O_3^{2-}$ (aq) $+ I_2$(aq) $\rightarrow 2I^-$(aq) $+ S_4O_6^{2-}$ (aq) (reactants ✓ products ✓)

ii. Redox reaction as both oxidation and reduction are taking place. ✓

b. When starch is added, a blue/black colour forms (due to the presence of iodine). At the end point, this colour disappears. ✓ This is because all the iodine (I_2) reacts with the thiosulfate ions forming iodide ions (I^-). ✓ (Iodide ions do not result in a blue/black colour with starch).

137.

a. Oxidation is shown by an increase in the oxidation number. Reduction is shown by a reduction in the oxidation number. ✓

b. i. Fe^{2+} (aq) → Fe^{3+} (aq) + e^-. ✓

   ii. Colour changes from pale green to brown / orange / yellow ✓

c. The two half-equations are: Fe^{2+} (aq) → Fe^{3+} (aq) + e^- and Br_2(aq) + $2e^-$ → $2Br^-$(aq) ✓

   Iron half-equations has to be doubled to balance the equations: $2Fe^{2+}$ (aq) → $2Fe^{3+}$ (aq) + $2e^-$ ✓

   This gives the overall ionic equation: $2Fe^{2+}$(aq) + Br_2(aq) → $2Br^-$(aq) + $2Fe^{3+}$(aq) (reactants ✓ products ✓)

   Iron ions: pale green to brown ✓. Bromine: orange to colourless ✓.

138. No of moles = mass (in grams) / molar mass ✓

139.

a. $K_2Cr_2O_7$ ✓

b. Iron has been oxidised ✓ from +2 to +3 ✓. Chromium has been reduced ✓ from +6 to +3 ✓.

   (Oxidation number of chromium in the dichromate ion is +6. This can be worked out because we know the oxidation number of combined oxygen is -2. Since there are seven oxygens present, this would mean oxygen is -14. As the ion has a charge of 2-, chromium in total would have to be +12. +12 -14 = -2. Because there are two chromiums present, each would have an oxidation number of +6.)

c. Moles of dichromate ions used = concentration x volume (in dm^3) = 0.0324 x 0.04230 = 1.371 x 10^{-3} ✓

   Looking at the molar ratios from the balanced equation, it is evident that 1 mole of dichromate ions reacted with 6 moles of Fe^{2+} ions. Therefore, moles of iron ions would be: (1.371 x 10^{-3}) x 6 = 8.223 x 10^{-3}. ✓

   Now find the amount in grams of iron that reacted:

   (Molar mass of Fe is 55.8)

   Mass(g) = moles x molar mass = 8.223 x 10^{-3} x 55.8 ✓ = 0.459g ✓

d. Percentage by mass = (0.459 / 2.357) x 100 ✓ = 19.5% ✓

140.

a. $C_6H_{14} + 9.5O_2$ → $6CO_2 + 7H_2O$ ✓✓

b.

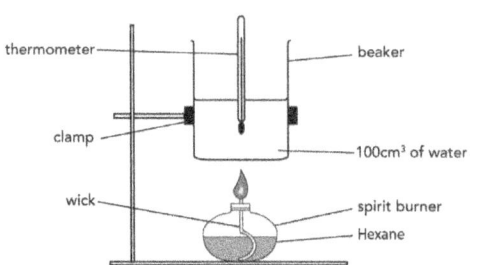

Clamp ✓
Thermometer ✓
Beaker of water ✓
Hexane in spirit burner ✓

c. $\Delta S = \Sigma S^\ominus_{products} - \Sigma S^\ominus_{reactants}$ ✓

  Change in entropy = $(214 \times 6 + 189 \times 7) - (295 + 205 \times 9.5)$ ✓

  Change in entropy = $2607 - 2242.5 = 364.5$ J K^{-1} mol^{-1} ✓

  $Q = mc\Delta T = 100 \times 4.18 \times 51.3$ ✓ $= -21443$ J ✓ (as heat is given out: exothermic : minus sign)

  Moles of hexane = $0.4502 / 86 = 0.005235$ moles ✓

  $\Delta H = 21443 / 0.005235 = 4100000$ J mol^{-1} = -4100 kJ mol^{-1} ✓

  $\Delta G = \Delta H - T\Delta S$ $0 = -4100 - T(0.3645)$ ✓ $T = -11248.3$ K $= -10975.3$ °C ✓.

141.

a. Magnesium carbonate ✓

b. $MgCO_3 + 2HCl \rightarrow MgCl_2 + CO_2 + H_2O$ (species ✓ balancing ✓)

c. i.

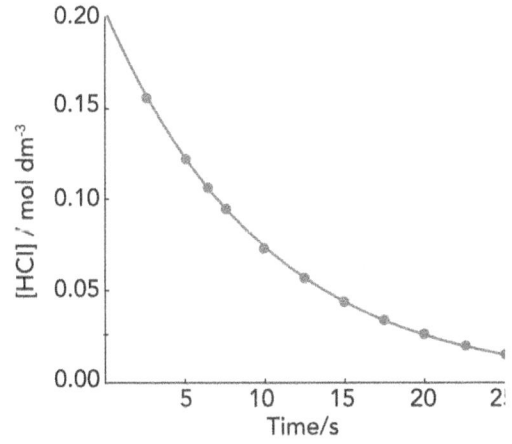

Axis labelled ✓
Points plotted correctly ✓
Points joined by curve ✓

ii. Moles = concentration x volume = $0.0600 \times 0.0200 = 0.00120$ moles ✓   (Use the graph to find out the concentration of HCl after 12s).

Mass = moles x molar mass = $0.00120 \times (1 + 35.5) = 0.00120 \times 36.5 = 0.0438$ g ✓   (Correct answer scores both marks)

iii.

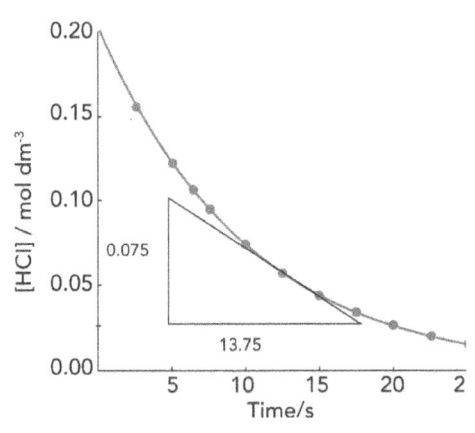

Triangle to the curve drawn ✓
Vertical and horizontal length measured ✓

ANSWERS — Physical Chemistry and Transition Elements

Rate of reaction = gradient = 0.075 / 13.75 = 0.00545 ✓ mol dm^{-3} s^{-1} ✓

iv. First order ✓ (Because successive half-lives are constant).

d. To ensure the term [Y]^y remains a constant throughout the experiment. ✓

e.

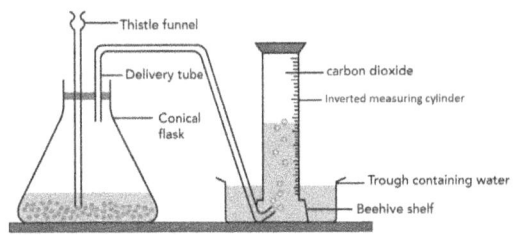

Conical flask sealed ✓
inverted measuring cylinder ✓
Trough with water ✓

142. The pH of the buffer solution = 4.16. [H$^+$(aq)] = 10$^{-4.16}$ = 6.92 × 10^{-5} mol dm^{-3} ✓.

K_a = [H$^+$(aq)][CH$_3$COO$^-$(aq)] / [CH$_3$COOH(aq)]

1.74 × 10^{-5} = 6.92 × 10^{-5} × [CH$_3$COO$^-$(aq)] / [CH$_3$COOH(aq)] ✓

1.74 × 10^{-5} / 6.92 × 10^{-5} = [CH$_3$COO$^-$(aq)] / [CH$_3$COOH(aq)]

0.251 = [CH$_3$COO$^-$(aq)] / [CH$_3$COOH(aq)] ✓

Since the student is using 5.25 cm^3 of sodium ethanoate, the volume of ethanoic acid required is 5.25 / 0.251 = 20.9 cm^3. (Answer ✓ Given to 2 or 3 s.f. ✓)

143.

a.

ln k	T / s	1 / T
-6.17	333	0.00300
-18.56	250	0.00400
-30.95	200	0.00500
-43.35	167	0.00600

All four correct ✓✓
Three correct ✓

b.

i.

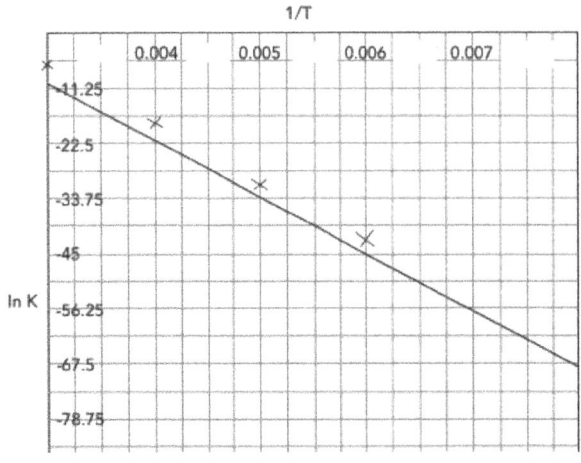

Axes labelled ✓
Points plotted ✓
Straight line graph ✓

ii.

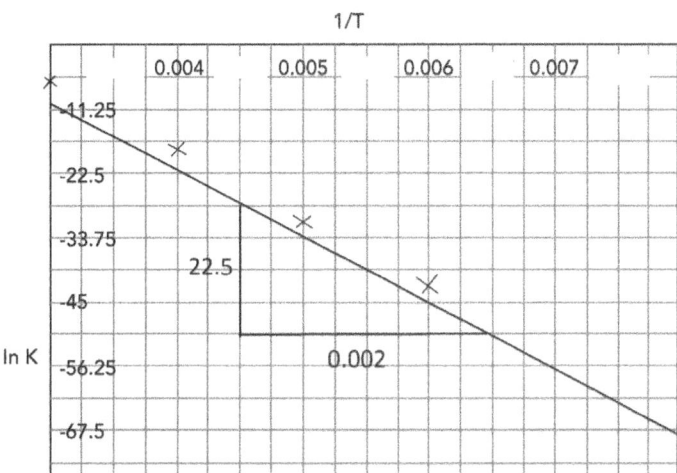

Gradient = y / x = 22.5 / 0.002 = -1.13 × 10⁴ ✓ (Note that gradient is negative as the line has a reducing "slope").

Gradient = - E_a / R ✓

-E_a = -1.13 × 10⁴ × 8.31 ✓ = -9.39 × 10⁴ ✓ J mol⁻¹ E_a = 9.39 × 10⁴ = 93.9 ✓ kJ mol⁻¹ ✓

(Answer given in J mol⁻¹ is also acceptable)

144.

a.

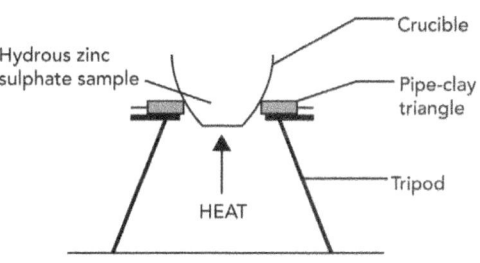

Heat source ✓
Sample in crucible ✓
Tripod and pipe-clay triangle ✓

b.

Mass of water lost = 18.5 - 12.7 = 5.8 g. ✓

Mass of anhydrous compound = 12.7 g (given in the question)

Molar mass of $ZnSO_4$ = 65.4 + 32.1 + 16 × 4 = 161.5 ✓

Molar ratio of $ZnSO_4$: H_2O is   12.7 / 161.5 :   5.8/18 ✓   = 0.0786 : 0.322   =   1:4

(Dividing 12.7g by 161.5 gives the moles of $ZnSO_4$ and dividing 5.8 by 18 gives the moles of H_2O)

Therefore, water of crystallisation, x = 4. ✓

145.

a. May not have the same activation energy.✓

b. i.

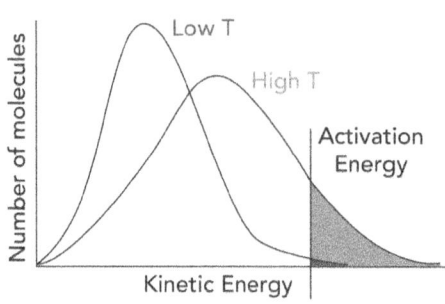

Graph moved to the right✓
Highest point lower than the lower temperature curve ✓

ii. At a higher temperature, molecules collide more frequently✓, leading to successful reactions as they exceed the activation energy✓. The proportion of molecules exceeding the activation energy is shown by the area under the graph✓ after the activation energy line.

c.

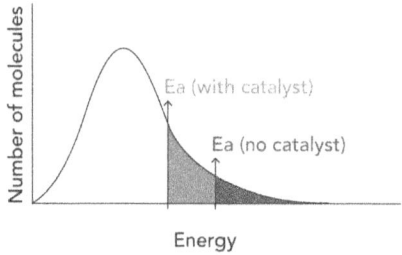

Activation energy line moved to the left in the presence of a catalyst✓

146.

a. As the molar ratio of HCl to NaOH is 1:1 in the equation, equal amounts in mol, of HCl and NaOH will react.

Place a known volume of HCl of known concentration✓ into the polystyrene cup✓. Add an equal amount or an excess of NaOH✓ to the cup, stir it and cover the cup with the polystyrene lid quickly✓.

The temperature of both HCl and NaOH should be the same before they are made to react✓.

Note the increase in temperature✓.

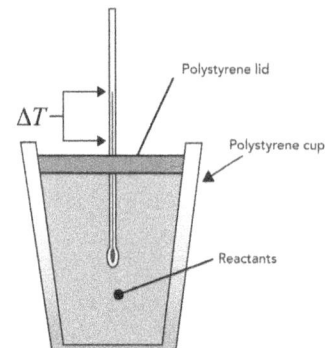

Reaction mixture sealed ✓
Correct placement of thermometer ✓

ANSWERS 5

Physical Chemistry and Transition Elements

b.

i. Q = mcΔT = (50.0 + 50.0✓) x 4.18 x (30.7-23.2✓) = 100 x 4.18 x 7.5 = 3135 J ✓ = -3.34 kJ (answer✓ 3 s.f.✓)

(Note that, we assume that the density of the solution is the same as the density of water: 1g / mL which is the same as 1 g / cm^3. As we had 50 cm^3 of the two solutions each, we get 100g as the mass).

ii. Calculate the numer of moles of HCl that reacted to work out the numer of moles of water formed:

Moles = concentration x volume = 1.00 x 0.0500 = 0.0500 mol of HCl = 0.0500 mol of H_2O✓ (the ratio of HCl : H_2O is 1:1).

3.34 kJ of energy given out per 0.0500 mol of water.

To find out energy given out per 1 mol: 3.34 / 0.0500 = 66.8 kJ mol^{-1}

Therefore, enthalpy change of neutralisation = -66.8 kJ mol^{-1} (answer✓ minus sign✓)

iii. No heat is lost to the surrounding✓. Density of the solution is the same as the density of water✓. Specific heat capacity of the solution is the same as water✓.

147.

At equilibrium n(C_6H_{12}) = 0.140 mol AND n(H_2) = 0.220 mol✓

	C_6H_6	$3H_2$	C_6H_{12}
Starting no. of moles	0.340	0.640	0
Change in moles	(0.340 - 0.200 =) -0.140	(0.140 x 3 =) -0.420	+0.140
Moles at equilibrium	0.200	(0.640 - 0.420) = 0.220	0.140

Explanation: Start by writing the amount in mol of each species. 0.340, 0.640 and 0.200 are given in the question. Since there is no C_6H_{12} at the start of the reaction, you always write 0 as the starting number of moles.

Then you can fill the 3rd row, 2nd column of the table, by substracting 0.200 from 0.340. This gives the change in the number of moles of C_6H_6. Using this, 3rd row, 3rd column can be worked out: since the ratio of C_6H_6 : $3H_2$ is 1:3 as shown by the equation, if there is a change of 0.140 for C_6H_6, there will be a change of 0.140 x 3 for H_2. Now substract 0.420 from 0.640 to give 0.220 as the equilibrium number of moles of H_2.

The change in moles of C_6H_{12} is 0.140 as the ratio of C_6H_6 : C_6H_{12} is 1:1. However, as C_6H_{12} is formed in the reaction, you would add 0 and 0.140 to give 0.140 as the equilibrium concentration of C_6H_{12}.

(We use "-" for the change in moles of the reactants as they are used up and "+" for change in moles of products as they are formed during the course of the reaction).

(Table not essential for marks - first two marks will be awarded for correct figures shown on the table)

p(C_6H_6) = (0.200 / 0.560) x 400 = 142 AND

$p(H_2) = (0.220 / 0.560) \times 400 = 157$ AND

$p(C_6H_{12}) = (0.140 / 0.560) \times 400 = 100$ ✓

(Note: It is very important that you enter these numbers correctly into your calculator. If you are not using the fractions function on your calculator, always use brackets as shown above. For example, if you just type 0.140 / 0.560 x 400 without the brackets, it will give a completely different and incorrect answer.)

$Kp = p(C_6H_{12}) / p(C_6H_6) \times p(H_2)^3 = 100 / 142 \times 157^3$ ✓ $= 1.82 \times 10^{-7}$ ✓

kPa / kPa x (kPa)³ = kPa / kPa x kPa³ = kPa / kPa⁴ = kPa⁻³ ✓

148.

Moles of hydrogen peroxide used = concentration x volume = 1 x 0.015 = 0.015 mol ✓

The ratio of $H_2O_2 : O_2$ in the equation is 2:1. Therefore, 0.015 mol of H_2O_2 would form 0.0075 mol ✓ of O_2 (theoretical yield).

Since the student was able to form 70 cm³ of oxygen, covert this into moles using moles = V (dm³) / 24 = 0.070 / 24 = 0.00292 mol of O_2 ✓ (actual yield).

Percentage yield = (0.00292 / 0.00750) x 100 = 38.9 %. (answer ✓ 3 s.f. ✓)

149.

a. ln k ✓

b.

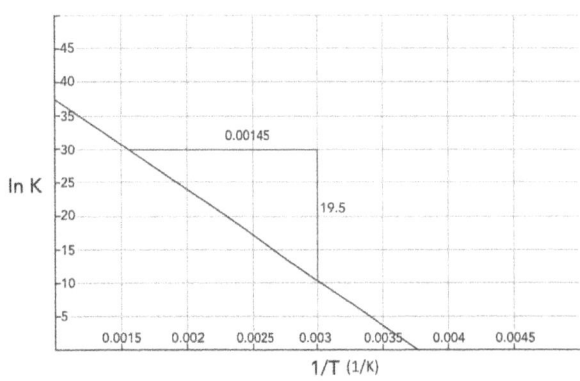

Gradient of the graph = 19.5 / 0.00145 = -13400 ✓

Gradient = $-E_a/R$ ✓ = -13400

$-E_a$ = -13400 x 8.314 = -111407.6 J mol⁻¹ ✓ E_a = 13400 x 8.31 = 111407.6 J mol⁻¹ ✓

y-intercept = ln A = 50✓. Therefore A = 5.18 x 10²¹ s⁻¹.✓

(To work out A when ln A is given, use the e function on your calculator. In this example, you would enter e⁵⁰ into the calculator and obtain the answer by pressing the equal button.)

$$k = Ae^{-\frac{E_a}{RT}}$$

40 °C = 313 K.

k = 5.18 x 10²¹ e ⁻¹¹¹⁴⁰⁷·⁶/⁸·³¹⁴ˣ³¹³ ✓ = 1322.9

k = 1320   (answer✓ 3 s.f.✓)

150.

a.

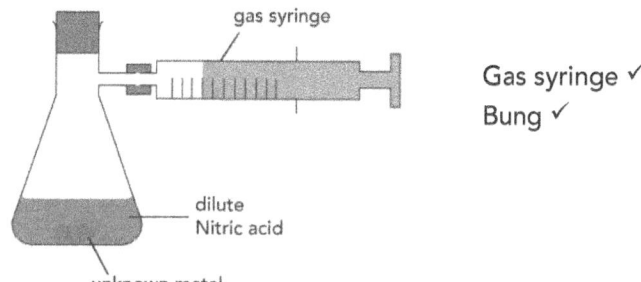

Gas syringe ✓
Bung ✓

Add the metal to the flask✓. Using a measuring cylinder, keep adding HNO₃ to the flask immedietly replacing the bung until no further reaction is observed✓. Measure the maximum volume of gas in the syringe✓.

(Note: The concentration of HNO₃ added is not important as long as it is in excess. If the moles of the acid is less than the moles of the metal, not all the metal will react, consequently affecting the volume of gas produced. It is the volume of gas formed that helps us in the identification of the metal.)

b. Moles of H₂ (g) = volume (dm³) / 24 = 0.0200 / 24 = 0.000833✓

The equation for the reaction:

(Say metal = X as we do not know its identity yet).

X(s) + 2HNO₃(aq) → X(NO₃)₂(aq) + H₂(g) ✓

The molar ratio of X : H₂ is 1:1. Therefore 0.000833 mol of H₂ formation needs 0.000833 mol of X✓.

moles = mass(g) / molar mass

0.000833 = 0.0200 / molar mass   (note that mass is given in mg in the question)

Molar mass = 0.0200 / 0.000833 = 24✓

Since a molar mass of 24 corresponds to the molar mass of magnesium, the metal is Mg✓.

151.

$$k = Ae^{-\frac{E_a}{RT}}$$

$k = Ae^{-55000 / 8.314 \times 303} = A3.297 \times 10^{-10}$ ✓ (Activation energy converted to J mol^{-1}✓. Temperature in K.✓)

$k = Ae^{-55000 / 8.314 \times 318} = A9.234 \times 10^{-10}$ ✓

$A9.234 \times 10^{-10} / A3.297 \times 10^{-10} = 2.80$ ✓

Therefore, the reaction at 45 °C would be 2.8 times faster than at 30 °C.

Note that A cancels out as it is constant. Remember, if the same number / term is present in the top and bottom of a fraction, it will cancel out.

152. Moles of CH_3CH_2COOH = 0.200 x 0.070 = 0.014 **AND**

Moles of $CH_3CH_2COO^-Na^+$ = 0.400 x 0.200 = 0.08 ✓

Total volume of the mixture = 200 + 70 cm^3 = 270 cm^3 = 0.270 dm^3 ✓.

Concentration of CH_3CH_2COOH = 0.014 / 0.270 = 0.05185 mol dm^{-3} AND

Concentration of $CH_3CH_2COO^-Na^+$ = 0.08 / 0.270 = 0.2963 mol dm^{-3} ✓

$[H^+(aq)] = K_a \times [HA(aq)] / [A^-(aq)]$ = 1.32 x 10^{-5} x 0.05185 / 0.2963✓ = 2.31 x 10^{-6} mol dm^{-3}✓.

pH = -log (2.31 x 10^{-6}) = 5.64 ✓

(Correct answer scores full marks).

153. Moles of precipitate **Y** = 2.30 / 103 = 0.02233 mol✓.

Using molar ratios, moles of complex ion = 0.02233✓. Mass of complex ion = 3.57g

Mass of transition metal in the complex ion: 0.325 x 3.57 = 1.16g✓. Rest of the molecule = 2.41g✓

Molecular mass of the ligands present = 2.41 / 0.02233 = 108✓.

Molar mass of transition metal = 1.16 / 0.02233 = 52 g mol^{-1}✓. Therefore Cr✓.

108 / 18 = 6✓. Therefore six water ligands are present. (Molar mass of water is 18).

Molar mass of OH in the precipitate = Molar mass of Y - Molar mass of transition metal = 103 - 52 = 51

51 / 17 (mass of OH) = 3 ✓. Therefore there should be 3 of OH in precipitate Y.

ANSWERS 5

Precipitate Y should be $Cr(OH)_3$ ✓

Colour of precipitate is green ✓. (Aqueous solution of Cr^{3+} reacts with aqueous sodium hydroxide to form a green precipitate. This confirms that six water ligands are present in complex ion M).

M: $[Cr(H_2O)_6]^{3+}$ ✓

Ionic equation between the complex ion and aqueous sodium hydroxide: $Cr^{3+}(aq) + 3OH^-(aq) \rightarrow Cr(OH)_3(s)$ (equation ✓ balacing ✓)

154.

a. For a strong acid, H^+ concentration = concentration of the acid. ✓

b. For a weak acid, $[H^+]_{equilibrium} \sim [A^-]_{equilibrium}$ ✓

c. $\Delta G = \Delta H - T\Delta S = -90 - (27 + 273) \times (-0.234) = -19.8$ kJ. $\Delta G < 0$. Therefore spontaneous. ✓

d. 1.07 V < 1.36 V. The half-cell with the lower $E^{\ominus}$ value provides the electrons and is oxidised in the process. ✓

e. Melting point can be measured using a melting point apparatus and compared with data book values. An impure sample does not melt at the same temperature as a pure sample. ✓

f. The electron that is being added is repelled by the 1- ion. Energy is needed to overcome this repulsion. ✓

155.

a.

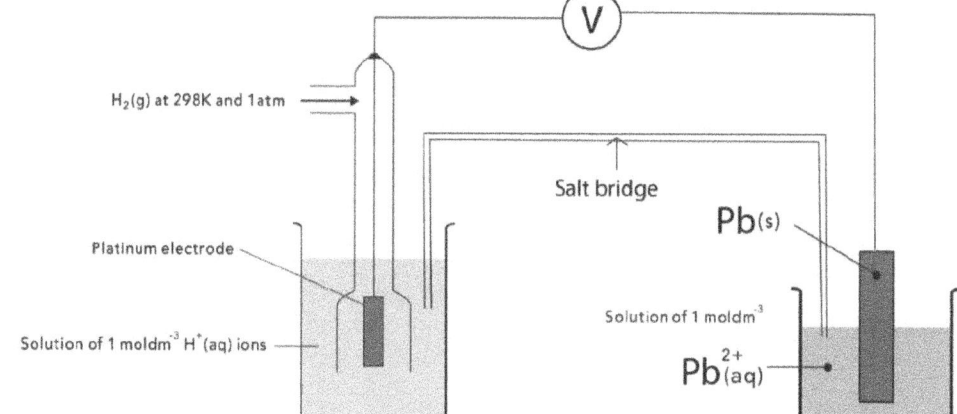

Half-cells (2 marks)

Pb(s) and 1 mol dm^{-3} Pb^{2+}(aq) ✓

1 mol dm^{-3} H$^+$(aq) **AND** H$_2$(g) at 298K and 1atm **AND** platinum electrode ✓

Complete circuit (1 mark)

Salt bridge **AND** voltmeter **AND** wires✓

Standard conditions (1 mark)

298 K / 25 °C **AND** 100 kPa / 101 kPa pressure✓

b. $E^\ominus{}_{cell}$ = $E^\ominus$ (positive terminal) - $E^\ominus$(negative terminal) = 2.87 - (-0.76) = 2.87 + 0.76 = 3.63 V✓

c. $F_2(g)$✓

d. Moves to the left✓

e. Zn^{2+}(aq)/ Zn(s) ✓

f. i. Zn^{2+} will be gaining electrons (equilibrium moving to the right - reduced) if the reaction was to be feasible and Cl^- would be losing electrons (oxidised). As the electrode potential of the reduction half-cell (-0.76) is smaller than that of the oxidation half-cell (1.36), the reaction would not be feasible.✓

ii. Actual conditions used during the reactions may be different from the standard conditions✓. Standard electrode potentials apply to aqueous equilibria✓ - the actual reactions may take place under different conditions.

156.

a.

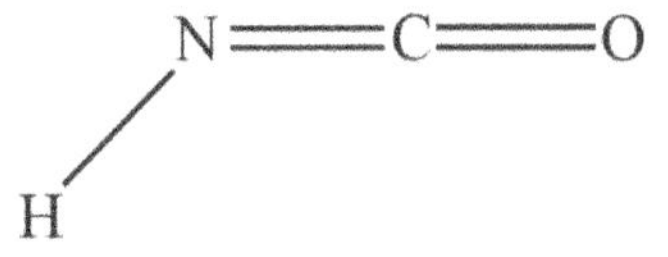

N≡C=O with H bonded to N ✓

b. i. HOCN(aq) + H_2O(l) ⇌ OCN^-(aq) + H_3O^+(aq)✓

HOCN **AND** OCN^-

H_2O **AND** H_3O^+✓

ii.

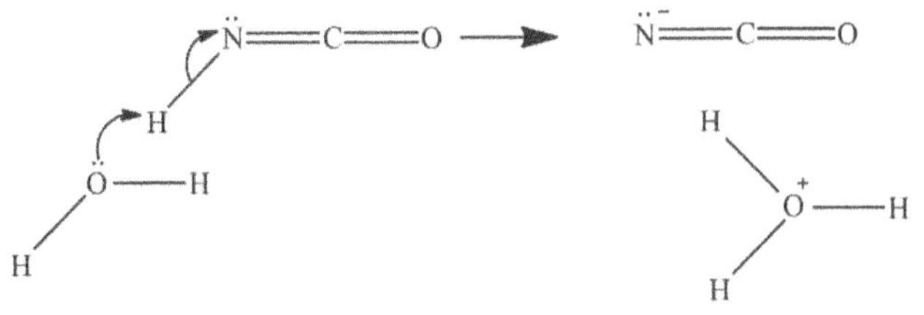

Correct reactants **AND** products **AND** correct positioning of + and - charges on products✓

Two correct curly arrows **AND** H_2O curly arrow starting from lone pair of O✓

c.

	HOCN	H₂O	OCN⁻	H₃O⁺
Starting no. of moles	1.0	5.00×10^{-2}	0	0
Change in moles	-3.00×10^{-3}	-3.00×10^{-3}	$+3.00 \times 10^{-3}$	$+3.00 \times 10^{-3}$
Moles at equilibrium	9.97×10^{-1}	4.70×10^{-2}	3.00×10^{-3}	3.00×10^{-3}

(Moles at equilibrium correct ✓)

$$Kc = \frac{[OCN^-][H_3O^+]}{[HOCN][H_2O]} = \frac{(0.003/V)(0.003/V)}{(0.997/V)(0.0470/V)} = \frac{0.003 \times 0.003}{0.997 \times 0.0470} \checkmark$$

$= 1.92 \times 10^{-4}$ ✓

No units ✓

(Note: the volume cancels out in the equilibrium expression. The volume is not given in the question. Even if it was provided, it would still not make any difference to the final answer).

Note: Award 4 marks to the correct answer and units (calculation not necessary). Award 3 marks to the correct answer without the units.

157.

a. $\Delta S = \Sigma S^\ominus_{products} - \Sigma S^\ominus_{reactants}$
   $15.2 = (2X + 3 \times 214) - (87.4 + 3 \times 198)$ ✓
   $15.2 = 2X - 39.4$
   $X = 27.3$. Standard entropy of Fe(s) = 27.3 J K⁻¹ mol⁻¹ ✓

b. $\Delta H = +824 + (111 \times 3) + 0 - (394 \times 3)$ ✓ = -25 kJ mol⁻¹ ✓

c. $\Delta G = \Delta H - T\Delta S = -25 - (3 + 273$ ✓$) \times 0.0152$ ✓ $= -29.2$ kJ mol⁻¹ ✓

   $\Delta G < 0$. Therefore reaction is feasible. ✓

   Note: Units of ΔS should be kJ K⁻¹ mol⁻¹ and temperature should be in K.

158.

a.

[Structure: 1,1,2-trichloroethene with two Cl atoms on one carbon and one Cl on the other carbon of C=C double bond] ✓

b. Titrate the mixture (trichloroethene) with 5.00×10^{-2} mol dm^{-3} NaMnO$_4$ ✓. As the trichloroethene reacts, the purple colour of MnO$_4^-$ will turn to the brownish black colour of MnO$_2$ and when all of the trichloroethene has reacted, the colour will stay purple ✓. Note the volume of NaMnO$_4$ required ✓.

c. $n(MnO_4^-) = 5.00 \times 10^{-2} \times 0.0345 = 1.725 \times 10^{-3}$ ✓

$n(C_2HCl_3) = 1.725 \times 10^{-3} / 2 = 8.625 \times 10^{-4}$ mol ✓

$n(ClCH_2CH_2Cl) = 3.45 \times 10^{-2} \times 0.025 = 8.625 \times 10^{-4}$ mol ✓

Ratio of $ClCH_2CH_2Cl : C_2HCl_3$ = 1:1 **AND** Therefore equation 1 is correct ✓

159.

a. Titration 1: Strong acid-weak base ✓. Titration 2: Weak acid-strong base ✓.

b. Titration 1: Bromocresol Green ✓ (Equivalence point is at pH = 5.3 ✓ and this is within the range of the indicator). Titration 2: Phenolphthalein ✓ (Equivalence point is at pH = 9.0 ✓ and this is within the range of the indicator).

c. Using a pH meter ✓ - the probe of the pH meter should be inserted into the conical flask and regular readings should be taken as the base is being added ✓.

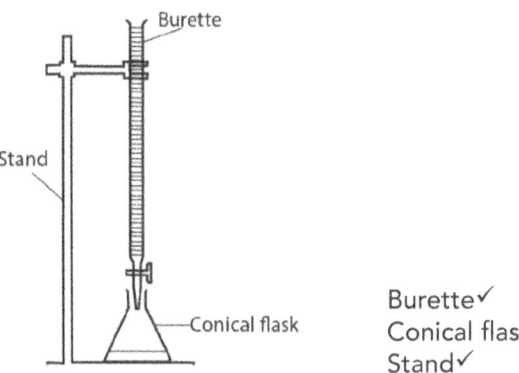

Burette ✓
Conical flask ✓
Stand ✓

d. pH = $-\log[H^+(aq)] = -\log(0.01) = 2$ ✓.

pH is 2 only in titration 1 ✓ when the volume of base added is 0.020 dm^3 ✓.

As the concentration of the base is 0.1 mol dm^{-3}, moles = $0.020 \times 0.1 = 2.0 \times 10^{-3}$ ✓.

160.

a. From the rate-determinig step, the rate equation can be worked out:

rate = $k[P]^2[S]$ ✓

$0.600 = k[P]^2 \times 0.500$ ✓

To determine k, a graph should be plotted using the data given in the table. Complete the table to give:

ln k	T/K	1/T
1.570	1140	8.770 x 10⁻⁴
1.770	1180	8.470 x 10⁻⁴
1.980	1220	8.200 x 10⁻⁴
2.180	1270	7.870 x 10⁻⁴

All four values correct ✓✓
Three correct ✓

Now plot a graph of ln k against 1/T.

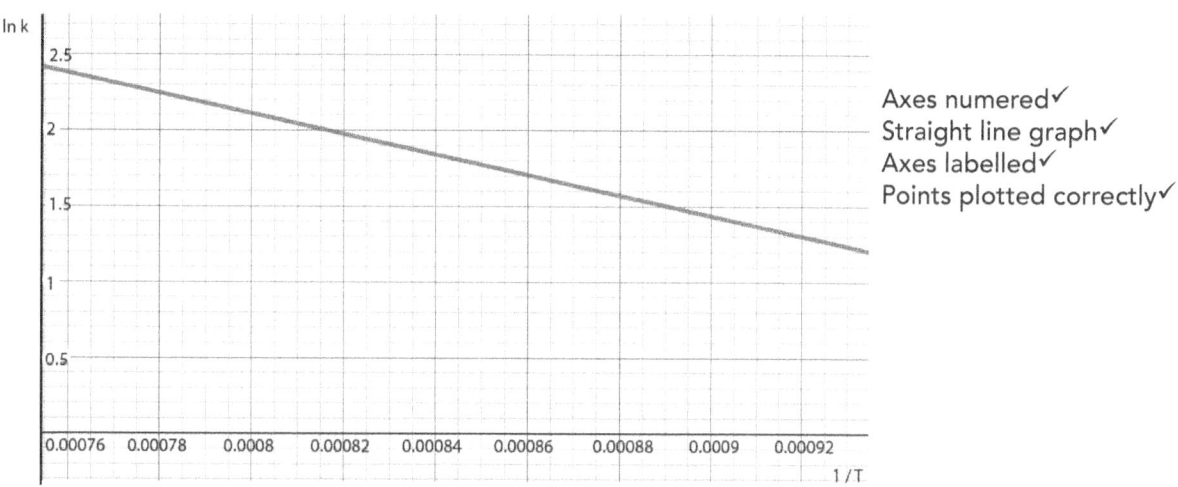

Axes numered ✓
Straight line graph ✓
Axes labelled ✓
Points plotted correctly ✓

The graph can be used to work out the rate of reaction when the temperature is 927 °C. In Kelvin, the temperature is equal to 1200 K ✓. 1/T = 1/1200 = 0.0008333 ✓. Find the ln k value corresponding to this figure using the graph. ln k = 1.9. Therefore $k = e^{1.9} = 6.7$ ✓.

$0.600 = 6.7[P]^2 \times 0.500$

[P] = 0.42 mol dm⁻³ ✓

b. Gradient = y / x = 0.75 / 0.00011 = -6.800 x 10³ ✓

Gradient = $-E_a / R$ ✓

$-E_a$ = -6.800 x 10³ x 8.314 ✓ = -5.654 x 10⁴ J mol⁻¹ E_a = 5.654 x 10⁴ ✓ J mol⁻¹

With a catalyst, the activation energy of reaction would be 0.72 x 5.654 x 10⁴ = 4.07 x 10⁴ ✓ J mol⁻¹

$k = 1810 \times e^{(-4.07 \times 10^4)/8.314 \times 1200} = 30.6$ ✓

Units = mol dm⁻³ s⁻¹ / (mol dm⁻³)² x mol dm⁻³ ✓ = mol⁻² dm⁶ s⁻¹ ✓

5 ANSWERS — Physical Chemistry and Transition Elements

Note: In questions like this, you may not get the exact value for the gradient as given in the mark scheme. However, provided your answers are within a certain range of the correct value, you will still be awarded the marks. If you have made a mistake in one of the initial steps in your working and you used the incorrect figures, you will still be awarded marks for the rest of your working: Error Carried Forward.

161.

a. Using the ideal gas equation,

Mass of 1 mole (g) = mass (g) × (RT/pV) = 0.108 × (8.314 × 298✓ / 98000 × 62 × 10^{-6}✓) = 44.0✓

Therefore, the gas should be carbon dioxide.✓

b. Adding barium nitrate to a carbonate will produce a white precipitate as well due to the formation of $BaCO_3$✓. The student should first identify the carbonate by adding dilute nitric acid and bubbling the gas formed through lime water✓. A white precipitate will confirm that it is a carbonate✓.

c. HCl is a strong acid.✓

$[HCl] = [H^+] = 10^{-0.9} = 0.1259$ mol dm^{-3} ✓

Moles of HCl present = 0.1259 × 0.05 = 6.295 × 10^{-3}✓

$2H^+ + CO_3^{2-} \rightarrow H_2O + CO_2$

Moles of CO_2, theoretical yield = 6.295 × 10^{-3} / 2 = 3.147 × 10^{-3}✓

Moles of CO_2, actual yield = mass / Mr = 0.108 / 44 = 2.454 × 10^{-3}✓

Percentage yield = (2.454 × 10^{-3}) / (3.147 × 10^{-3}) × 100 = 78.0%✓

162.

a. ✓✓✓

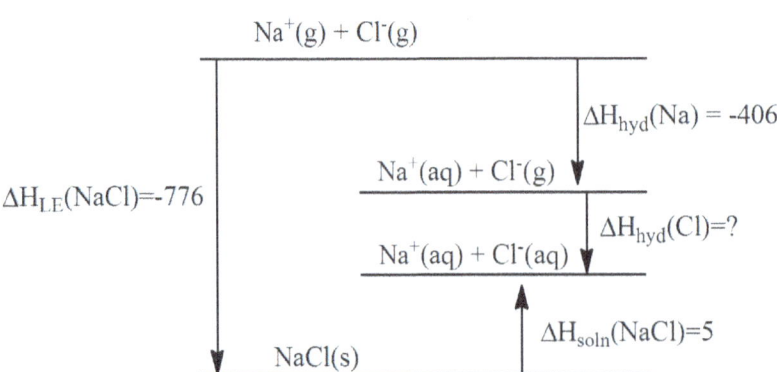

Lose marks for missing:
state symbols
labels
arrow wrong way

Sum of clockwise enthalpy changes = sum of anticlockwise enthalpy changes

-406 + $\Delta H_{hyd}(Cl)$ = -776 + 5✓ $\Delta H_{hyd}(Cl)$ = -365 kJ mol^{-1}✓

b. Amount of NaCl, in mol = 7 / 58.5 = 0.1197 mol✓

   Heat absorbed = 0.1197 x 5 = 0.5985 kJ = 598.5 J✓

   Heat absorbed = mcΔT = 30 x 4.18 x ΔT

   598.5 = 30 x 4.18 x ΔT✓

   ΔT = 4.773 K✓   Final temperature = 296 - 4.8 = 291 K✓

163.

   a. Fe(OH)$_2$(s) is oxidised to Fe(OH)$_3$(s) due to the presence of oxygen in the air.✓

   b. The 3+ charge of Cr^{3+} is balanced by the 3- charge of the three OH$^-$ groups.✓

   c. Scandium does not form at least one ion with a partially filled d sub-shell.✓

   d. NO$_2$ is an electron-withdrawing group with a 3-directing effect.✓

   e. The oxygen ligands of the iron in haemoglobin are substituted by CO ligands irreversibly.✓

164.

   a. [Co(H$_2$O)$_6$]$^{2+}$(aq) + 4Cl$^-$(aq) ⇌ [CoCl$_4$]$^{2-}$(aq) + 6H$_2$O(l) (reactants ✓ products✓)

   b. As the concentration of H$_2$O increases, the system opposes the change✓ by moving the equilibrium position to the left✓. The blue solution turns into a pink solution.✓

   c.

   $$\left[\begin{array}{c} Cl \\ | \\ Cl-Co \cdots Cl \\ \quad \searrow \\ \quad\quad Cl \end{array}\right]^{2-}$$

   Tetrahedral shape✓
   Rest of the structure✓

   d. i. $$K_c = \frac{[[CoCl_4]^{2-}]}{[[Co(H_2O)_6]^{2+}][Cl^-]^4}$$

   Top of the expression✓
   Bottom of the expression✓

   (only aqueous solutions are included in the expression because water has a constant concentration)

   ii. $$K_c = \frac{[\text{mol dm}^{-3}]}{[\text{mol dm}^{-3}][\text{mol dm}^{-3}]^4}$$

   Top of the expression✓
   Bottom of the expression✓

   K_c = mol^{-4} dm^{12}✓

5 ANSWERS — Physical Chemistry and Transition Elements

165.

  a.

    i. Dark blue solution turns into a green precipitate which dissolves with excess ammonia to form a purple solution. ✓

    ii. The charge of the Cr^{3+} ion is 3+ and H_2O ligands are neutral. ✓

    iii. Octahedral ✓

  b.

    i. +2. ✓ H_2O ligands are neutral and OH ligands have a 1- charge each. The 2+ charge of the Fe^{2+} ion is balanced by the 2- charge from the two OH ligands. ✓

    ii. Green precipitate (solid) ✓

  c.

    i. $[Fe(H_2O)_6]^{3+}$ ✓

    ii. Aqueous ammonia or aqueous sodium hydroxide. ✓

166.

  a. Start by writing the balanced equation for the reaction: $2NO_2 + F_2 \rightarrow 2NO_2F$ ✓

    RDS: $NO_2 + F_2 \rightarrow NO_2F + F$ (reactants ✓ products ✓)

    $NO_2 + F \rightarrow NO_2F$ ✓

    Note that in order to work out the RDS, the rate equation has to be taken into account. Anything which appears in the rate equation will always appear in the RDS. Ie. The reactant side of the RDS was worked out using the rate equation. The product side can then be deduced using the reactants. The F which appears on the right side of the first equation and the left side of the second equation, cancels out to give the overall balanced equation.

  b. 2 ✓ (1 + 1 = 2. Both NO_2 and F_2 are first order reactants which is worked out by looking at the rate equation)

  c. Using the graph, when the rate of reaction is 0.5 mol dm^{-3} s^{-1}, the concentration of NO_2 is 0.125 mol dm^{-3} ✓.

    rate = $k[NO_2][F_2]$ $0.5 = 38(0.125)[F_2]$ ✓ $[F_2] = 0.105$ mol dm^{-3} ✓

    Concentration = moles / volume.   moles = 0.105 × 0.3 = 0.0315 mole ✓

    F_2 molar mass = 38   mass required = 0.0315 × 38 ✓ = 1.2 g ✓

167.

a.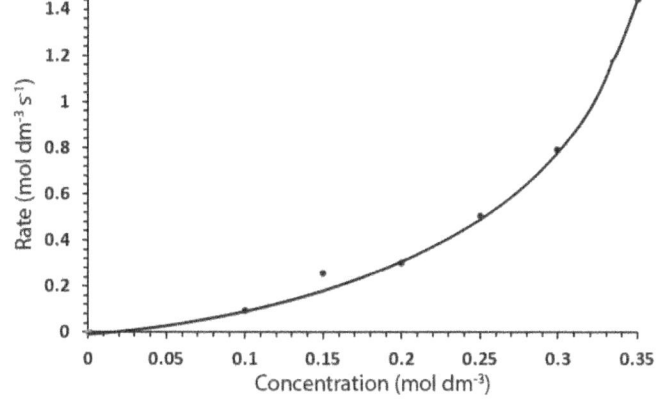

Point plotted correctly ✓
Curve of best-fit ✓
Axes labelled ✓

Order: Second order. ✓

b. Insert the probe of a pH meter and measure the pH of the solution ✓. $[H^+] = 10^{-pH}$ ✓.

c. Using different concentrations of X would lead to a different set of results and would not allow determination of the order with respect to nitric acid ✓ unless the order with respect to X is zero ✓.

d. **M:** NH_3 ✓ **X:** $(NH_4)_2CO_3$ ✓

168.

a. (Calculate A based on the values given in the question and then use that to determine the rate constant at 315 K. The value of the gas constant, R, is given on the data sheet).

$k = Ae^{-E_a/RT}$

$1.02 \times 10^{-10} = Ae^{-112000 ✓/8.314 \times 290}$ ✓

$1.02 \times 10^{-10} = A \times 6.697 \times 10^{-21}$ $A = (1.02 \times 10^{-10}) / (6.697 \times 10^{-21}) = 1.523 \times 10^{10}$ ✓

$k = Ae^{-E_a/RT} = 1.523 \times 10^{10} \times e^{-112000/8.314 \times 315}$ ✓ $= 4.07 \times 10^{-9}$ ✓ s^{-1}.

b. rate = $k[NO_2]$ ✓. The units of the rate constant is s^{-1}.

As the order with respect to NO_2 is 1 ✓, the units of k would be = mol dm^{-3} s^{-1} / mol dm^{-3} = s^{-1} ✓.

169.

a. **Equation 1:** Using the rate equation, the left-hand side (LHS) of the rate-determining step can be worked out. Looking at the reactants, it is obvious that H_2O is produced. As HOI does not appear in the overall equation, HOI should appear on the LHS of one of the remaining two equations, so it can cancel out.

Equation 2: It is likely that HOI is a reactant in the second equation. H_2O has to be a product because the overall equation has two water molecules on the right-hand side (RHS). Therefore another water molecule has to be formed in either the second or the third equation. It cannot be the third because, the product of the third equation needs to contain I. This is because the LHS of the third equation will definitely contain I as the overall equation has 3 I's on the LHS. To balance, a H^+ needs to be added on to the LHS of the second equation.

Equation 3: The products side of the overall equation contains an I_3^-. As I_3^- hasn't appeared in the products side of any of the other two equations, it should be produced in the third. I_2 does not appear on the products side of the overall equation, but it appears on the second equation. So this needs to be cancelled out. If we use I_2 in the reactants side of the third equation, it would cancel out. Now an I^- needs to be added as the other remaining reactant to balance the equation. This makes sense as the overall equation has three I^- on the reactants side. Now each of the three equations has one I^- on the left side giving a total of three I^-.

RDS✓: $H_2O_2(aq) + I^-(aq) + H^+(aq) \rightarrow H_2O(l) + HOI(aq)$

$HOI(aq) + I^-(aq) + H^+(aq) \rightarrow H_2O(l) + I_2(aq)$

$I^-(aq) + I_2(aq) \rightarrow I_3^-(aq)$ (all 10 missing species correct ✓✓✓✓✓ lose one mark for each incorrect)

b. rate = $k[H_2O_2][I^-][H^+]$

$$k = \frac{\text{rate}}{[H_2O_2][I^-][H^+]} = \frac{\text{mol dm}^{-3}\text{s}^{-1}}{(\text{mol dm}^{-3})(\text{mol dm}^{-3})(\text{mol dm}^{-3})} \checkmark = \frac{\text{s}^{-1}}{(\text{mol dm}^{-3})^2} = \frac{\text{s}^{-1}}{\text{mol}^2 \text{dm}^{-6}}$$

Units = $\text{mol}^{-2} \text{dm}^6 \text{s}^{-1}$ ✓.

170.

a. $H_2(g) + ICl(g) \rightarrow HI(g) + HCl(g)$

   $HI(g) + ICl(g) \rightarrow I_2(g) + HCl(g)$ (✓✓✓✓ lose a mark for each incorrect species)

b. i. rate = $k[H_2][ICl]$ ✓✓   (First equation is the RDS as its the slowest step)

   ii. moles = 0.480 / 24 = 0.0200 moles✓   (Moles of a gas = volume in dm^3 / 24)

   $$k = \frac{0.250}{0.0200 \times 0.0200} = \frac{0.250}{0.000400} = 625 \checkmark\checkmark$$

   iii.
   $$k = \frac{\text{rate}}{[H_2][ICl]} = \frac{\text{mol dm}^{-3}\text{s}^{-1}}{(\text{mol dm}^{-3})(\text{mol dm}^{-3})} = \frac{\text{s}^{-1}}{\text{mol dm}^{-3}} \checkmark$$

   Units = $\text{dm}^3 \text{mol}^{-1} \text{s}^{-1}$ ✓

c. 2 ✓   (The order with respect to H_2 is 1 and that of ICl is 1. 1+1 = 2)

171.

a. i.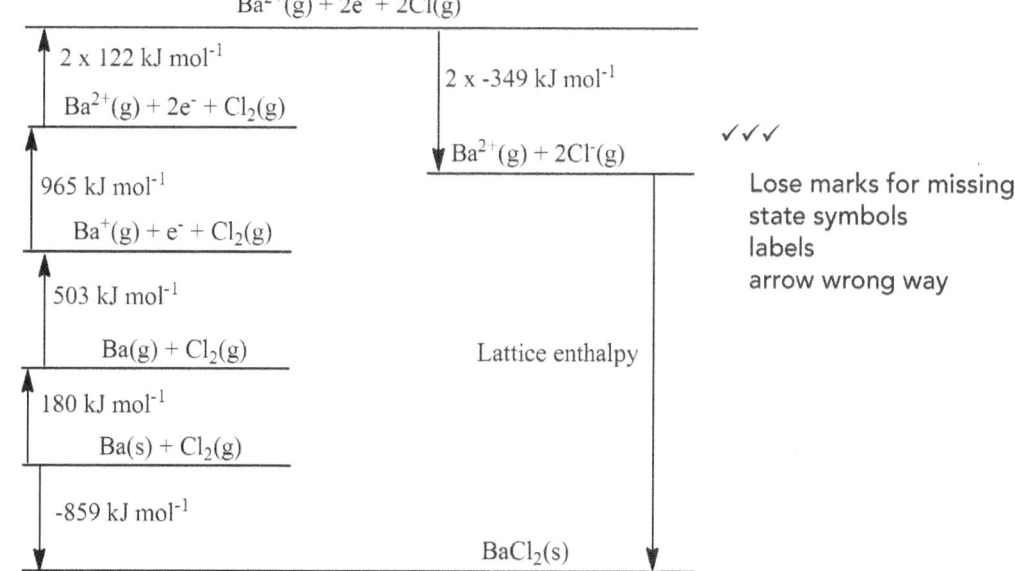

✓✓✓

Lose marks for missing:
state symbols
labels
arrow wrong way

ii. Sum of clockwise enthalpy changes = sum of anticlockwise enthalpy changes

-859 = 180 + 503 + 965 + (2 x 122) + (2 x -349) + LE ✓

-859 = 1194 + LE   LE = -859 - 1194 = -2053 ✓ kJ mol^{-1}.

b. Test for sulfate ions ✓. When $BaCl_2$ is added to a sample containing sulfate ions, a white precipitate of $BaSO_4$ is observed ✓

c. It is higher than that of $BaCl_2$ as O^{2-} ions have a smaller ionic radius and higher ionic charge than Cl^- ✓. Therefore there is a stronger electrostatic force of attraction between Ba^{2+} and O^{2-} ions than between Ba^{2+} and Cl^- ions ✓.

172.

a. i. $Na_2S_2O_3(aq) + 2HCl(aq) \rightarrow 2NaCl(aq) + SO_2(aq) + S(s) + H_2O(l)$ (reactants ✓ products ✓ balancing ✓)

ii. $S_2O_3^{2-}(aq) + 2H^+(aq) \rightarrow S(s) + SO_2(g) + H_2O(l)$ (reactants ✓ products ✓)

b. Draw a cross on a piece of paper ✓. Place the conical flask on top of the cross. Add sodium thiosulfate of known volume and concentration ✓. Now add the acid to the flask while starting the timer at the same time ✓. Swirl the flask. Record the time taken for the cross to disappear ✓. Repeat with sodium thiosulfate solutions of different concentrations but same volume ✓. The concentration and volume of HCl used each time should be kept the same. ✓

c. i.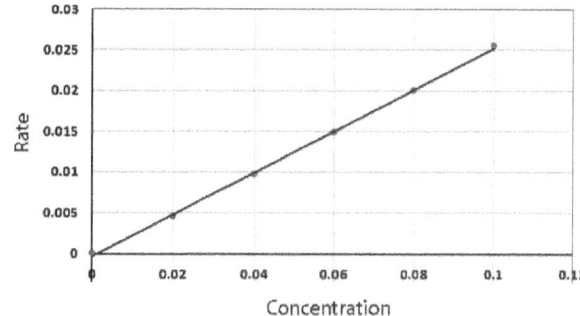

Points plotted correctly ✓
Line of best-fit ✓
Axes labelled ✓

First order✓. Rate is proportional to the concentration.

ii. Temperature✓ (or any other valid factor)

173.

a. Co^{3+}✓. Has the most negative $E^\ominus$ value ✓(when you reverse the last equation in the table). A reducing agent donates electrons✓. (Note that the strongest reducing agent will have the most negative $E^\ominus$ value, as it oxidises itself)

b. i. Standard cell potential = Standard electrode potential of positive terminal - standard electrode potential of negative terminal
= 1.19 - (-0.86)✓ = 2.05 V✓

ii. Note that the half-cell with the most negative $E^\ominus$ supplies the electrons. It is the iron half-cell in this case and to reflect this, the electrons have been written on the right side of the half-equation. This has to be done with all questions like this before the overall equation can be worked out (unless you can do it mentally).

$2Fe(OH)_2(s) + 2OH^-(aq) \rightleftharpoons Fe_2O_3(s) + 3H_2O(l) + 2e^-$
$ClO_2(g) + H^+ + e^- \rightleftharpoons HClO_2(aq)$

The electrons are not balanced. Multiply the second equation by two, so the electrons are $2e^-$ on both equations.

$2Fe(OH)_2(s) + 2OH^-(aq) \rightleftharpoons Fe_2O_3(s) + 3H_2O(l) + 2e^-$
$\mathbf{2}ClO_2(g) + \mathbf{2}H^+ + \mathbf{2}e^- \rightleftharpoons \mathbf{2}HClO_2(aq)$

If anything appears in the left side of one equation and the right side of the other, they can be cancelled out. Now add the equations together.

$2Fe(OH)_2(s) + + 2OH^-(aq) + 2ClO_2(g) + 2H^+ \rightarrow Fe_2O_3(s) + 3H_2O(l) + 2HClO_2(aq)$
(reactants✓ products✓ balancing✓)

iii. ClO_2 electrode✓ as it has the largest $E^\ominus$ value✓.

c. Co✓. The oxidation number goes down from +4 in CoO_2 to +3 in Co^{3+}✓.

(Note the formula CoO_2. We know oxygen is 2-. So the only way we can get CoO_2 is if Co is 4+. This way, the 2- of each oxygen will cancel out 4+ of Co to give CoO_2)

174.

a. Iron oxidised✓ from Fe^{2+} to Fe^{3+}. Oxidation number increases from +2 to +3✓. Chromium has been reduced✓ from +6 in $Cr_2O_7^{2-}$ to +3 in Cr^{3+}✓.

b. $Cr_2O_7^{2-}$ has an orange colour✓. Cr^{3+} has a green colour✓. Colorimetry✓ can be used to determine the concentration of Cr^{3+} as it is formed - small samples of the mixture is taken and analysed at regular time intervals✓. A graph of concentration against time can be plotted using this information✓. The gradient of the graph at any given point will be equal to the reaction rate at that point in time✓.

c. Moles of $Cr_2O_7^{2-}$ = 0.0350 x 0.0100 = 0.000350 moles✓

Ratio of $Cr_2O_7^{2-}$: Fe^{2+} = 1:6✓ (using the equation given in the question).

Moles of Fe^{2+} = 6 x 0.000350 = 0.00210 moles✓

Original moles in 250 cm³ = 0.00210 x 10 = 0.021 moles✓. Mass of $FeSO_4.7H_2O$ = 0.00210 x 277.9✓ = 5.84g✓

175.

a. 2✓P + Q → W + H Looking at the graph, the concentraion decreases very rapidly initially and then the rate of decrease goes down✓ as you would expect from a second order reactant✓. (Q is a first order reactant as the coefficient of Q is 1 in the rate-determining step. So the graph does not belong to Q).

b. Intermediete✓. (Even though it is produced in the rate-determining step, it will be used up in a subsequent step. We can conclude it is an intermediete as it does not appear in the overall equation).

c. To determine the initial rate of reaction, a tangent needs to be drawn to the curve at t=0. Then the gradient of this tangent needs to be found.

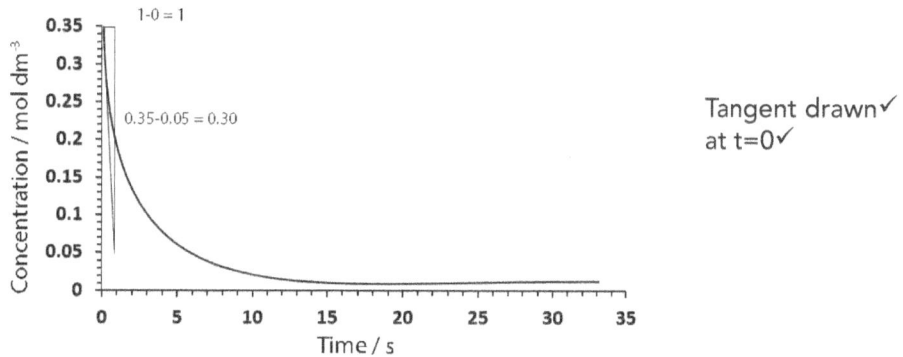

Gradient = y / x = 0.30 / 1✓ = 0.30 ✓ mol dm⁻³ s⁻¹✓. (You may get slightly different answers. In the exams, mark schemes will accept answers within a certain range).

d. Rate equation: rate = $k[P]^2[Q]$✓

0.30 = $k(0.35)^2(0.35)$✓
0.30 = k x 0.042875   k = 7.00✓

Units: mol dm⁻³ s⁻¹ = k (mol dm⁻³)²(mol dm⁻³) ✓   mol dm⁻³ s⁻¹ = k x mol³ dm⁻⁹
k = mol⁻² dm⁶ s⁻¹✓

(We have already worked out the initial rate of reaction to be 0.30 mol dm⁻³ s⁻¹. We can use this information along with the initial concentrations of the two reactants to work out the rate constant. The initial concentration of P is given on the graph as 0.35 mol dm⁻³ and that of Q is given in the question, again, as 0.35 mol dm⁻³).

176.

a.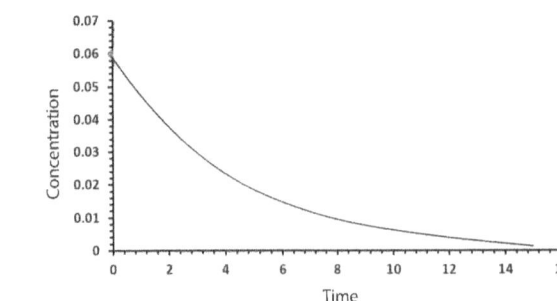

Points plotted ✓
Curve of best fit ✓
Axes labelled ✓

b.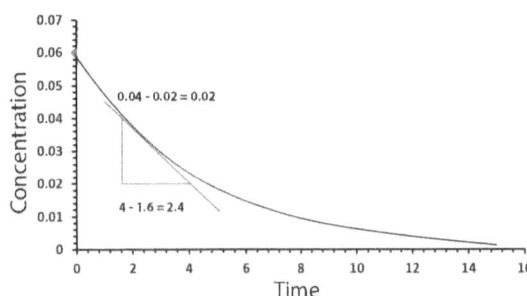

Tangent at 2.4s ✓
Y length worked out ✓
X length worked out ✓

Rate of reaction = Gradient = y / x = 0.02 / 2.4 ✓ = 0.0083 ✓ mol dm^{-3} s^{-1}.

c.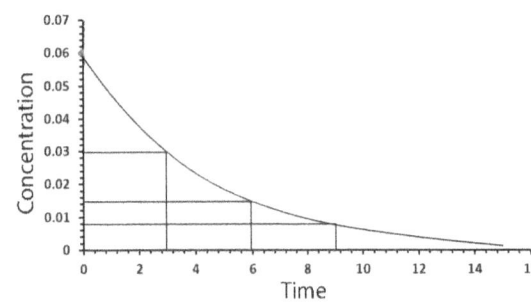

Line at 0.03s going to x-axis ✓
Line at 0.015s ✓
Line at 0.0075s ✓

Half-life is 3 seconds ✓.

d. Rate constant, k = (ln 2) / half-life = (ln 2) / 3 ✓ = 0.23 ✓ s^{-1} ✓

(Question states this is a first order reaction. Therefore the rate equation will look like: rate = k[reactant])

(When you enter ln 2 to your calculator, it is best to type it as (ln 2) / 3. If you type ln 2/3, you will get the wrong answer).

e. rate = k[reactant] ✓   At 2.4 seconds, reaction rate was found to be 0.0083 mol dm^{-3} s^{-1} and the concentration of the reactant was 0.034 mol dm^{-3} ✓.

Therefore: 0.23 x 0.034 ✓ = 0.0078 ✓

(you may get slightly different answers)

177.

a. Fe ✓   for Haber process: $N_2 + 3H_2 \rightleftharpoons 2NH_3$ ✓

   MnO_2 ✓   for decomposition of H_2O_2: $2H_2O_2 \rightarrow 2H_2O + O_2$ ✓

b. i. Cis-platin ✓

ii. Trans-platin ✓   cis / trans isomerism ✓   (Reject E/Z isomerism which is only used with alkenes).

178.

a. i.

ii. Octahedral ✓.

b. +3 ✓. H_2O ligands are neutral / no charge ✓. C_2O_4 is 2-. As there are two of these ligands, there will be an overall 4- charge from all the ligands ✓. The net charge of the entire complex ion is 1-. Therefore chromium should be Cr^{3+}.

c. $C_2O_4^{2-}$ is a bidentate ligand ✓. H_2O is a monodentate ligand ✓.

179.

a. i.

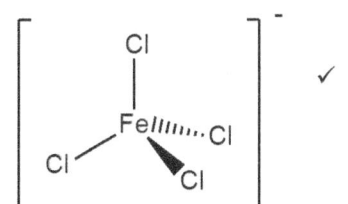

ii. 109.5°✓. There are four bonding pairs and no lone pairs of electrons✓. Electron pairs repel and get as far apart as possible✓.

iii. Tetrahedral.✓

iv. Square planar✓.

b. +3✓. Each chlorine ligand has a 1- charge✓ leading to a total 4- charge from the chlorine atoms. As the overall charge of the ion is 1-, iron must have a charge of 3-.

180.

a.

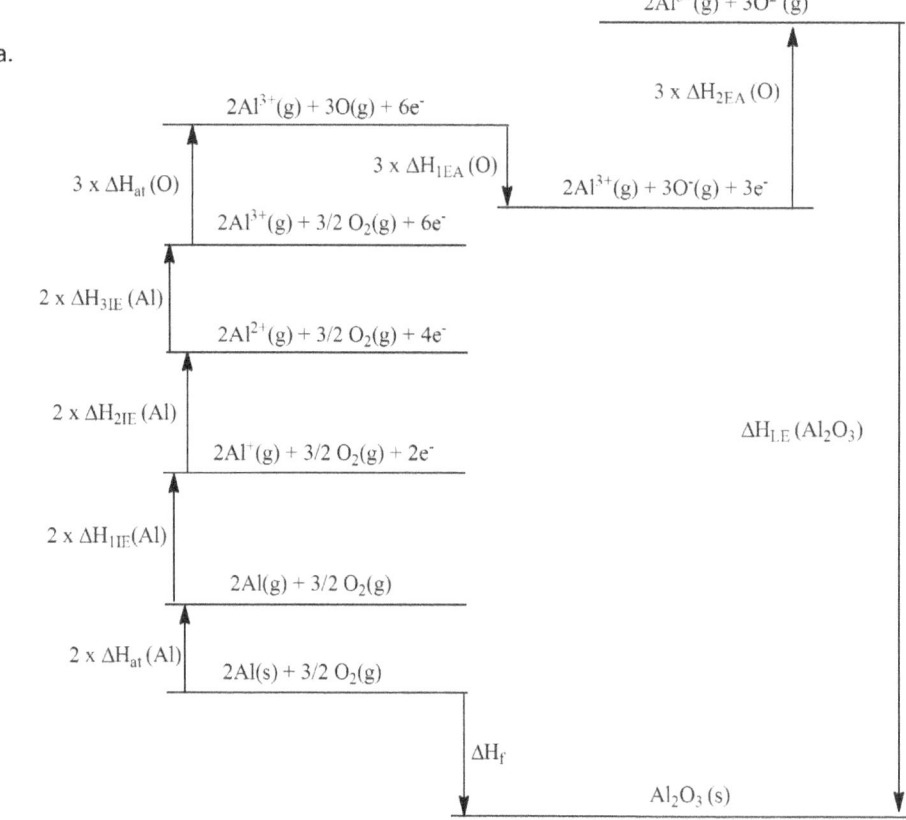

✓✓✓✓
Lose marks for :
incorrect state symbols
incorrect labels
arrow wrong direction
incorrect multiples

b. Sum of clockwise enthalpy changes = sum of anti-clockwise enthalpy changes

OR Enthalpy change of formation = sum of all other enthalpy changes (this may be an useful way to think about it, if you find the terms "clockwise" and "anti-clockwise" confusing).

$-1676 = 2 \times 326 + 2 \times 578 + 2 \times \Delta H_{2IE}$ (Al) $+ 2 \times 2745 + 3 \times 249 + 3 \times (-142) + 3 \times 844 - 15461$ ✓✓

$-1676 = 2 \times \Delta H_{2IE}$ (Al) $- 5310$

$+5310 - 1676 = 2 \times \Delta H_{2IE}$ (Al)

$2 \times \Delta H_{2IE}$ (Al) $= +3634$ ✓

ΔH_{2IE} (Al) $= 1817$ kJ mol^{-1} ✓.

(Note that there are 2 of aluminium and 3 of oxygen in aluminium oxide. Therefore all enthalpy changes relating to aluminium have to be multiplied by 2 and those of oxygen have to be multiplied by 3.)

181.

a. SO_4^{2-} : Add dilute hydrochloric acid ✓ (to remove any carbonate ions which might also produce a white precipitate) followed by barium chloride solution✓. If sulfate ions are present a white precipitate will be observed✓. $Ba^{2+}(aq) + SO_4^{2-}(aq) \rightarrow BaSO_4(s)$✓

b. Cu^{2+} : Add aqueous sodium hydroxide.✓ Pale blue solution will turn to a blue precipitate✓ if Cu^{2+} is present. $Cu^{2+}(aq) + 2OH^-(aq) \rightarrow Cu(OH)_2(s)$✓

c. Br^- : Add dilute nitric acid✓ (reacts with and removes any other ions that might also form a precipitate with silver nitrate) followed by silver nitrate solution✓. If bromide ions are present, a cream precipitate will be observed.✓ $Ag^+(aq) + Br^-(aq) \rightarrow AgBr(s)$✓. This precipitate dissolves in concentrated ammonia to give a colourless solution✓.

d. Mn^{2+} : Add aqueous sodium hydroxide✓. Pink solution will turn to a white / cream precipitate if Mn^{2+} ions are present✓. $Mn^{2+}(aq) + 2OH^-(aq) \rightarrow Mn(OH)_2(s)$✓

e. NH_4^+ : Add aqueous sodium hydroxide and warm the mixture✓. If ammonium ions are present, a (pungent-smelling) gas which turns damp red litmus paper blue will be produced✓. $NH_4^+(aq) + OH^-(aq) \rightarrow NH_3(g) + H_2O(l)$✓

f. I^- : Add dilute nitric acid✓ (reacts with and removes any other ions that might also form a precipitate with silver nitrate) followed by silver nitrate solution✓. If iodide ions are present, a pale yellow precipitate will be observed✓. $Ag^+(aq) + I^-(aq) \rightarrow AgI(s)$✓. This precipitate is insoluble in ammonia solution of any concentration✓.

182.

a. i. Purple✓ to almost colourless at the end-point✓.

ii. Moles of sodium oxalate = 0.340 / 134✓ = 0.002537✓

Ratio of $Na_2C_2O_4$: $KMnO_4$ is 5:2. Therefore moles of potassium permanganate = 0.002537 x (2/5) = 0.001015 moles✓

Concentration = moles / volume = 0.001015 / 0.0273 = 0.0372 mol dm^{-3}✓

iii. $5C_2O_4^-(aq) + 2MnO_4^-(aq) + 16H^+(aq) \rightarrow 2Mn^{2+}(aq) + 10CO_2(g) + 8H_2O(l)$ (reactants✓ products✓)

b.

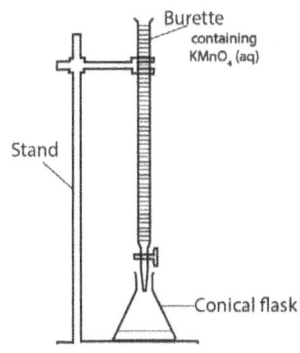

Burette✓
Labels✓
Conical flask✓
Stand✓

5 ANSWERS — Physical Chemistry and Transition Elements

183.

a. $K_a = [H^+]^2 / [HA]$ $1.77 \times 10^{-5} = [H^+]^2 / 0.203$ ✓ $[H^+]^2 = 3.593 \times 10^{-6}$

   $[H^+] = 1.90 \times 10^{-3}$ ✓   pH = 2.7 ✓

b. $CH_3COOH + NaOH \rightarrow CH_3COONa + H_2O$

   Moles of CH_3COOH = 0.203 x 0.025 = 5.08×10^{-3} ✓   Molar ratios of CH_3COOH:NaOH is 1:1.

   Moles of NaOH = 5.08×10^{-3} ✓   Volume required = 5.08×10^{-3} / 0.250 = 0.0203 dm^3 = 20.3 cm^3 ✓

c. i.

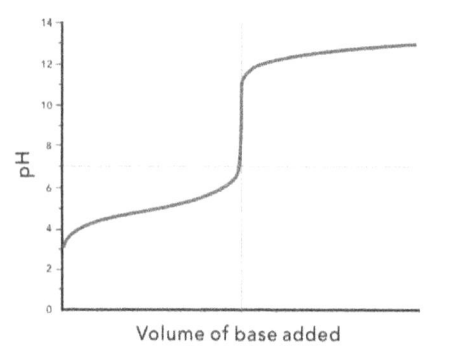

   ii. Using a pH meter. ✓

d. Because the pH change is too gradual closer to the equivalence point. ✓

184.

a. (Calculate A based on the values given in the question and then use that to determine the rate constant at 27 °C. The value of the gas constant, R, is given on the data sheet).

   $k = Ae^{-E_a/RT}$

   $1.03 \times 10^{-3} = Ae^{-110000 ✓ / 8.314 \times 313}$ ✓   (Note that units of E_a in the formula is J mol^{-1} and not kJ mol^{-1})

   $1.03 \times 10^{-3} = A \times 4.386 \times 10^{-19}$ $A = (1.03 \times 10^{-3}) / (4.386 \times 10^{-19}) = 2.348 \times 10^{15}$ ✓

   $k = Ae^{-E_a/RT} = 2.348 \times 10^{15} \times e^{-110000/8.314 \times 300} = 1.65 \times 10^{-4}$ ✓ mol dm^{-3} s^{-1}.

b. At 27 °C the rate constant is 1.65×10^{-4}. If it increases by 1.5 times, it would be 2.475×10^{-4} ✓.

   $\ln k = -E_a/RT + \ln A$ $\ln 2.475 \times 10^{-4} = (-E_a / 8.314 \times 300) + \ln 2.348 \times 10^{15}$ ✓

   $-E_a = (\ln 2.475 \times 10^{-4} - \ln 2.348 \times 10^{15}) \times 8.314 \times 300 = -8.83 \times 10^4$ J mol^{-1} ✓

   E_a = 88.3 kJ mol^{-1} ✓

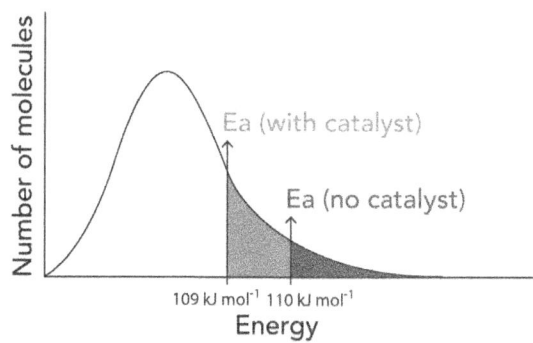

Activation energies labelled ✓
Correct shape of curve ✓

185.

a. i. Pb (s) + 2Cr^{3+} (aq) → Pb^{2+} (aq) + 2Cr^{2+} (aq) (species✓ balancing✓)

ii. Standard cell potential = Standard electrode potential of positive terminal - standard electrode potential of negative terminal

Note that in this case, since it is solid lead that is reacting with the Cr^{3+} ions, the lead equilibrium will be moving to the left (losing electrons) and the chromium equilibrium will be moving to the right, gaining electrons. The equilibrium which gains the electrons will be the positive terminal. In this part of the question, the student is not constructing an electrochemical cell. If this was the case, you would take the positive terminal to be the half-cell with E$^\ominus$ = -0.13 V and the negative terminal to be the half-cell with E$^\ominus$ = -0.41 V and in such an electrochemical cell a different reaction would be taking place.

Standard cell potential = -0.41 - (-0.13) = -0.28 V.✓ The reaction is not feasible as the standard cell potential is negative / less than 0.✓ (The feasibility could have been determined without the need for any calculation. Since the lead equilibrium's electrode potential is higher than that of chromium, common sense tells us it cannot lose electrons. It would have to lose electrons if solid lead were to react with Cr^{3+} ions.)

Another acceptable answer is: chromium equilibrium is reduced as it gains electrons. As the reduction half-cell does not have the greater electrode potential, the reaction is not feasible. -0.13 > -0.41.

b. i. Pb^{2+} (aq) + 2Cr^{2+} (aq) → Pb (s) + 2Cr^{3+} (aq) (species✓ balancing✓)

ii. Standard cell potential = -0.13 - (-0.41) = +0.28 V.✓   (In this case, the chromium equilibrium loses the electrons and hence acts as the negative electrode. When an electrochemical cell is created, the equilibrium with smaller electrode potential automatically becomes the negative terminal. Ie: it is not assigned by anyone).

186. The ethanoic acid would react with the calcium hydroxide to form a salt (calcium ethanoate). The equation for the reaction: 2CH$_3$COOH + Ca(OH)$_2$ → (CH$_3$COO$^-$)$_2$Ca^{2+} + 2H$_2$O (species✓ balancing✓)

Moles of ethanoic acid = 1 x 0.05 = 0.05✓. Moles of Ca(OH)$_2$ = 0.30 x 0.02 = 0.006✓. Molar ratio of CH$_3$COOH : Ca(OH)$_2$ in the equation is 2:1. Therefore 0.012 moles of ethanoic acid is required to react with 0.006 moles of calcium hydroxide✓. Ethanoic acid is in excess.

This leaves 0.038 moles of ethanoic acid unreacted. 0.006 moles of (CH$_3$COO$^-$)$_2$Ca^{2+} is formed from the reaction.

ANSWERS — Physical Chemistry and Transition Elements

As the total volume of the mixture is 1.5 dm³, you do not need to work out the concentrations of the remainig acid and the salt. Using the concentrations or the moles will give you the same answer because the volumes cancel out.

$[H^+(aq)] = K_a \times [HA(aq)] / [A^-(aq)] = 1.74 \times 10^{-5} \times 0.038 / 0.006$ ✓ $= 1.102 \times 10^{-4}$ mol dm⁻³ ✓.

pH = -log (1.102×10^{-4}) = 3.96 ✓

187.

a. $CH_3COO^-Na^+$ ✓

b. i. $K_a = 10^{-pKa} = 10^{-4.76} = 1.74 \times 10^{-5}$ mol dm⁻³ ✓.

$[H^+] = 10^{-pH} = 3.16 \times 10^{-6}$ ✓

$$K_a = \frac{[CH_3COO^-][H^+]}{[CH_3COOH]}$$

$$1.74 \times 10^{-5} = \frac{[CH_3COO^-] \times 3.16 \times 10^{-6}}{[CH_3COOH]}$$ ✓

$$\frac{[CH_3COO^-]}{[CH_3COOH]} = \frac{1.74 \times 10^{-5}}{3.16 \times 10^{-6}} = \frac{5.5}{1}$$ ✓

Therefore the ratio of **ethanoic acid to sodium ethanoate is 1: 5.5** ✓

ii. Moles of sodium ethanoate = 0.0200 x 0.06 = 0.0012 moles ✓.

Using the ratio of 1:5.5, we would need 0.0012/5.5 = 2.18 x 10⁻⁴ moles ✓ of ethanoic acid.

188.

a. $K_c = [R]^2/[P]^2[Q]$ ✓   mol⁻¹ dm³ ✓

b. i. $K_c = (0.27/0.75)^2 / (0.25/0.75)^2 \times (0.10 / 0.75)$ ✓ = 8.75 ✓ mol⁻¹ dm³

ii. Increase ✓. There are more moles of gas on the left-hand side (3) and fewer on the right-hand side (2) ✓. System will oppose the increase in pressure by moving in the direction with fewer number of gas particles ✓. (Note that decreasing the volume of a gas system increases its pressure. See the ideal pressure gas equation you learnt in your year 1 of the course).

c. Y ✓ is higher. The forward reaction is exothermic ✓ (-70 kJ mol⁻¹). If K_c has decreased, it means the equilibrium has opposed the increase in temperature by moving to the left in the endothermic direction ✓.

189.

a. NH_4Br ✓

b. ΔS is positive ✓. ΔG is negative OR TΔS>ΔH ✓.

c. i. Moles of NH_4Br = 5 / 97.9 = 0.05107 moles ✓. Heat absorbed = 16.78 x 0.05107 = 0.857 kJ mol⁻¹ ✓.

Q = mcΔT   857 = 70 x 4.2 x ΔT ✓   ΔT = 2.91 ✓   Final temperature = 25 - 2.91 = 22.1 °C ✓

Physical Chemistry and Transition Elements — ANSWERS 5

(Note: Q is measured in J mol⁻¹ and not kJ mol⁻¹. Hence why you should times 0.857 by 1000 to get 857. Final temperature is determined by initial-change because the reaction is endothermic. If the enthalpy of the solution was exothermic then it would be final = initial + change)

ii. Specific heat capacity of the ammonium bromide solution is different to that of water✓. Standard enthalpy change of solution takes place under standard conditions but this reaction does not✓.

190.

   a. 7.45 / 74.6 = 0.100 moles✓ of KCl

   Q = mcΔT = 90 x 4.18 x (27.5 - 23) = 1690✓

   1690 / 0.100 = 16900 J = 16.9✓ kJ mol⁻¹

   b. Some heat is lost to the surrounding✓

191. (This question assumes knowledge that phenol is a weak acid - Module 6)

   a. K_a = 10^{-pKa} = 10$^{-9.80}$ = 1.58 x 10⁻¹⁰ ✓

   [H⁺]² = K_a x [C_6H_5OH]✓   [H⁺] = 3.37 x 10⁻⁶ ✓   pH = 5.47 ✓

   b. The equilibrium concentration of the acid is approximately equal to the undissociated concentration.✓
   [H⁺]~[A⁻] so [H⁺][A⁻]~[H⁺]²

192. KOH is a strong base. Therefore [OH⁻(aq)] = [KOH(aq)] = 0.082 mol dm⁻³

   K_w = [H⁺(aq)][OH⁻(aq)] = 1.00 x 10⁻¹⁴ ✓   [H⁺(aq)] = K_w / [OH⁻(aq)] = 1.00 x 10⁻¹⁴ / 0.082 =
   1.22 x 10⁻¹³ ✓ mol dm⁻³   pH = -log (1.22 x 10⁻¹³) = 12.9 ✓

193.

   a. K_c = [NH_4^+] [OH⁻]✓ / [NH_3]✓   (Note that H_2O is omitted as it has a constant concentration of 55.56 mol dm⁻³).

   b. K_c = (0.06 x 0.7) / 0.03 = 1.4✓ mol dm⁻³ ✓

194.

   a. When oxygen reacts with iron (II) hydroxide, the oxygen half-cell will be moving from left to right. As it does so, oxygen gains electrons and is therefore reduced✓. As the reduction half-cell has the greater electrode potential, (0.41 > -056✓), the reaction will be feasible✓.

   b. 4Fe(OH)$_2$ + O$_2$ + 2H$_2$O → 4Fe(OH)$_3$ (reactants✓ products✓ balancing✓)

195.

   i. k = Ae$^{-Ea/RT}$ = 30.2e$^{(-(-5)/315 \times 8.314)}$ = 30.3✓   ii. k = Ae$^{-Ea/RT}$ = 24.5e$^{(-6/298 \times 8.314)}$ = 24.4✓

   iii. k = Ae$^{-Ea/RT}$ = 25.5e$^{(-8.03/302 \times 8.314)}$ = 25.4✓   iv. k = Ae$^{-Ea/RT}$ = 29.6e$^{(-(-2.03)/300 \times 8.314)}$ = 29.6✓

   Note : units of T is K and E_a is J mol⁻¹. Be careful of the minus sign in the formula.

ANSWERS

196.

i. $\ln k = -E_a/RT + \ln A$ Rearrange to give $E_a = -(\ln k - \ln A) \times RT = -(\ln 78 - \ln 30.2) \times 8.314 \times 288 = -2271.99$ J mol^{-1} = -2.27 kJ mol^{-1} ✓

ii. $E_a = -(\ln k - \ln A) \times RT = -(\ln 58 - \ln 16.5) \times 8.314 \times 270 = -2821.87$ J mol^{-1} = -2.82 kJ mol^{-1} ✓

197.

a. $\ln k = -E_a/RT + \ln A$ Rearrange to give $T = -E_a / R \times (\ln k - \ln A) = 3.05 / 8.314 \times (\ln 60 - \ln 17)$ ✓ $= 0.291$ K = -273 °C ✓

b. $T = -E_a / R \times (\ln k - \ln A) = 2500 / 8.314 \times (\ln 60 - \ln 35)$ ✓ $= 558$ K = 285 °C. ✓

c. $T = -E_a / R \times (\ln k - \ln A) = -6.02 / 8.314 \times (\ln 26.3 - \ln 29.6)$ ✓ $= 6.13$ K ✓

198.

a. $CuSO_4$ ✓

b. i. $Cu(s) + H_2SO_4(aq) \rightarrow CuSO_4(aq) + H_2(g)$ ✓

ii. Effervescence (bubble given off) ✓

199. $MnO_4^-(aq) + 8H^+(aq) + 5Fe^{2+}(aq) \rightarrow Mn^{2+}(aq) + 5Fe^{3+}(aq) + 4H_2O(l)$ (equation ✓ balancing ✓)

Moles of $Fe^{2+} = 0.03 \times 0.02 = 6.00 \times 10^{-4}$ ✓   Moles of MnO_4^- required $= 6.00 \times 10^{-4} / 5 = 1.20 \times 10^{-4}$ ✓

Concentration of $MnO_4^- = 1.20 \times 10^{-4} / 0.0253 = 4.74 \times 10^{-3}$ ✓ mol dm^{-3}

200.

a. The temperature values have to be converted to K.

k ($\times 10^{12}$)	$\ln k$	T / °C	T / K	$1 / T$
1.963	28.31	-73.00	200.0	0.005000
2.102	28.37	-123.0	150.0	0.006667
2.409	28.51	-173.0	100.0	0.01000
3.627	28.92	-223.0	50.00	0.02000

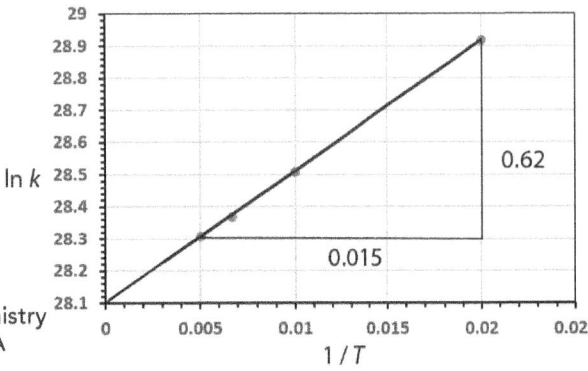

$1/T$ values calculated ✓
$\ln k$ values calculated ✓

Axes labelled correctly ✓
Points plotted correct ✓
Line of best fit ✓

ANSWERS 5

Physical Chemistry and Transition Elements

Gradient = 0.62 / 0.015 = 41.333✓   Gradient = - E_a/R = 41.333   - E_a = 41.333 x 8.314 = 343.64✓

E_a = -343.64 J mol^{-1} ✓

b. $k = Ae^{-E_a/RT}$ = 1.60 x 10^{12} e$^{-(-343.64)/8.314 \times (25+273)}$ ✓ = 1.84 x 10^{12} ✓

Answers to MCQ

1. B

    The effect on the reaction rate of changing the concentration of a reactant depends on the order of that reactant. For example, doubling the concentration of a zero order reactant will have no effect on the reaction rate whereas doubling the concentration of a first order reactant will double the reaction rate.

2. B

3. A

4. C

5. B

6. D

7. D

8. A

9. D

10. C

11. A

    K_a is the acid dissociation constant and its units will always be mol dm^{-3}.
    K_a = [H$^+$(aq)][A$^-$(aq)] / [HA(aq)]
    On the contrary, the units of the equilibrium constant (K_c) will differ from one equilibrium to the next. Likewise, the units of the rate constant (k) will differ from one reaction to the next.

12. A

13. C

14. D

    A buffer solution will not prevent pH changes. It will only minimise any change in the pH.

15. A

    In the first and second ionisation energies, an electron is lost. This electron (negatively charged) has to

ANSWERS

overcome the attraction from the nucleus (positive due to the protons inside). This needs energy and therefore both first and second ionisation energies are endothermic.

16. **B**

    Ions of opposite charge with smaller radii are attracted more towards each other. Similarly, ions of opposite charge with higher charges experience greater attraction towards each other.

17. **A**

    Entropy (s) is always positive. However, note that the *change* in entropy (ΔS) can be positive or negative.

18. **B**

    When an atom gains or loses electrons, its number of protons and electrons are not balanced any longer. If it loses 2 electrons, there will be 2 more protons than the number of electrons and the ion would have a charge of 2+.

19. **C**

20. **D**

    Ions are carried in the salt bridge between the two half-cells. Electrons are carried in the wire.

21. **C**

    When a transition element loses electron(s), it forms an ion with an incomplete d sub-shell. (Metals tend to lose electrons; non-metals tend to gain electrons)

22. **B**

23. **A**

    Cr and Cu are two exceptions when it comes to the electron configurations of the transition metals

24. **C**

25. **B**

    A precipitate does not dissolve in water and is given the "s" state symbol whereas solutions have the "aq" state symbol. s: solid. aq: aqueous.

26. **A**

27. **A**

    The ligands form coordinate bonds with the central transition metal ion. A coordinate bond is one in which both electrons in the shared pair are provided by one atom. In this example, each water ligand donates a pair of electrons to the central metal ion.

28. **D**

29. **B**

30. **C**

31. **D**

ANSWERS 5

Physical Chemistry and Transition Elements

32. D

33. A

34. A

35. B

    A hexadentate ligand is able to donate 6 pairs of electrons to the central metal ion.

36. B

    A common bidentate ligand is ethane-1,2-diamine often referred to as "en" for short. It donates two PAIRS of electrons to the central metal ion.

37. C

38. B

    It is very difficult to store hydrogen as it is a gas and has to be stored under pressure as a liquid.

39. A

40. D

    The diagram shows a hydrogen half-cell. (Standard electrode potentials are measured by connecting a half-cell to the standard hydrogen half-cell).

41. C

    Among the four possible options, a and b mention calcium and magnesium, both of which form 2+ ions. Formation of 2+ ions would need both the first and second ionisation energies. However, the Born-Haber cycle only shows the first ionisation energy. This rules out options a and b.
    Option d is lithium oxide, which consists of O^{2-} ions. Formation of 2- ions require both the first and second electron affinity. However, the Born-Haber cycle only shows the first electron affinity. This rules out option D.

42. A

    Chloride ions have a smaller ionic radius than bromide ions and are able to attract water molecules more strongly.

43. B

44. A HCl is a strong acid. NaOH is a strong base / alkali.

45. D

46. C

47. A Note that one mole of $Ca(OH)_2$ would release two moles of OH^- ions. $[OH^-] = 2[Ca(OH)_2]$.

48. A

49. D (A) shows no change, both (B) and (D) has a change of 5 and (D) has a change of 3.

50. D (ΔG when calculated isn't negative for any of the three reactions).

51. C (Use the formula $Q = mc\Delta T$. Specific heat capacity of water, c, is given in the data sheet as 4.18 J g^{-1} K^{-1}. If option C was to be a possible value, the change in temperature would have to be 110 °C. This is not possible as water boils at 100 °C. So this apparatus is limited to those combustion reactions where the temperature of water would not exceed 100 °C.)

52. B $[OH^-] = 3[Sc(OH)_3]$

53. D The rate equation is rate = $k[A]^2[B]$. Order with respect to A is 2 and that of B is 1. Therefore, the overall order is 3. The overall reaction may have other reactants and not just A and B. The overall reaction may consist of more than one step and not just the rate-determining step. The other steps may have different reactants which may be carried on to the overall reaction.

54. A $[HA]_{equilibrium} \sim [HA]_{undissociated}$ may not be valid for "stronger" weak acids as the acid dissociates to some extent.

55. C A buffer solution may be made from a weak acid and a strong alkali. However, the weak acid has to be in excess. This way, the **salt** of the weak acid will be formed from the reaction: $CH_3COOH + NaOH \rightarrow CH_3COONa + H_2O$: and there will still be some weak acid left. In option C, there is 1.25×10^{-3} mol dm^{-3} of the weak acid (not in excess) and 1.40×10^{-3} mol dm^{-3} of the strong alkali (NaOH).

56. B. The gas produced in test 1 is NH_3. Test 2 is for halide ions.

57. B Four of the ligands are the same. Two of the other two ligands are the same.

58. D In (1), bromine is being reduced as Br_2 gains electrons to become $2Br^-$. As the reduction half-cell (bromine) has the greater electrode potential, the reaction will be feasible. (3) In an electrochemical cell, the half-cell with the smaller electrode potential loses the electron (oxidised).

59. C When roughly equal amounts of reactants and products are present at equilibrium, K_c = 1. As K_c is 52.1 at 720K, it means the reaction is product favoured as $K_c > 1$.

60. D There is a big jump from 4th to 5th. This suggests the 5th electron is in a shell closer to the nucleus. Si has four outer shell electrons (its in the fourth group of the periodic table).

61. D $k = (\ln 2)$ / half-life   Therefore $k = 0.02773$ rate = $0.02773 \times 0.05 = 1.39 \times 10^{-3}$.

62. C (1) No indicator is suitable for weak acid / weak base titrations. (2) An indicator is considered to be a weak acid. (3) There are 1.26×10^{-3} moles of each.

63. C The electrode has to be rinsed with deionized water between different solutions.

64. A Sum of clockwise enthalpy changes = sum of anticlockwise. -910 + 45.6 = X - 390   X = -910 + 45.6 + 390 = -474.4

65. C Alcohols have low volatility (higher boiling points) compared to alkanes because the OH group forms hydrogen bonds.

Physical Chemistry and Transition Elements — ANSWERS 5

66. D (1) Some are tetrahedral (2) Some show *cis-trans* isomerism (3) Not 3 but 6 : bidentate ligands.

67. D $K_a = 10^{-pK_a}$ (K_a of a stronger acid is bigger than that of a weaker acid and would therefore have a more negative pK_a)

68. B There are 0.0105 moles of $[Cu(H_2O)_6]^{2+}$ and 0.0425 moles of NaOH. The molar ratio of $[Cu(H_2O)_6]^{2+}$: NaOH is 1:2 in the reaction equation. Therefore 0.0105 x 2 = 0.021 moles of NaOH will be required. This means 0.0425 - 0.021 = 0.0215 moles of NaOH will be in excess. 0.0215 x 40 = 0.86 g of NaOH.

69. A 5.30 / 160 = 0.03313 moles of $[Cr(H_2O)_6]^{3+}$ are present which would require 0.03313 x 6 = 0.1988 moles of NH_3 to fully react. Molar ratio of $[Cr(H_2O)_6]^{3+}$: NH_3 is 1:6 in the reaction equation. 0.1988 x 17 = 3.38 g of NH_3 required. $[Cr(NH_3)_6]^{3+}$ is the resulting complex which has only one type of ligand. The green precipitate which forms dissolves in excess of ammonia to form a purple solution.

70. B In the K_c and K_p expression, products are written as the numerator. Note that solid and liquid concentrations are omitted from K_c and K_p expressions. If H_2O was in the liquid form, it would have been omitted. But in this case, it is in a gaseous form.

71. B The graph crosses the y-axis at 27. ln A = 27. Therefore $A = e^{27} = 5.3 \times 10^{11}$

ANSWERS
Organic Chemistry and Analysis

MODULE 6

1.
   a. All six carbon-carbon bonds in benzene are the same length.✓ Enthalpy change of hydrogenation of benzene is less than expected.✓ Benzene does not react as readily with Bromine as it should if it had 3 C=C double bonds, as alkenes would do with cyclohexa-1,3,5-triene.✓

   b. 152 kJ mol^{-1}.✓ Enthalpy change of hydrogenation of cyclohexa-1,3,5-triene is -360 kJ mol^{-1} and that of benzene is -208 kJ mol^{-1}. The difference in these enthalpy changes, and the resulting increased stability is known as delocalisation energy.✓

   c. Benzene is a cyclic hydrocarbon with six carbon and six hydrogen atoms arranged in a planar hexagonal ring.✓ The shape around each carbon is trigonal planar with a bond angle of 120^0.✓ Three out of four electrons in the outer electron layer of carbon bond to two other carbon atoms and one hydrogen atom. These are sigma bonds. The fourth electron is in a 2p orbital above and below✓ the plane of carbon atoms overlapping with other 2p electrons✓ of other carbon atoms, forming a delocalised π cloud of electrons across the ring. These p electrons are delocalised over the whole structure.✓ (Delocalised electrons do not belong to any one single carbon).

2.
   a.

   Two curly arrows✓✓
   Intermediate✓
   Products✓

   b. $H_2SO_4 + HNO_3 \rightarrow H_2O + NO_2^+ + HSO_4^-$. ✓

   $H^+ + HSO_4^- \rightarrow H_2SO_4$.✓

   c. A nitrating mixture is first prepared in a round-bottom flask by mixing together concentrated nitric acid and concentrated sulfuric acid✓, while carefully cooling the mixture in a beaker containing cold water✓. Benzene is then added carefully to this mixture, making sure the temperature is kept below 50°C.✓ After all the benzene has been added, a reflux condenser is fitted to the flask and the mixture is heated to 50°C in a water bath.✓ (If the mixture gets any hotter, more than one nitro group may be substituted onto the benzene ring).

3.
   a. Amide✓

   b. Secondary amide✓ (The N atom is bonded to two carbon atoms).

   c. $CH_3(CH_2)_2COOH$✓ and $NH_2CH_2CH_3$✓

   d. React ethylamine✓ with butanoyl chloride✓

   $CH_3CH_2NH_2 + CH_3(CH_2)_2COCl \rightarrow CH_3(CH_2)_2CONHCH_2CH_3 + HCl$ (reactants✓ products✓)

   e. Hydrogen bonds form between the lone pair of electrons on the nitrogens and hydrogens on the water molecules.✓

ANSWERS 6

Organic Chemistry and Analysis

4. a
   i. Molecules with the same structural formula✓ different arrangement of atoms in space✓

   ii. Cis-trans / geometric / E/Z isomerism ✓

   iii. Optical isomerism. Compounds with optical isomers have chiral centres; there are four different atoms or groups attached to a carbon.✓ Optical isomers are non-superimposable mirrors images of each other.✓

b.
   i. Butane.✓ Nickel. ✓

   ii. $1s^2\ 2s^2\ 2p^6\ 3s^2\ 3p^6\ 4s^2\ 3d^8$✓

   iii. Transition metal.✓

c. The spectrum of butan-2-ol would have a peak around 3200-3600 cm^{-1} corresponding to the OH alcohol group✓

   but-2-ene would not have any peak at this range✓

5.
   a. i. $Br_2 + C_6H_6 \rightarrow C_6H_5Br + HBr$✓

   ii.
   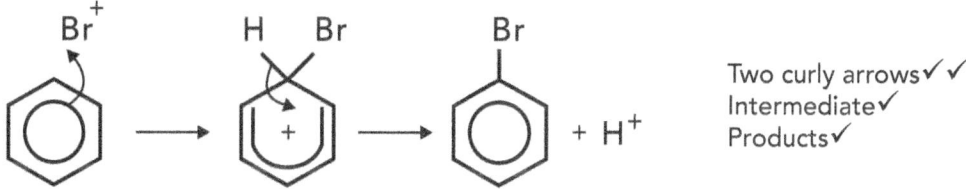

   Two curly arrows✓✓
   Intermediate✓
   Products✓

   $Br_2 + FeBr_3 \rightarrow Br^+ + FeBr_4^-$✓ (Equation depends on which halogen carrier is used).

   $FeBr_4^- + H^+ \rightarrow FeBr_3 + HBr$✓

   iii. $AlBr_3$ or $FeBr_3$✓ (both required for answer)

   iv. $AlCl_3$ or $FeCl_3$✓ (both required for answer). The halogen carrier helps in the formation of the electrophile Cl^+✓ Cl_2 is non-polar whereas Cl^+ will attract an electron pair from benzene. It is a stronger electrophile.✓

   b. A reaction where an atom or group of atoms is replaced by a different atom or group of atoms.✓

   c. Substitution reactions maintain the stability of the benzene ring✓, whereas addition reactions disrupt that stability and the delocalisation✓.

6. R_f values:

   Component A = 1.3/6.2 = 0.21✓, component B = 2.7/6.2 = 0.44✓.

ANSWERS
Organic Chemistry and Analysis

7.
   a. Molecular formula of butan-2-ol is $C_4H_{10}O$. Therefore formula mass / molar mass = (12x4) + (1x10) + 16 = 74. ✓

   b. Four peaks. ✓

   c. Four peaks. ✓

8.
   a. Atomic radii increases down the group. ✓ The larger the atomic radius, the weaker the electrostatic force of attraction from the nucleus ✓ on an electron from outside the atom. As you go down the group, the number of electron shells increases which increases the amount of shielding from the nucleus and further decreases attraction to the nucleus. ✓ (Remember, when halogens react, they <u>accept</u> electrons as they form ions. For <u>group 1</u> elements, it is the opposite: reactivity increases down the group. In their case, they have to <u>lose</u> the electron from their last shell.)

   b. Trigonal planar. ✓ 120°. ✓

   c. $Br_2 + FeBr_3 \rightarrow Br^+ + FeBr_4^-$. ✓

   d. Benzene reacts with bromine at room temperature and pressure in the presence of a halogen carrier such as $AlBr_3$ or $FeBr_3$. The halogen carrier helps in the formation of the electrophile Br^+. ✓

      $C_6H_6 + Br_2 \rightarrow C_6H_5Br + HBr$. ✓

   e. Electrophilic substitution. ✓

9. R_f values:
Component X = 0.8/4.0 = 0.2 ✓, component Y = 1.2/4.0 = 0.3 ✓, component Z = 2.0/4.0 = 0.5 ✓.

10.
   a. Cyclohexa-1,3,5-triene. ✓

   b. Electrophilic substitution. ✓

   c. If the mixture gets hotter than 50°C, more than one nitro group may be substituted onto the benzene ring. ✓

   d. $HNO_3 + H_2SO_4 \rightarrow NO_2^+ + HSO_4^- + H_2O$. ✓

11.
   a. Electron pair acceptor. ✓ (Different textbooks may use other definitions. However, this answer will be credited in the exam).

   b. A substitution reaction where an electrophile is attracted to an electron-rich centre or atom, where it forms a new covalent bond after accepting a pair of electrons. Another group then leaves the molecule,

Organic Chemistry and Analysis — ANSWERS 6

reforming the electron rich centre and the electrophile has replaced this group. ✓ (This is in contrast to electrophilic addition where the electron rich centre is not reformed)

12. a.

   i. Electrophilic substitution. ✓

   ii. Movement of a pair of electrons. ✓

   iii. Because it accepts a pair of electrons. ✓

   iv. It is reformed at the end. ✓

   b. Electrophilic addition. Orange to colourless. ✓

13.

   a. Electrons are delocalised in benzene and does not have sufficient electron density to induce a dipole in the Br-Br bond and attract Br_2 ✓. Br_2 is non-polar whereas Br^+ will attract an electron pair from benzene: Br^+ is a stronger electrophile. ✓

   b. A reaction where a reactant is added to an unsaturated molecule making a saturated molecule. ✓

   c. In cyclohexene the π electrons are localised ✓ above and below the two carbon atoms in the double bond, giving it a high π electron density ✓ unlike benzene where the π electrons are delocalised ✓ over all six carbon atoms. Therefore, cyclohexene can polarise the halogen sufficiently and attract it allowing the reaction to take place ✓.

14.

   a. A group of organic compounds in which a hydroxyl group (OH) is connected directly to a benzene ring. ✓

   b. i. $C_6H_5OH + 3Br_2 \rightarrow C_6H_2Br_3OH + 3HBr$. ✓ Orange colour disappears forming a white precipitate. ✓

   ii. The lone pair of electrons on the oxygen atom is drawn into the benzene ring, creating a region of high electron density ✓, activating the ring. This polarises bromine molecules ✓, attracting them more strongly towards the ring.

   c. The lone pairs of electrons on the O atoms in the OH groups forms hydrogen bonds ✓ with the H atoms on the water molecules ✓.

   d. $C_6H_5OH + NaOH \rightarrow C_6H_5O^-Na^+ + H_2O$. ✓

ANSWERS
Organic Chemistry and Analysis

15.

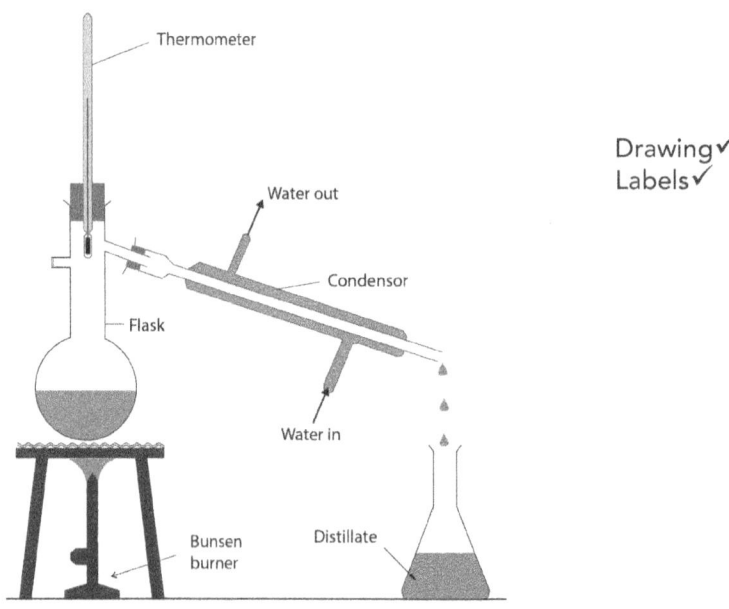

Two curly arrows ✓✓
Intermediate ✓
Dipole on Br_2 ✓

16.
   a. Butanoic acid. ✓ $C_4H_8O_2$. ✓

   b.

   Drawing ✓
   Labels ✓

   (Diagram: distillation apparatus labelled Thermometer, Water out, Condensor, Flask, Water in, Bunsen burner, Distillate)

   c. Potassium dichromate and sulfuric acid. ✓ (Both required for answer)

17. $CH_3CH_2CH_2OH + 2[O] \rightarrow CH_3CH_2COOH + H_2O$. (Species ✓ Balancing ✓)

18. The ability of an atom to attract a shared pair of electrons in a covalent bond. ✓

19. Aqueous sodium hydroxide is added to aqueous silver nitrate until brown precipitate (silver oxide) is formed. ✓ Dilute aqueous ammonia is then added to this, enough to make the precipitate just dissolve. ✓

20. $FeBr_3 + Br_2 \rightarrow FeBr_4^- + Br^+$. ✓

    $H^+ + FeBr_4^- \rightarrow FeBr_3 + HBr$. ✓ (Note that $FeBr_3$ is regenerated: not used up).

21.
   a. The lone pair of electrons on the oxygen atom is drawn into the benzene ring. ✓
   b. Prevents further oxidation to a carboxylic acid. ✓

Organic Chemistry and Analysis

ANSWERS 6

22.

   a. Acidified potassium dichromate. ($H^+/Cr_2O_7^{2-}$).✓

   b. $CH_3CH_2CH_2CH_2OH + [O] \rightarrow CH_3CH_2CH_2CHO + H_2O$. (Reactants✓ Products✓)

   c. To prevent the aldehyde from being oxidised further to a carboxylic acid.✓

   d. When Brady's reagent✓ (solution of 2,4- dinitrophenylhydrazine in a mixture of methanol and sulfuric acid) is added to an aldehyde or ketone, a yellow or orange precipitate✓ is formed.

23.

   a. Butan-2-one.✓ Ketones.✓

   b. $CH_3CH_2CH(OH)CH_3 + [O] \rightarrow CH_3CH_2COCH_3 + H_2O$. (Reactants✓ Products✓)

   c. Secondary alcohol.✓

   d. Tertiary alcohols.✓

24.

   a. Sodium tetrahydridoborate (III), $NaBH_4$.✓

   b. $CH_3CH_2CH_2CHO + 2[H] \rightarrow CH_3CH_2CH_2CH_2OH$.✓✓

   c. $CH_3CH_2COCH_3 + 2[H] \rightarrow CH_3CH_2CH(OH)CH_3$.✓✓

   d. Nucleophilic addition reaction.✓

25. **Confirm the carbonyl bond**: add 2,4-dinitrophenylhydrazine / 2,4-DNPH✓; Orange / yellow precipitate.✓

    OR

    Infrared spectrum✓; has a peak around 1630 - 1820 cm^{-1}.✓

    Identify the ketone: add Tollens' reagent✓. Aldehyde will cause a silver mirror✓. Ketone results in no reaction.

    OR

    Proton N.M.R spectrum; aldehyde spectrum has peak at 9-10✓. ketone spectrum has no peak at 9-10✓.

    Confirm the identity of formaldehyde: Use the 2,4-dinitrophenylhydrazone derivative; recrystallise the derivative to purify✓; measure its melting point✓; compare with data book - should match with that of formaldehyde✓;

26. Electron pair donor✓. (An atom or a group of atoms that is attracted to an electron-deficient centre donating a pair of electrons forming a covalent bond).

ANSWERS

27.
  a. $CH_3CH_2CH_2CHO + [O] \rightarrow CH_3CH_2CH_2COOH$. (Reactants✓ Products✓)

  b. Acidified potassium dichromate.✓ Colour changes from orange to green.✓

  c. Butanoic acid.✓

  d. *First find the amount of moles in 132 g and then divide it by 2 (2000 cm³ is equal to 2 dm³) to find the concentration.*

  Moles = mass (in grams) / molar mass = 132 / 88 = 1.5 moles of butanoic acid✓

  Concentration = moles / volume (in dm³) = 1.5 / 2 = 0.75 mol dm⁻³✓

28. $2C_6H_5OH + 2Na \rightarrow 2C_6H_5O^-Na^+ + H_2$. (Species✓ Balancing✓)

29. The 2,4-dinitrophenylhydrazone derivative is purified by filtering and recrystallising✓, its melting point is measured✓ and compared with data table values to identify the aldehyde or ketone✓.

30. $CH_3CH_2COOH + Na \rightarrow CH_3CH_2COO^-Na^+ + 1/2H_2$.✓ Sodium propanoate and hydrogen.✓.

31. Aromatic amines can be prepared by reducing nitrobenzene or a nitroarene✓ using a mixture of tin and concentrated hydrochloric acid✓. The reagents should be heated under reflux✓, after which the excess hydrochloric acid has to be neutralised.

32.
  a. Esters are used in perfumes and flavourings.✓

  b. Reacting a carboxylic acid with an alcohol in the presence of an acid catalyst.✓ Or, gently heating an acid anhydride with an alcohol.✓ Reacting an alcohol with an acyl chloride ✓ (except methanoyl chloride).

  ci. Methanol ✓ with either butanoic acid / butanoic anhydride / butanoyl chloride.✓

  cii. Propan-1-ol ✓ with either methanoic acid / methanoic anhydride.✓ (Note: methanoyl chloride is not stable and breaks down into carbon monoxide and hydrogen chloride)

  d. -COO-✓

33. Propanoic anhydride✓ can be gently heated with pentan-1-ol✓ to make pentyl propanoate.

34. A reaction where a chemical compound is broken down into two compounds using water or hydroxide ions.✓

35.
  a. $NaBH_4$.✓

  b. Add 2,4-dinitrophenylhydrazine. A yellow/orange precipitate indicates it is a carbonyl compound.✓
  Add Tollens' reagent to the precipitate. A "silver mirror" will confirm it is an aldehyde.✓

  c. i. 4-bromopentan-1-ol.✓

  ii. Nucleophilic addition.✓

Organic Chemistry and Analysis — ANSWERS 6

36.

a. Ester has to be heated under reflux with dilute sulfuric acid or dilute hydrochloric acid.✓ While the acid acts as the catalyst, water breaks down the ester. This reaction will form a carboxylic acid and an alcohol.✓ The type of carboxylic acid and alcohol formed will depend on the ester used. This is a reversible reaction.

b. Aqueous sodium hydroxide or aqueous potassium hydroxide has to be refluxed with the ester.✓ This reaction will form the sodium salt of the carboxylic acid and an alcohol.✓ The type of carboxylic acid and alcohol formed will depend on the ester used. It is a non-reversible reaction.

37.

a. Ethanoic acid.✓

b. React with a metal or a metal carbonate.✓ Effervescence will be observed.✓

c. $Li_2CO_3 + 2CH_3COOH \rightarrow 2CH_3COO^-Li^+ + CO_2 + H_2O$. (correct species✓ balancing✓) Lithium ethanoate.✓

38.

a. Amines are derivatives of ammonia where one or more H atoms of ammonia have been replaced.✓

b. The lone pair of electrons on the nitrogen atom can accept a proton forming a dative covalent bond.✓ Proton acceptors are bases.

c. $CH_3CH_2OH + NH_3 \rightarrow CH_3CH_2NH_2 + H_2O$ ✓✓

d. $K_a = 10^{-pK_a} = 10^{-10.7}$ ✓ $= 1.995 \times 10^{-11}$ ✓

e. A stronger acid would have a larger K_a than a weaker acid. ✓

39. In a saturated fatty acid, there are no C=C double bonds. In unsaturated fatty acids, there are C=C double bonds.✓

40. Butanoic acid and ethanol✓ may be reacted together to form ethyl butanoate. Concentrated sulfuric acid can be used as the acid catalyst.✓

41.

a. The OH's in 4-hydroxybenzoic acid form hydrogen bonds with water molecules.✓

Dipoles✓
Dashed line✓
Lone pairs✓

(Tip: dotted lines represent hydrogen bonds).

b. Carboxylic acid and phenol. (Note: do not allow "alcohol").✓

c. 5✓

ANSWERS — Organic Chemistry and Analysis

42. A silver-grey solid / "silver mirror" ✓ is formed when Tollens' reagent ✓ is added to aldehydes. No reaction is observed with ketones.

43. O-H on the carboxylic acids form hydrogen bonds ✓ with water molecules ✓.

44. $CH_3CH_2CH_2NH_2 + HNO_3 \rightarrow CH_3CH_2CH_2NH_3^+NO_3^-$. ✓ Propylammonium nitrate. ✓

45.
   a. 1-bromobutane is warmed gently ✓ with an excess of ammonia ✓ using ethanol as the solvent ✓.
   b. Nucleophilic substitution. ✓
   c. $CH_3CH_2CH_2CH_2Br + NH_3 \rightarrow CH_3CH_2CH_2CH_2NH_2 + HBr$. ✓
   d. Butylamine and hydrogen bromide. ✓

46.
   a. $CH_3CH_2CH_2COOH + NaOH \rightarrow CH_3CH_2CH_2COO^-Na^+ + H_2O$ ✓ ✓ (One mark awarded for the correct formula of sodium butanoate.)

   Sodium butanoate. ✓

   b. Alcohols form hydrogen bonds ✓ with each other due to the presence of the OH group. Hydrogen bonds are stronger than London forces and hence take more energy to break. ✓ Alcohols can also form dipole-dipole interactions. (Note than van der Waals forces are present in alkanes as well.)

   c. Statement is not correct. ✓ When a substance changes state, the intramolecular forces (forces within a molecule - in this case covalent bonds) are not broken. It is the intermolecular forces (forces between molecules - in this case London forces) that are broken. ✓

47. $RCH(NH_2)COOH$. ✓

48.
   a. Addition polymer. ✓
   b. [Structure of styrene: phenyl group attached to $C=C$ with H atoms] ✓

   c. Calculate the R_f value ✓; Compare the R_f values with those from known compounds ✓.
   d. Compounds with similar properties have similar R_f values so they are difficult to distinguish from each other. ✓

ANSWERS 6

Organic Chemistry and Analysis

49.

    a. Different R groups result in different isoelectric points. ✓

    b. The NH$_2$ group of an amino acid is able to accept a proton acting as a base ✓ and the COOH group of an amino acid is able to donate a proton acting as an acid. ✓

50.

    a.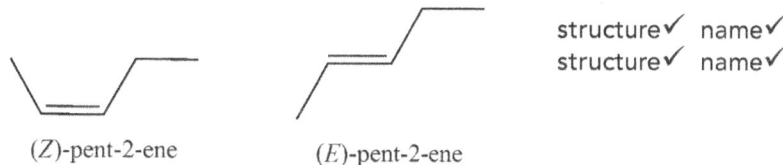

    CH$_2$CH$_3$ group shown ✓
    Rest of the structure ✓

    b. i. The C=C bond in small unsaturated monomers is broken and they join together in a long chain. The polymer is the only product. ✓

       ii. Alkenes. ✓

    c. A reaction where monomers are joined together with the elimination of a small molecule such as water. ✓

    d. In addition polymerisation, the only product is the polymer: no other products are formed. Only one functional group is needed (eg. C=C). ✓ Condensation polymers are made from one or two types of monomers and the monomers must have two functional groups (eg. COOH and OH). A small molecule such as water is also produced. ✓

    e. Stereoisomerism / cis-trans isomerism / E-Z isomerism ✓.

    (Z)-pent-2-ene   (E)-pent-2-ene

    structure ✓ name ✓
    structure ✓ name ✓

    Note that cis-trans is a type of E-Z isomerism where one of the atoms bonded to each carbon is the same - hydrogen in this example. E-Z isomerism only exists in alkenes. However, cis-trans isomerism can also occur in transition metal complex ions for example.

51.

    a. In nature, only one optical isomer is made. Only this may interact with a given enzyme due to the stereospecific nature of enzymes. ✓

    b. Stereoisomers that are non-superimposable mirror images of each other. ✓

    c. Chemically optical isomers behave in the same way. However, biologically, usually only one optical isomer is active. ✓

52. The polypeptide or protein is hydrolysed by heating with an aqueous acid or aqueous alkali. ✓ In acid hydrolysis, amino acids formed are positively charged salts. In alkaline hydrolysis, amino acids are formed in the form of their sodium salt. ✓

53.

a.

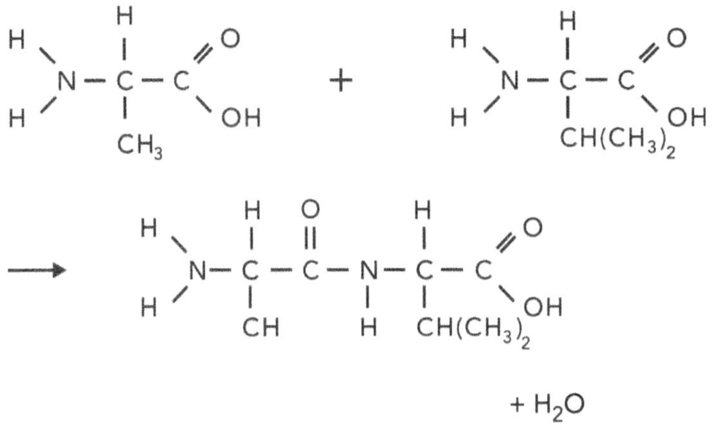

Correct reactants✓
Correct products✓

b. Condensation ✓

54.

a. i. $AlBr_3 + Br_2 \rightarrow AlBr_4^- + Br^+$✓

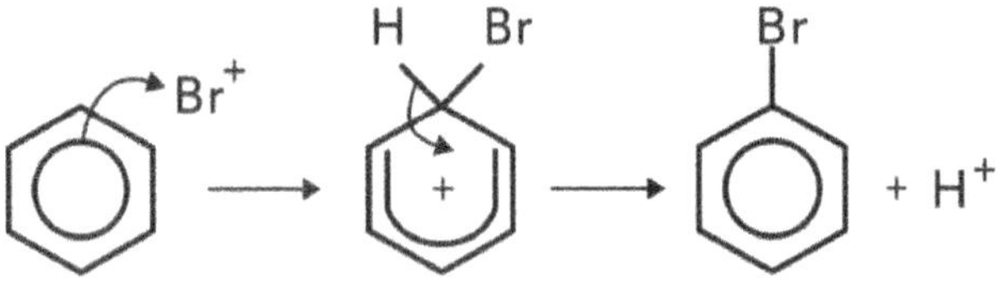

Curly arrows✓✓
Intermediate✓

$H^+ + AlBr_4^- \rightarrow AlBr_3 + HBr$✓

ii. Electrophilic substitution.✓

b. The lone pair of electrons on the O in phenol is delocalised into the ring✓; activating the ring;

Resulting in increased electron density; (There is sufficient electron density to) polarise a Br_2 molecule.✓

(Explanation: The high electron density of phenol's ring structure causes charges to be induced in the bromine molecule).

55.

a. Structural isomerism✓

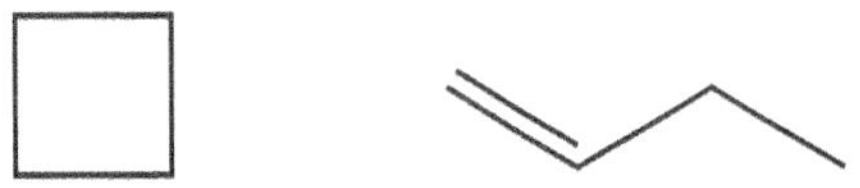

✓✓

cyclobutane but-1-ene

Organic Chemistry and Analysis — ANSWERS 6

b. i. Species with the same structural formula✓ but with a different arrangement of atoms in space✓.

   ii. Optical isomers and geometric (E/Z) isomers.✓

c. i. A carbon atom that is attached to four different atoms or groups of atoms.✓

   ii. Optical isomerism✓

56.

a. i ✓ (structure: benzene ring with NH$_2$ and CH$_3$ in ortho positions)

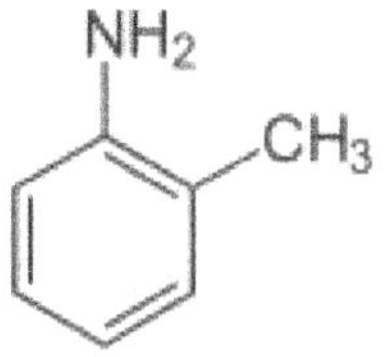

   ii. 3-amino-1-methylbenzene✓ and 4-amino-1-methylbenzene✓.

b. i. The lone pair of electrons on the nitrogen atom can accept a proton forming a dative covalent bond. Proton acceptors are bases. ✓

   ii. $(C_6H_4CH_3NH_3^+)_2 \, SO_4^{2-}$ ✓

57.

a. Optical isomers are stereoisomers that are non-superimposable mirror images of each other.✓

b. Low drug doses can be used✓; risks from undesirable side effects are decreased✓.

c. ✓ (structure: chiral carbon with CH$_3$, H, HO, COOH)

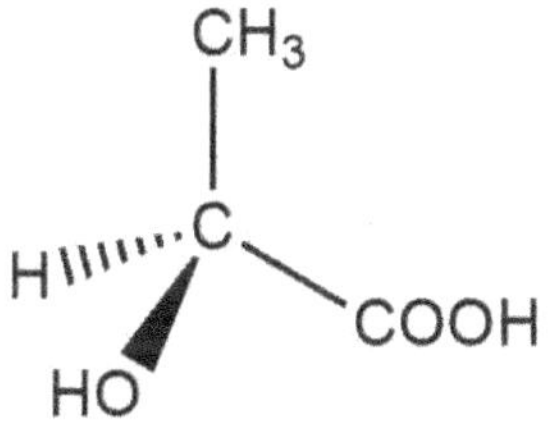

58. Peptide linkages are found in polypeptides and proteins. Between two amino acids is a peptide linkage / amide bond.✓

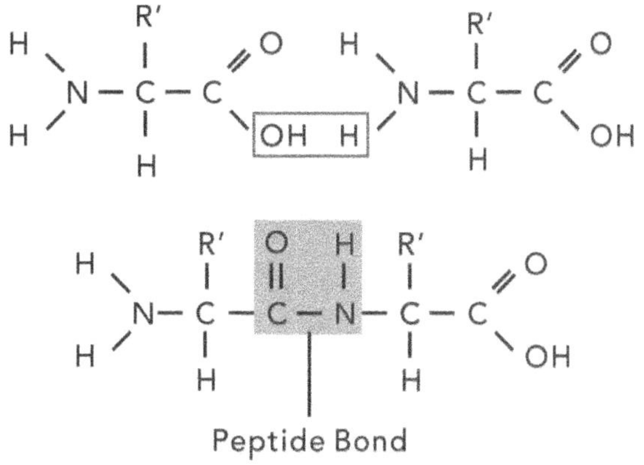

(diagram showing two amino acids condensing to form a dipeptide with the peptide bond highlighted)

59. At a pH of 1, the amino acid acts as base accepting a proton from the acid forming a positive ion.✓ At a pH of 13, the amino acid acts as an acid donating a proton to the hydroxide ion, forming a negative ion.✓ (At a low pH, the amine group gains a proton. At a high pH, the carboxylic acid group loses a proton).

6 ANSWERS — Organic Chemistry and Analysis

60.

a. RCH(NH$_2$)COOH ✓

b. i.

At PH 1

$H_3N^+ - \underset{CH_3}{\underset{|}{\overset{COOH}{\overset{|}{C}}}} - H$

At PH 12

$H_2N - \underset{CH_3}{\underset{|}{\overset{COO^-}{\overset{|}{C}}}} - H$

✓✓

ii. Peptide bond (amide bond) ✓

Structure showing: H$_2$N—CH(CH$_3$)—C(=O)—NH—CH(CH$_3$)—C(=O)—OH ✓

61. Polyamides are polymers where the monomer units consist of carboxyl groups (-COOH) and amine groups (-NH$_2$). ✓

62.

a.

Structure of 1,2-dichloro-2-phenylethene: Cl and H on one carbon, C=C, Cl and phenyl on the other carbon. ✓

b. Addition polymerisation. ✓

63.

a. It has a chiral centre (a carbon atom attached to four different atoms or groups of atoms). ✓

b.

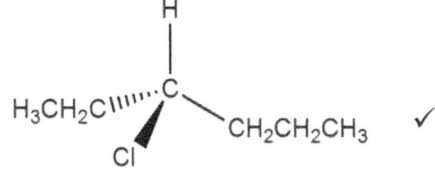 ✓

c. Catalysts provide an alternative route for the reaction to occur. The new path has a lower activation energy. ✓

Organic Chemistry and Analysis — ANSWERS 6

(Activation energy is the minimum energy needed for a reaction to occur).

64.

a. Moles = mass (g) / molar mass

Moles = 0.57 / 140.5 ✓ = 0.0041 = 4.1 x 10^{-3} ✓ (Two s.f. ✓ standard form ✓)

b. Percentage yield = (actual yield / theoretical yield) x 100 ✓

= (0.57 / 0.70) x 100 = 81% ✓

(Working the percentage yield in moles or grams would give the same answer.)

65.

a. Condensation polymerisation ✓

b. Proteins. ✓

66. Polyesters can be hydrolysed using hot aqueous alkali or hot aqueous acid. ✓ Following hydrolysis using hot aqueous sodium hydroxide, sodium salt of carboxylic acid group is formed together with hydroxyl group. ✓ Following acid hydrolysis, monomer units from which the polyester was originally formed is produced (the corresponding alcohol and carboxylic acid).

67.

a.

$$\begin{array}{c} H \\ \diagdown \\ C = C \\ \diagup \diagdown \\ H Cl \end{array} \begin{array}{c} H \\ \diagup \\ \end{array}$$ ✓

b. Addition polymerisation. ✓

c. The monomer contains a double bond. ✓ However, on one carbon, both the atoms bonded to it is the same (both are hydron atoms). For E/Z isomerism to be present, there should be two different atoms / groups of atoms attached to each carbon. ✓

68.

a. Add bromine water to the sample. ✓ Orange colour changes to a white precipitate if phenol is present. ✓ 2,4,6-tribromophenol. ✓

b. Moles = mass(g) / molar mass

Mass(g) = moles x molar mass = 0.800 x 94.1 = 75.3g ✓

Concentration (g dm^{-3}) = amount (g) / volume (dm^3) = 75.3 / 2.3 = 32.7 g dm^{-3} ✓. (Correct answer scores full marks.)

c. Because the OH ✓ group of phenol forms hydrogen bonds ✓ with water.

ANSWERS

Organic Chemistry and Analysis

69. Polyamides can be hydrolysed by hot aqueous alkali or hot aqueous acid.✓ In acid hydrolysis, carboxylic acid group and the ammonium salt of the amine group is formed.✓ In base hydrolysis, sodium salt of carboxylic acid is formed together with the amine group.

70. If something is photodegradable, it will break down in the presence of light.✓ The C=O bond found in some polymers absorbs radiation (UV) and is hydrolysed in the presence of light. This causes the polymer to break down (degrade).

71.
   a. Benzene does not decolourise bromine water, the alkene does✓

   b. The electrons in the p-orbitals✓ of the carbon atoms in benzene overlap with each other forming a delocalised π cloud of electrons above and below the plane of the carbon atoms✓. Delocalised means the electrons are "not local" to any given carbon and are therefore common to the entire structure.

   c. Phenylamine.✓

   d. $C_6H_5NH_2$.✓

   e. Nitrobenzene is heated under reflux✓ with a mixture of tin and concentrated hydrochloric acid✓ to form phenylamine. Reduction✓ (Tin and hydrochloric acid act as the reducing agent.)

72.
   a. $C_8H_8ClNO_2$.✓

   b. Five peaks✓ as there are five carbon environments✓. (As this is a symmetrical molecule, carbon atoms on either side which are at the same level are in the same enviornment.)

   ✓

73.
   a. Nitrobenzene.✓

   b. $C_6H_5NO_2$.✓

   c. Nitrobenzene is heated✓ under reflux✓ with a mixture of tin (Sn) and concentrated hydrochloric acid (HCl)✓ to form phenylamine.

   d. HCl✓ H_2SO_4✓

   e. NO_2^+✓

Organic Chemistry and Analysis — ANSWERS 6

74.

a. 2-aminobutan-2-ol. ✓

b. $CH_3C(NH_2)(OH)CH_2CH_3$. ✓

c. One. ✓ NH_2, OH, CH_3 and CH_2CH_3. ✓

d.

[Two mirror-image stereochemical structures of 2-aminobutan-2-ol, each with NH₂, OH, CH₃ and CH₂CH₃ groups around a central carbon.] ✓✓

e. When manufacturing a pharmaceutical. ✓

75.

a. It is an analytical technique separating components in a mixture. ✓

b. Thin-layer chromatography ✓ and gas chromatography. ✓

c. Mobile phase: the phase that moves in chromatography. ✓ (For example, the mobile phase is often a fluid in thin layer chromatography - it moves up the plate carrying with it the different components and separating them in the process). Stationary phase: the phase that does not move (stationary) in chromatography. ✓ (For example, the stationary phase is a solid in thin layer chromatography - it allows the mobile phase to move through it).

76.

a. A sample of the substance to be analyzed is placed on the chromatography plate towards the bottom and the plate is then placed in contact with a solvent ✓. The sample line should be above the solvent and not in contact. The plate and the solvent should be contained in a closed tank, covered on top until the solvent front reaches the top of the plate. Finally, the plate is removed and the position of the solvent front is marked. ✓ The solvent should then be allowed to evaporate and the plate can be analysed. The distance travelled by each component is measured. ✓

b. Stationary phase is the chromatographic plate. ✓ Mobile phase is the solvent. ✓

c. The solid stationary phase of thin-layer chromatography separate by adsorption ✓. The molecules of the sample bind to the solid plate at its surface. The stronger the adsorption, the more the components are slowed down.

77.

a.

1-bromo-3-methylbenzene   1,2-dimethylbenzene ✓✓

b. 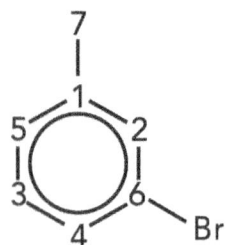 ✓✓

c. There are seven carbon environments in 1-bromo-3-methylbenzene✓ and only four in 1,2-dimethylbenzene✓. Therefore there will be seven peaks in the spectrum of 1-bromo-3-methylbenzene (hence spectrum "a") and only four in the spectrum of 1,2-dimethylbenzene (hence "b")✓.

d. Five.✓

78. R_f value = distance moved by a component / distance moved by the solvent front.✓

79. There are two peaks in the spectrum and therefore the compound should have two proton environments, excluding the proton environment in the OH group. The quartet at 4.2 ppm should have been produced from a proton environment attached to a carbon that is bonded to Br.✓ The adjacent C should be attached to three protons (n+1 rule).✓ The doublet at 1.8ppm indicates that the adjacent C should be attached to only one proton.✓ Based on this information the compound should be $CH_3CHBrCOOH$.✓ ($CH_3CHBrCH_3$ has two proton environments, but its spectrum would have a multiplet and CH_3CH_2COOH does not have Br / the splitting pattern would form a triplet and a quartet).

80.

a. It is the time taken for a component to move from the column inlet to the detector in gas chromatography.✓

b. Gas chromatography-mass spectrometry (GC-MS) combines the two techniques giving a more powerful analytical tool than gas chromatography alone. Components are initially separated by gas chromatography✓ and these components are then detected using mass spectrometry. Mass spectra are compared with spectral databases for identification.✓ A mass spectrum is unique for a given compound, allowing accurate identification of each component in the sample.

81. 1,2,4-tribromobenzene.✓

82. Both these compounds have 4 carbon environments and therefore 4 peaks.✓ Since the carbon environments are almost identical, the chemical shifts of these peaks are almost the same. Carbon-13 NMR does not produce splitting patterns in the peaks, so no further information about the nature of the peaks is available.

83.

a. Tetramethylsilane (TMS) is used as the standard for chemical shift measurements.✓ When NMR spectroscopy is carried out, a small amount of TMS is often added. TMS leads to a peak on the NMR spectrum, which is used as reference in chemical shift measurements. As TMS is chemically unreactive and volatile, it is easy to separate from the sample after NMR spectroscopy is carried out.

b. Chemical shift compares the frequency of an NMR absorption with the frequency of the TMS reference peak at 0 ppm.✓

c. Number of different carbon environments and the types of carbon environments.✓

84. Deuterated solvents✓. Example: $CDCl_3$✓

85. Pentan-1-ol.

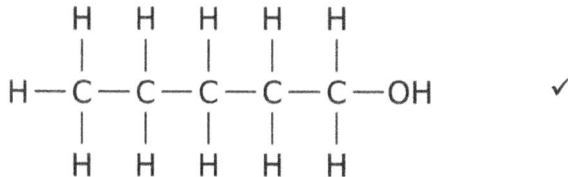

2,2-dimethylpropan-1-ol

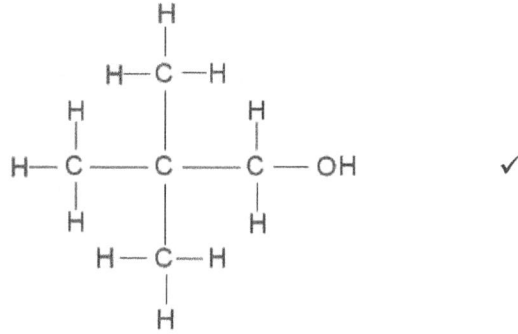

Pentan-1-ol and 2,2-dimethylpropanol. Pentan-1-ol has 5 different carbon environments and therefore 5 peaks in its C-13 NMR spectrum.✓ 2,2-dimethylpropanol has 3 different carbon environments and therefore 3 peaks in its C-13 NMR spectrum.✓ Both the spectra of these compounds will have a peak around 70 ppm as a result of the carbon atom in C-OH.

86.
   a. $(CH_3CH_2CO)_2O$ + $CH_3CH(OH)CH_2CH_3$ → $CH_3CH_2COOCH(CH_3)(C_2H_5)$ + CH_3CH_2COOH
      (reactants✓ products✓)

87. 4 different carbon environments and therefore 4 peaks. **CH_3**-C (0-50 ppm), **C**=O (160-220 ppm), **C**-O (50-90 ppm), **CH_3**-C (0-50 ppm).✓ Ethyl ethanoate.✓

88.
   a. Ester✓
   b. First order✓

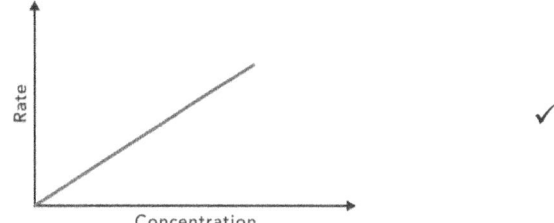

   c. Sodium ethanoate.✓ $CH_3COO^-Na^+$.✓

89. 6 different carbon environments and therefore 6 peaks.✓ C-O (50-90 ppm), C-Br (20-50 ppm) and 4 x C-C (0-50 ppm).✓ 5-bromohexan-2-ol.✓ Hydroxyl functional group (alcohol) OR haloalkane✓

ANSWERS
Organic Chemistry and Analysis

90.

a.

b. The m/z value of the molecular ion peak is equal to the molar mass of the compound being analysed.✓ The fragmentation patterns in the mass spectrum of a compound helps in the identification of individual groups of atoms present in that compound. The m/z value of a given fragmentation peak is equal to the molar mass of the respective group of atoms✓. Mass spectrum of the compound is compared with spectral databases allowing identification.✓

[1,3,5-tribromobenzene structure]

91. $CH_3CH(OH)CH_2CH_3$ (Butan-2-ol) has 4 carbon environments and therefore 4 peaks.✓ $CH_3COHCH_3CH_3$ (2-methylpropan-2-ol) has 2 carbon environments and therefore 2 peaks. Peaks at 70 ppm is due to C-O.✓

92. Ethanoic acid   Propanoic acid

[Structural formulas of ethanoic acid and propanoic acid]   ✓One mark for both formulae

CH_3COOH has two carbon environments and therefore two peaks in its spectrum.✓ CH_3CH_2COOH has three carbon environments and therefore three peaks in its spectrum.✓ The carbon-13 NMR shown is that of CH_3COOH.

93. Similar compounds often have similar retention times.✓ Unknown compounds do not have reference retention times for comparison.✓

94. 2,2-dimethylpropan-1-ol.✓ This compound has 3 different carbon environments and therefore its spectrum will have 3 peaks.✓ The peak at 71 ppm in the spectrum is a result of the carbon atom in C-OH bond. The peaks at 15ppm and 32ppm are a result of the carbon atoms in the C-C bonds.✓

95. 2,2-dimethylbutane (a)✓

[Structural formula of 2,2-dimethylbutane]   ✓

This compound has 4 different carbon environments and therefore 4 peaks in its spectrum.✓ All the peaks will be in the range 10-50ppm as all the carbon atoms are of the C-C bonds. There will be no peak closer to 70ppm.✓

Organic Chemistry and Analysis — ANSWERS 6

96. *(In questions like this, it is best to draw the compounds before answering the question).*

    2-methylpropane has two proton environments. H_9C_3-C protons will produce a peak between 0-2ppm and will be a doublet as the adjacent carbon has one proton (n+1 rule).✓ HC-C_3H_9 protons will produce a peak around 0-2ppm and will be a multiplet as the adjacent carbons have 9 protons.✓ Therefore, this will be spectrum (a).✓

    2,2-dimethylpropanal has two proton environments and its spectrum will have two peaks. H_9C_3-C protons will produce a peak between 0-2 ppm and will be a singlet as the adjacent carbon has no protons (n+1 rule).✓ HCO protons will produce a peak around 9-10ppm and it will be a singlet as the adjacent carbon has no protons.✓ Therefore, this will be spectrum (c).✓

    Methylpropanoate has three proton environments and its spectrum will have three peaks. H_3C-C protons will produce a peak between 0-2ppm and will be a triplet as the adjacent carbon has two protons (n+1 rule).✓ H_2C-C=O protons will produce a peak between 2-3 ppm and it will be a quartet as the adjacent carbon has three protons.✓ The H_3C-O will produce a peak between 3-4.5ppm and it will be a singlet as there are no adjacent carbons and therefore protons.✓ Therefore, this will be spectrum (b).

97. $CH_3CH_2CH(CH_2CH_3)COOH$ has 4 carbon environments and therefore 4 peaks in its spectrum.✓ $CH_3CH_2CH_2COOCH_2CH_3$ has 6 carbon environments and therefore 6 peaks in its spectrum. In both the spectra, there is a peak around 180 ppm as a result of the carbon atom in O-C=O.✓

98. The molecular weight is 86. As it is a ketone, it will have the carbonyl functional group C=O with a molecular weight of 28 (12+14). 86-28 = 58.✓ This "58" will consist of carbon and hydrogen atoms. 4 carbon atoms can fit into this leaving space for 10 hydrogen atoms. This way the compound will have 5 C atoms in total, 10 H atoms and 1 O atom. Molecular formula: $C_5H_{10}O$.✓ In the proton NMR spectrum, there are three proton environments as there are three peaks. The heptet at 2.6ppm should be as a result of the proton in O=CCH. The adjacent carbon should have 6 protons and therefore this part of the compound should have the formula $(CH_3)_2CHCO$.✓ The singlet at 2.1ppm should be the result of the proton in O=CCH. There will be no protons on the adjacent carbon (n+1 rule).✓ The doublet at 1.1 ppm should be the result of $(CH_3)_2CH$ protons. There will only be one proton on the adjacent carbon. The compound is 3-methylbutan-2-one.✓

$$CH_3-\underset{\underset{H}{|}}{\overset{\overset{CH_3}{|}}{C}}-\overset{\overset{O}{\|}}{C}-CH_3$$

99.
   a. Some signals in a proton NMR spectrum may be split into specific patterns. Spin-spin coupling specify the number of hydrogen atoms on the adjacent carbon through the n+1 rule.✓

   b. A proton NMR spectrum is run, as would normally be done. A small amount of D_2O is then added to the sample and shaken.✓ Another proton NMR spectrum of this mixture is taken. Any peaks due to protons in -OH and -NH would have disappeared. Deuterium in D_2O exchanges with H found in -OH and -NH.✓

100.
   a. The peak with the highest m/z value indicates the mass of the compound.✓

   b. A molecular ion is formed when an electron is removed from a molecule forming a positive molecular ion✓. The peak responsible for the molecular ion is referred to as the molecular ion peak✓. (The m/z value of the molecular ion peak is equal to the molar mass of the molecule).

ANSWERS — Organic Chemistry and Analysis

101. Mass spectrometry can be used to find the molecular mass of a compound and the percentage by mass of each element.✓ Infrared spectroscopy tells us what bonds are present in a certain molecule and the possible functional groups. Carbon-13 NMR spectroscopy is used to find out the number and types of carbon environments in a given molecule.✓ Proton NMR spectroscopy gives information about the number and types of proton environments, proportions of protons in each environment and the number of protons on adjacent carbon atoms.✓

102. There are two proton environments: two peaks.✓ HC-I will be a heptet between 3-4.5ppm as the adjacent carbon atoms have 6 protons. $(H_3C)_2$-C will be a doublet as the adjacent carbon has only one proton (n+1 rule).✓

103.

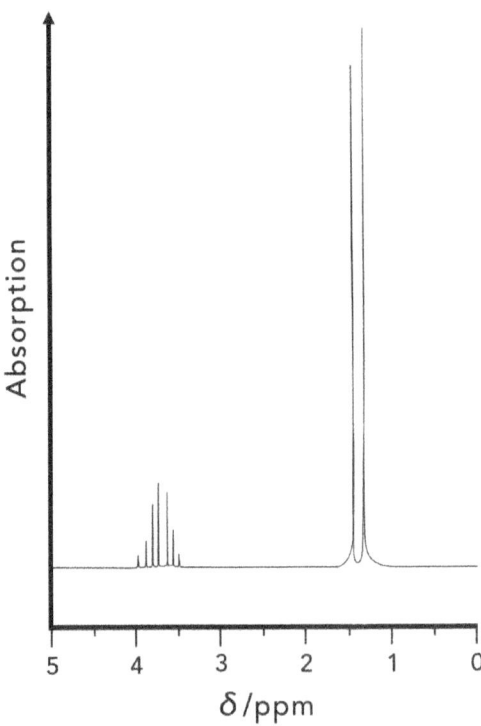

a. n+1 rule is used to work out the number of equivalent protons on the adjacent carbon to a given carbon. If the adjacent carbon has n equivalent protons, number of peaks in the splitting pattern will be n+1.✓

b. Peaks due to -OH and -NH protons are difficult to identify and also make it difficult to identify other peaks in the spectrum. Using D_2O solves this problem as it makes peaks resulting from -OH and -NH disappear. This works because deuterium in D_2O exchanges with H in OH and NH groups.✓

c. Number of proton environments✓ in the given compound, types of proton environments✓, proportion of protons in each environment✓ and proton NMR spectra also tell us the number of protons that are bonded to the adjacent carbons.✓ (all four✓✓✓ three✓✓ two✓).

104.

a. The time taken for the substance to travel from the injector through to the detector and leave the column✓

b. Similar compounds have similar retention times.✓ Unknown compounds have no reference retention times for comparison.✓

ANSWERS 6

Organic Chemistry and Analysis

c.

[Structure: 3-chlorobutan-2-one with C=O and Cl substituents] ✓

δ = 1.6 ppm, doublet, 3H, H_3C-R. Doublet, so adjacent C should have one proton (n+1 rule). ✓

δ = 2.4 ppm, singlet, 3H, H_3C-C=O. Singlet, so adjacent C should have no protons (n+1 rule). ✓ This means the compound cannot be an aldehyde as the C of C=O, should be bonded to another C.

δ = 4.2 ppm, quartet, 1H HC-Cl. Quartet, so adjacent C should have three protons (n+1 rule). ✓

105. There are three proton environments: three peaks ✓. H_3C-CH_2 protons will split into a quartet as the adjacent carbon has three protons. ✓ H_3C-CH_2 protons will split into a triplet as the adjacent carbon has two protons ✓. $(H_3C)_2$-CCl protons will show as a singlet as the adjacent carbon has no protons. ✓

106.

(Note that in questions like these, the reasoning scores most marks. Therefore, it is important you state your observations in your answer, even though they may be obvious to you.)

The mass spectrum shows the compound has a molecular weight of 116. ✓

Proton-NMR spectrum.

δ = 1.0 ppm, singlet, 9H, H_9C-R. Singlet, so adjacent C should have no protons (n+1 rule). ✓

δ = 2.2 ppm, singlet, 2H, H_2C-C=O. Singlet, so adjacent C should have no protons (n+1 rule). ✓

δ = 11.9 ppm, 1H, COOH- carboxylic acid. ✓

Supporting statements.

116 / 58 = 2   So molecular formula: $C_6H_{12}O_2$ ✓

Compound: ✓

[Structure: 3,3-dimethylbutanoic acid]

Carbon-13 NMR has 4 peaks, so the compound will have 4 C environments ✓

3 x R-CH peaks and 1 C=O peak ✓

107. The C-NMR spectrum shows 4 peaks, which is equivalent to 4 different C environments. ✓ As the compound has 6 C atoms in total, there should be 2 pairs of equivalent C environments. There are 3 peaks in the proton NMR spectrum which is equivalent to 3 different proton environments. The heptet at 1.8ppm

suggest the adjacent carbon has 6 protons (n+1 rule).✓ This part of the compound will therefore have the formula (CH$_3$)$_2$-CH. The singlet at 1.5ppm suggests that the adjacent carbon is not bonded to any H atoms. Therefore, this should be bonded to the Cl.✓ The doublet at 1 ppm suggests the adjacent C is bonded to 1 H atom. This should be the result of the protons in (CH$_3$)$_2$-CH.✓ The compound should be 2-chloro-2,3-dimethylbutane.✓

108.

a.

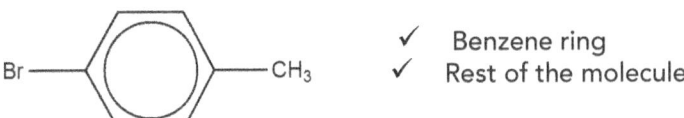

✓ Benzene ring
✓ Rest of the molecule

b. Five Peaks.✓

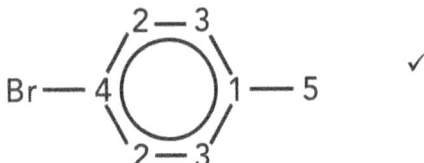

✓

c. 1-bromo-4-methylbenzene.✓ 2,2-dimethylbutane will only have four peaks in its carbon-13 NMR spectrum.✓

109.

a. Cyclopentane.✓

b. Spectrum (a).✓ Because cyclopentane has only one proton environment and therefore only one peak in its proton NMR spectrum.✓ (There is no spin-spin coupling because each proton is in the same environment as the adjacent protons and so produces the exact same signal)

(It also has only carbon environment and therefore its carbon-13 NMR spectrum would only have one peak. In the question, (a) is a proton NMR spectrum and (b) is carbon-13 NMR spectrum. This is evident from the x-axis of each graph - proton NMR spectra have a smaller scale.)

c.

✓

110.

a.

i. Propanone✓

ii.

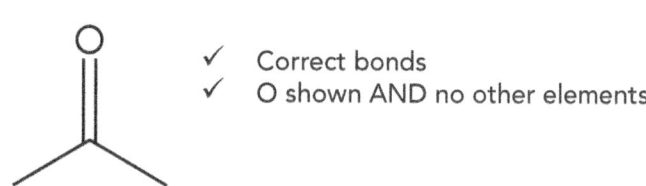

✓ Correct bonds
✓ O shown AND no other elements

b.

i.

Dipoles across C=O bond ✓
Curly arrows drawn correctly ✓
Lone pair of electrons on C and O ✓
Final product ✓
CN shown with - and lone pair ✓

ii. 2-hydroxy-2-methylpropanenitrile ✓

iii. Nucleophilic addition ✓

111.

a. Heptanoic acid ✓

b. Heptanal ✓

c. React the aldehyde with acidified potassium dichromate ✓. Heat the reaction mixture under reflux (this should either be stated or illustrated using the diagram).

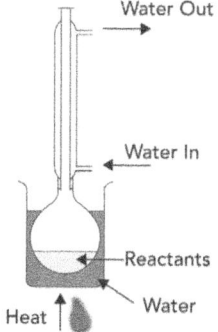

Reflux condensor ✓
Heat source - water bath not essential ✓
Round bottom flask ✓
Water in / out labelled ✓

6 ANSWERS — Organic Chemistry and Analysis

d.

i.

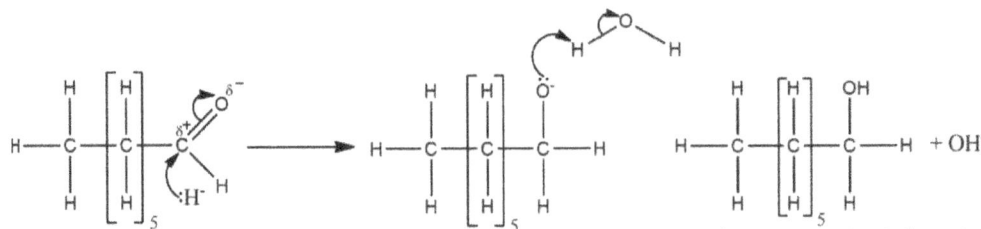

Dipoles across C=O bond ✓
Curly arrow from C=O to O and dipole ✓
Curly arrow from H⁻ to C ✓
Intermediete ✓

ii. Nucleophilic addition ✓

112.

a.

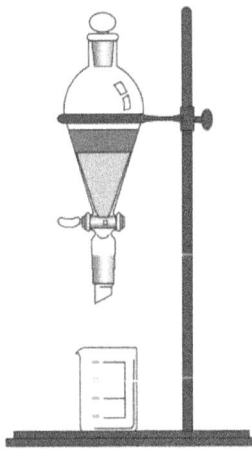

Separating funnel ✓
Beaker to collect liquid ✓

b. React the unknown sample with sodium carbonate ✓ and collect the gas produced ✓. Bubble this gas through lime water ✓. If the lime water turns cloudy, the sample is a carboxylic acid ✓.

113.

a.

i. Carbonyl ✓ and Nitrile ✓.

ii. React with 2,4-DNP. ✓ Yellow or orange precipitate ✓ will be observed due to the presence of the carbonyl function group.

b. Product from reaction 1:

Carboxylic acid group ✓
Rest of the molecule ✓

Product from reaction 2:

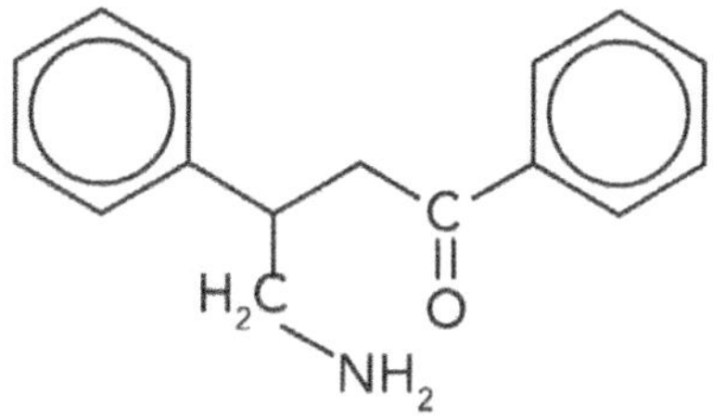

[Structure: 1,3-diphenyl-propan-1-one with CH₂NH₂ group on the carbon between phenyl and C=O]

Amino group ✓
Rest of the molecule ✓

c.

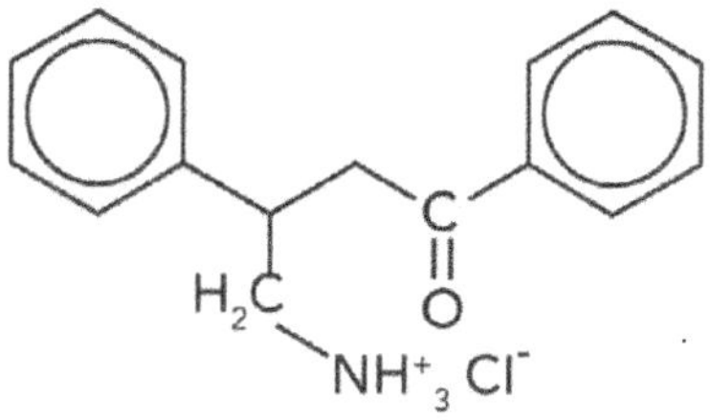

[Structure: same as above but with NH₃⁺Cl⁻ instead of NH₂] ✓

114.

a. Reaction 1

[Structure: 4-ethylacetophenone] ✓

Reaction 2

[Structure: 4-ethyltoluene] ✓

Allow substitution of CH₃ group at other positions.

b. i. HCl ✓   ii. Electrophilic substitution ✓

c.

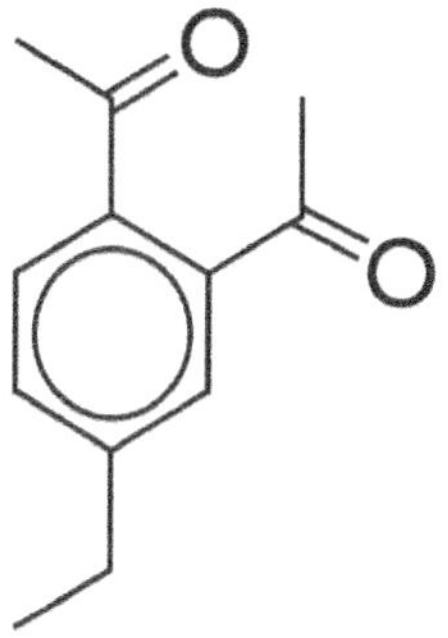

[Structure: benzene ring with two acetyl groups (1,2) and ethyl group at 4-position] ✓

Allow other valid structures

d. No reaction ✓   (2,4-DNP does not react with acyl chlorides).

ANSWERS — Organic Chemistry and Analysis

115. P =

¹H NMR spectrum

δ = 1.2 ppm, triplet, 3H CH3-CH2

δ = 2.3 ppm, quartet, 2H CH2-C=O

δ = 6.3 ppm, singlet, 2H N-H2

Supporting statements

δ = 6.3 ppm lost after D_2O, indicating N-H. (Since 2 protons are responsible for that peak, it rules out OH).

δ = 180 ppm in ¹³C NMR but no δ = 9-10 or 10-12 ppm in ¹H NMR so not aldehyde or ketone.

$M_r(NH_2COCH_3CH_3)$ = 73

For 5 marks: Structure correct **AND** all ¹H NMR signals analysed with at least two supporting statements.

For 3-4 marks: Structure has correct molecular formula **AND** C=O **AND** NH_2 but in incorrect positions **AND** at least two ¹H NMR signals analysed with at least one or two supporting statements.

For 1-2 marks: Structure has correct molecular formula **AND** C=O **OR** NH_2 but in incorrect positions **AND** at least one ¹H NMR signals analysed with no or one supporting statements.

116.

a. Intramolecular forces are stronger than intermolecular forces. ✓

b. There are two chiral centres in $CH_3CHBrCHClCH_3$. ✓

c. R_f value of different components vary and can be compared to data table values for identification. ✓

d. Deuterium in D_2O exchanges with H in OH and NH. ✓

e. 2.6 / 140.5 = 1.85 × 10⁻² mol **AND**

   1.85 × 10⁻² × 6.02 × 10²³ = 1.11 × 10²² ✓

f. Electrophilic substitution reactions maintain the stability of the benzene ring. ✓

g. Moles (of any gas) = V (in dm³) / 24

   24 x 0.05 = 1.2 dm³ **AND**

   Mass (g) of CO_2 = 44 x 0.05 = 2.2 g **AND** Mass (g) of NH_3 = 17 x 0.05 = 0.85 g ✓

h.
$$\frac{\frac{(400 \times 10^3) \times (2 \times 10^{-3})}{8.314 \times 25}}{\frac{(200 \times 10^3) \times (2 \times 10^{-3})}{8.314 \times 100}} = 8 \checkmark$$

(Allow alternative working)

(2 dm³ has been written as 2×10^{-3} in the above calculation as the SI units of V in PV = nRT is m³)

117.

a. Aldehyde ✓
   Rest of molecule ✓

b. HCN ✓

c. Curly arrow from C=O bond to O **AND** dipole ✓
   Curly arrow from CN to C ✓
   Intermediate **AND** curly arrow from O⁻ to H ✓

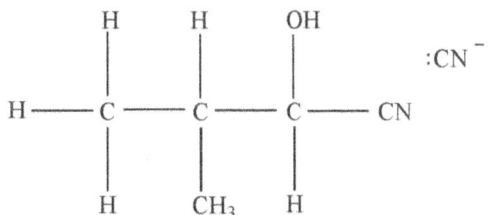

d. To monitor gases causing air pollution✓ and in modern breathalysers to measure ethanol in breath✓.

e. Moles of compound A used = 4.32 / 72 = 0.06 mol✓. (Molar ratio of the species are 1:1:1 in equation)

Say moles of compound C that needs to be produced = X

X / 0.06 = 0.80

X = 0.80 x 0.06 = 0.048 moles✓. (Therefore 0.048 moles of compound C needs to be produced for the percentage yield to be 80%).

Mass, in grams = 0.048 x 99 = 4.75 g✓. Therefore, if 4.75 g of compound C is formed, percentage yield = 80%.

(Notes: 72 is the molar mass of compound A. 99 is the molar mass of compound C. 80% = 0.80 as a decimal).

118. **Structure**

Mass spectrum

Molecular ion peak at m/z = 116, therefore M_r = 116 g mol^{-1}.

^{1}H NMR spectrum

δ = 1.2 ppm, triplet, 3H, CH_3-R

δ = 1.4 ppm, doublet, 6H, $(CH_3)_2$-R

δ = 2.3 ppm, quartet, 2H CH_2-C=O

δ = 4.1 ppm, multiplet, 1H HC-O

Supporting statements

δ = 172 ppm in ^{13}C NMR but no δ = 9-10 or 10-12 ppm in ^{1}H NMR so not aldehyde or ketone.

δ = 70 ppm in ^{13}C NMR, C-O

(Since C-O bond is present and as the compound is not an aldehyde or ketone, it must be an ester).

$M_r(C_3H_6O)$ = 58   116 / 58 = 2

Organic Chemistry and Analysis — ANSWERS

For 5-6 marks : Structure correct **AND** all ^{1}H NMR signals analysed with at least two supporting statements.

For 3-4 marks : Structure has correct molecular formula **AND** COO but in incorrect position **AND** at least two ^{1}H NMR signals analysed with at least one or two supporting statements.

For 1-2 marks : Structure has correct molecular formula **AND** at least one ^{1}H NMR signals analysed with no or one supporting statements.

119.

a. i. W:

[Structure showing: H-C(H)(H)-C(H)(H)-C(H)(H)-C(=O)-O-C(H)(H)-C(H)(H)-C(H)(H)-H] ✓

ii. Extract the product mixture. Shake and leave to settle in a separating funnel✓. Separate the layers by tapping off✓. Add some anhydrous $MgSO_4$ or $CaCl_2$ to the organic layer in a dry conical flask✓. Distil the organic layer✓

b. i. $CH_3CH_2CH_2COOH + CH_3CH_2CH_2OH \rightarrow CH_3CH_2CH_2COOCH_2CH_2CH_3 + H_2O$ (Species✓ Balancing✓)

ii. $n(CH_3CH_2CH_2COOH) = 4.00 / 88 = 0.04545$ ✓

Molar ratio of $CH_3CH_2CH_2COOH : CH_3CH_2CH_2OH$ is 1: 1 from the balanced equation. Therefore 0.04545 moles of butanoic acid would require 0.04545 moles of propanol✓.

Mass in grams of propanol required = 0.04545 × 60 = 2.73g ✓

iii. Concentration = moles / volume (dm^3)

5.00×10^{-2} = moles / 0.01

Moles = $5.00 \times 10^{-2} \times 0.01 = 5.00 \times 10^{-4}$ ✓

Mass (g) = $5.00 \times 10^{-4} \times 130 = 6.50 \times 10^{-2}$ g ✓

120.

a. i.

Product from step 1: ✓

[Benzene ring with CHO group and NO$_2$ group (meta-substituted)]

Reactant for step 1: ✓

[Benzene ring with CHO group]

6 ANSWERS — Organic Chemistry and Analysis

ii. Step 1: Mixture of concentrated nitric acid and concentrated sulfuric acid ✓.

Step 2: Mixture of tin and concentrated hydrochloric acid. ✓

b. i.

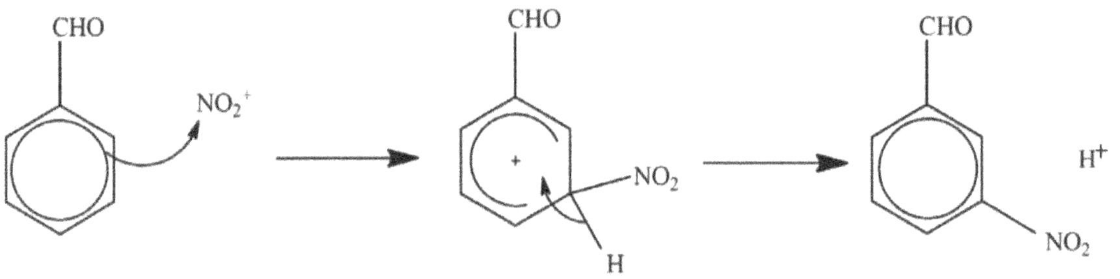

[Mechanism diagram: benzaldehyde + NO_2^+ → arenium ion intermediate with + charge and H shown → Compound A (3-nitrobenzaldehyde) + H^+]

Compound A

Curly arrow benzene ring to NO_2^+ ✓
Curly arrow from H bond to centre AND + charge shown ✓
H^+ shown AND NO_2 in correct position ✓

ii. Electrophilic substitution ✓

c. Yellow / orange precipitate ✓ when 2,4-DNP is added to compound A ✓.

121.

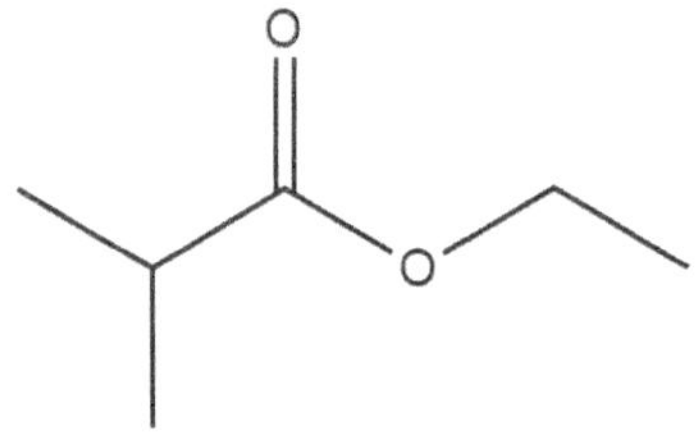

[Structure: ethyl 2-methylpropanoate / ethyl isobutyrate]

Molecular ion peak at m/z = 116, therefore M_r = 116 g mol^{-1}.

^{1}H NMR spectrum

δ = 1.2 ppm, doublet, 6H, $(CH_3)_2$-R

δ = 1.3 ppm, triplet, 3H, CH_3-R

δ = 2.7 ppm, multiplet, 1H HC-C=O

δ = 4.2 ppm, quartet, 2H H_2C-O

Supporting statements

	C	H	O
	62.1%	10.3%	27.6%
Divide by molar mass	5.175	10.3	1.725
Divide by the smallest no. (ie. 1.725)	3	6	1

*Therefore the empirical formula: C_3H_6O

*$M_r(C_3H_6O) = 58$ $116 / 58 = 2$ Hence molecular formula: $C_6H_{12}O_2$

For 5-6 marks: Structure correct **AND** all ^1H NMR signals analysed with at least two supporting statements.

For 3-4 marks: Structure has correct molecular formula **AND** COO but in incorrect position **AND** at least two ^1H NMR signals analysed with at least one or two supporting statements.

For 1-2 marks: Structure has correct molecular formula **AND** at least one ^1H NMR signals analysed with no or one supporting statements.

Note: This is an example of a question which is marked by the level of response. 6 marks are available for such questions.

122.

a. Forms a carbon to carbon bond with a benzene ring.✓

b. i. Alkylation: ✓   Nitration: ✓   Hydrogenation: ✓   Oxidation: ✓

ii. React with a mixture of propanoyl chloride✓ and aluminium chloride (catalyst)✓. Heat the mixture to about 60 °C for about 3 minutes.✓

iii. H_2O ✓

6 ANSWERS Organic Chemistry and Analysis

iv.

C₆H₅COOH + SOCl₂ → C₆H₅COCl + SO₂ + HCl

reactants ✓ products ✓

123.

a. Ethyl cyanide. ✓

b. Warm the compound with sodium hydroxide ✓ solution in a mixture of ethanol and water. Then add dilute nitric acid. Now add silver nitrate solution ✓. A cream precipitate ✓ will be observed if the sample is bromoethane.

c. i. $CH_3CH_2Br + KCN \rightarrow CH_3CH_2CN + KBr$ (reactants ✓ products ✓)

ii.
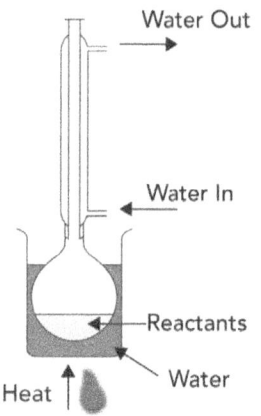

Reflux condensor ✓
Heat source - water bath not essential ✓
Round bottom flask ✓
Water in / out labelled ✓

iii.
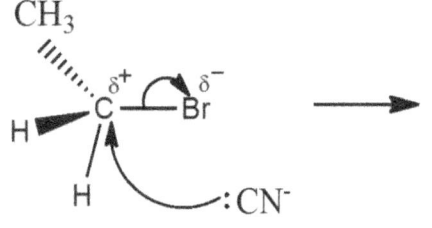

Lone pair on CN⁻ ✓
dipoles labelled correctly ✓
Correct curly arrows ✓
Br⁻ ion ✓

iv. Nucleophilic substitution. ✓

d. To ensure even heating of the mixture. ✓

124.

a.

One mark for each optical isomer ✓✓✓✓

[Four stereoisomer structures shown with wedge/dash notation of 2-bromopentane enantiomers]

b. Five ✓

c. i. **Step 1**: Heat 1-bromopentane under reflux with a solution of sodium / potassium cyanide in ethanol. ✓

$CH_3CH_2CH_2CH_2CH_2Br + NaCN \rightarrow CH_3CH_2CH_2CH_2CH_2CN + NaBr$ ✓

Step 2: Heat the nitrile under reflux with dilute hydrochloric acid. ✓

$CH_3CH_2CH_2CH_2CH_2CN + 2H_2O + HCl \rightarrow CH_3CH_2CH_2CH_2CH_2COOH + NH_4Cl$ ✓

ii. Nucleophilic substitution ✓

iii.

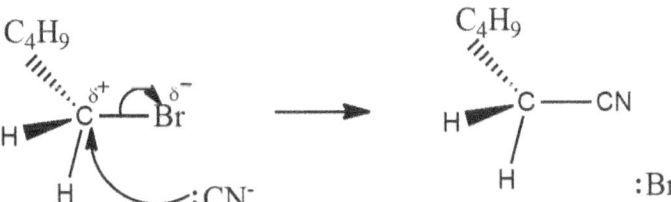

Lone pair on CN⁻ ✓
dipoles labelled correctly ✓
Correct curly arrows ✓
Br⁻ ion ✓

iv. 71: $CH_3CH_2CH_2CH_2CH_2^+$ ✓   59: CH_2COOH^+ ✓

125.

a. $RCH(NH_2)COOH$ ✓

b. Serine: CH_2OH ✓   aspartic acid: CH_2COOH ✓

c.

[Structure showing aspartic acid with protonated amine (NH₃⁺) and nitrate counter-ion (NO₃⁻), both carboxylic acid groups as -OH] ✓

d. i.

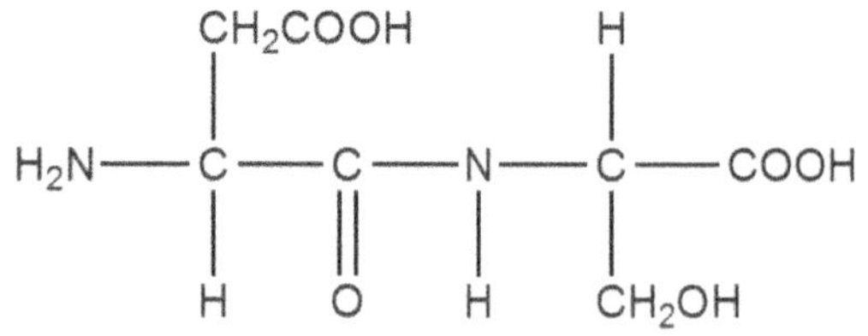

[Structure 1: H₂N—C(CH₂COOH)(H)—C(=O)—N(H)—C(H)(CH₂OH)—COOH]
Peptide linkage ✓
Rest of the molecule ✓

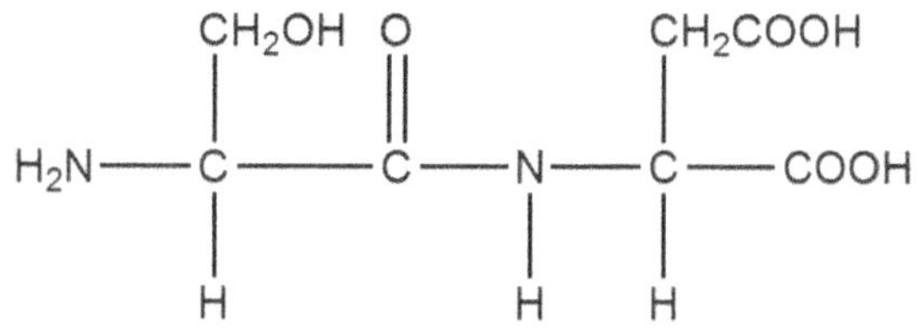

[Structure 2: H₂N—C(CH₂OH)(H)—C(=O)—N(H)—C(H)(CH₂COOH)—COOH]
Peptide linkage ✓
Rest of the molecule ✓

ii.

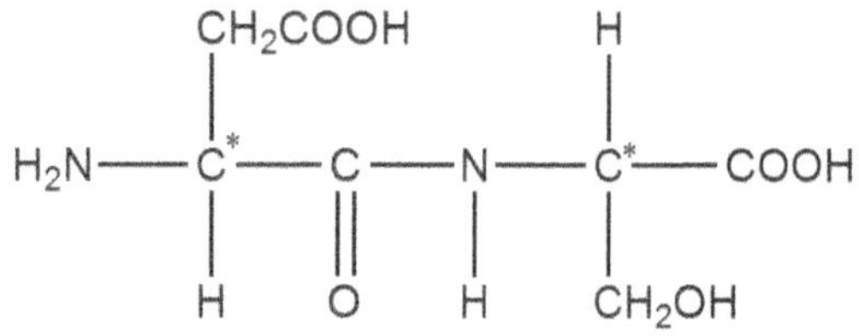

[Structure: H₂N—C*(CH₂COOH)(H)—C(=O)—N(H)—C*(H)(CH₂OH)—COOH] ✓✓

iii. Four ✓

126.

a.

Major organic product ✓
Rest of the equation ✓

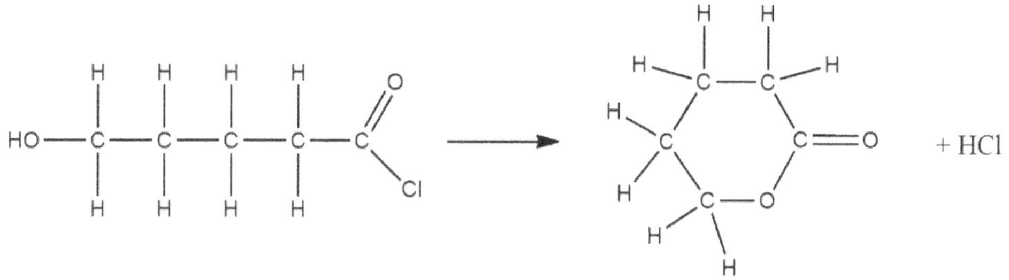

HO—CH₂—CH₂—CH₂—CH₂—C(=O)Cl → [cyclic lactone: 6-membered ring with O and C=O] + HCl

b.

[Repeating unit: —CH₂—CH₂—CH₂—CH₂—C(=O)—O—] ✓

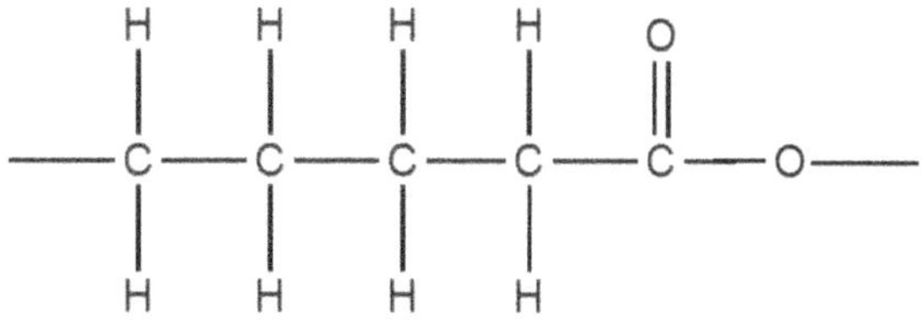

Polyester ✓   Condensation polymerisation ✓

c. **Step 1:** React 5-hydroxypentanoyl chloride with benzene in the presence of an aluminium chloride catalyst ✓

C_6H_6 + $HOCH_2CH_2CH_2CH_2COCl$ → $HOCH_2CH_2CH_2CH_2COC_6H_5$ + HCl ✓

Step 2: React the resulting compound under reflux with acidified potassium dichromate ✓

$HOCH_2CH_2CH_2CH_2COC_6H_5$ + 2[O] → $HOOCCH_2CH_2CH_2COC_6H_5$ + H_2O ✓

(OR vice-versa)

ANSWERS 6

127.

(structure: methylmalonic acid — HOOC–CH(CH₃)–COOH drawn with central C bonded to H and to a CH₃ group, flanked by two COOH groups)

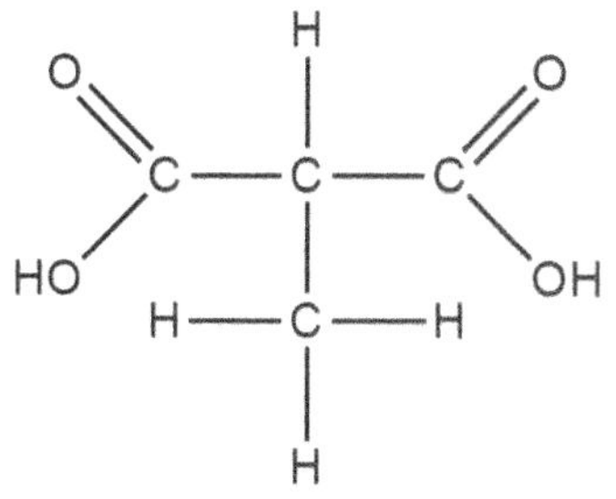

Mass spectrum

Molecular ion peak at m/z = 118, therefore M_r = 118 g mol^{-1}.

¹H NMR spectrum

δ = 2.8 ppm, quartet, 1H, HC-C=O

δ = 0.9 ppm, doublet, 3H, CH$_3$-CH

Supporting statements

	C	H	O
	40.7	5.08	54.2
Divide by molar mass	3.39	5.08	3.39
Divide by the smallest no. (ie. 3.39)	1	1.5	1
x2	2	3	2
Empirical formula: C$_2$H$_3$O$_2$			

δ = 172 ppm in ^{13}C NMR but no δ = 9-10 in ^{1}H NMR so not aldehyde. As C-O peaks not present, not an ester.

M_r(C$_2$H$_3$O$_2$) = 59   118 / 59 = 2. Therefore molecular formula: C$_4$H$_6$O$_4$

For 5-6 marks: Structure correct AND all ^{1}H NMR signals analysed with at least two supporting statements.

For 3-4 marks: Structure has correct molecular formula AND two COOH groups but in incorrect positions AND at least one ^{1}H NMR signals analysed with at least one or two supporting statements.

For 1-2 marks: Structure has correct molecular formula AND at least one ^{1}H NMR signals analysed with no or one supporting statements.

128.

a. **Mixture 1**

R_f value = distance travelling by spot / distance travelled by the solvent front

P: 2.50 / 5.52 = 0.45 (Tyrosine)✓   Q: 2.2/5.52 = 0.40 (Cysteine)✓

Mixture 2:

X: 1.90/4.74 = 0.4 (Cysteine)✓   Y: 3.46/4.74 = 0.73 (Leucine)✓   Z: 1.28/4.74=0.27 (Serine)✓

b. R_f value = distance travelling by spot / distance travelled by the solvent front

Distance travelled by aspartic acid = 0.24 x 7.23 = 1.74 cm✓

c. Easier to perform - does not require specialist equipment.✓

d. Concentration of components in the mixture.✓

129.

a. Peak at 10.77 min = Malathion. Peak at 11.49 min = Dicofol. Peak at 19.15 min = Tetradifon. Peak at 26.83 min = Azoxystrobin. (all four correct ✓✓✓ three correct✓✓ two correct✓)

Now plot a calibration curve using the data provided.

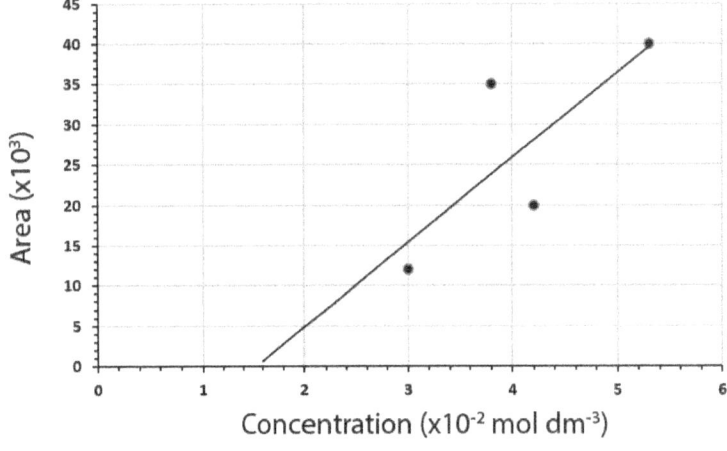

Axes labelled correctly ✓
Points correct✓
Line of best fit ✓

The calibration curve can now be used to work out the concentration of Malathion in the mixture.

Malathion = 0.039 mol dm^{-3} ✓

b. 37000 ✓   (2 sf✓)

130.

a. **Step 1:** $CH_3CH_2CH_2OH + 2[O] \rightarrow CH_3CH_2COOH + H_2O$. (Species✓ Balancing✓)

Heat the reaction mixture under reflux✓.

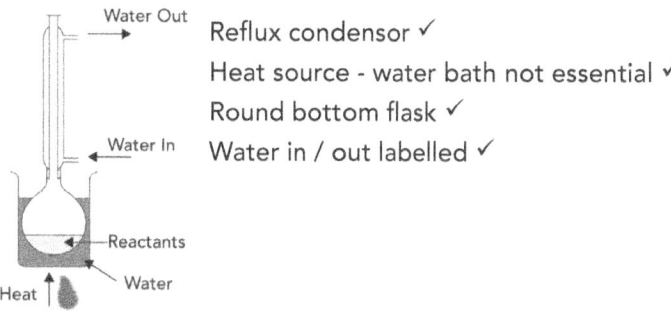

Reflux condensor ✓
Heat source - water bath not essential ✓
Round bottom flask ✓
Water in / out labelled ✓

Step 2: $CH_3CH_2COOH + SOCl_2 \rightarrow CH_3CH_2COCl + SO_2 + HCl$ (Species✓ Balancing✓)

b.

$$H-\underset{\underset{H}{|}}{\overset{\overset{H}{|}}{C}}-\underset{\underset{H}{|}}{\overset{\overset{H}{|}}{C}}-\overset{\overset{O}{\|}}{C}-O-\overset{\overset{O}{\|}}{\underset{\underset{}{\|}}{S}}-Cl$$

CH_3CH_2COO ✓
Rest of the structure ✓

131.

The two amino acids making up the dipeptide are serine✓ and valine.✓

It cannot be mixture 2 as mixture 2 has both serine and valine.✓

Peak integration values indicate the ratio of concentrations of the components in the mixture (or the statement implied)✓

Simplifying the peak integration values gives the ratio 12:27:110✓

(This can be worked out by dividing the integration values by 2000 as it is the smallest numer to give the following- 1: 2.25 :9.15. Multiplying these numbers by 12 gives the number as a ratio in whole numbers as 12:27:110).

Therefore it should either be mixture 3 or 4.✓

It cannot be mixture 4 as an amino acid with retention time of 34.0 is not one of the possible options.✓

It is mixture 3✓. The amino acids are threonine, glycine and valine.✓

132.

a. i. Butyl pentanoate✓

ANSWERS Organic Chemistry and Analysis

   ii. Ester ✓

b.

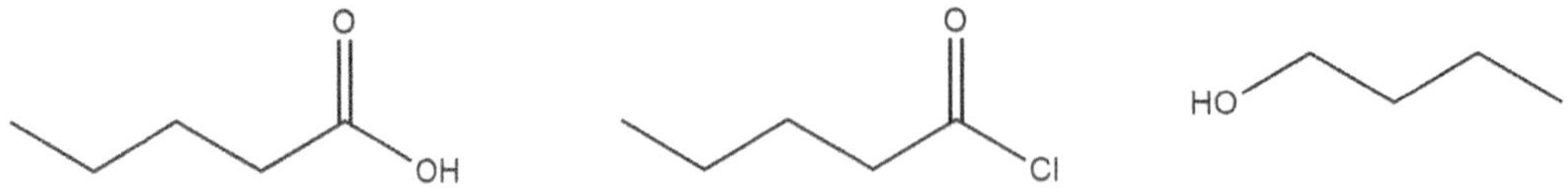

P: Pentanoic acid

Q: Pentanoyl chloride

R: Butanol

Three names ✓✓   Two names ✓
Three structures ✓✓   Two structures ✓
Name and structure ✓

c. i. $CH_3CH_2CH_2CH_2COOCH_2CH_2CH_2CH_3 + KOH \rightarrow CH_3CH_2CH_2CH_2COOK + HOCH_2CH_2CH_2CH_3$

(potassium salt ✓ butanol ✓)

   ii. Hydrolysis (alkaline) ✓

133.

a. Alkene ✓ and phenol ✓

b. i.

Major product ✓   Rest of the equation ✓

+ NaOH ⟶ ... + H_2O

   ii. No reaction ✓.

c. i.

✓

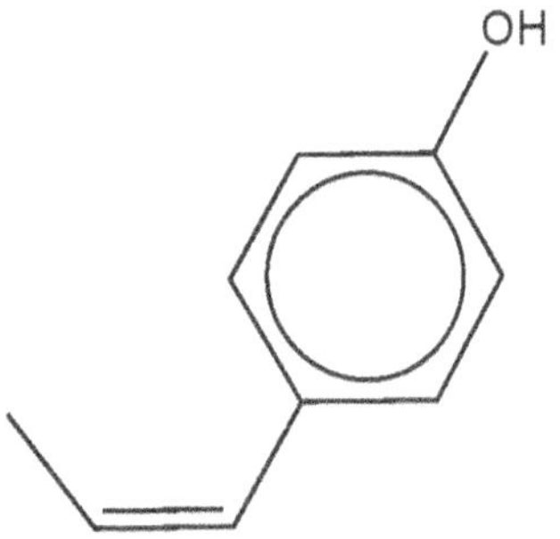

   ii. E/Z isomerism ✓

   iii. Z isomer ✓. The phenol group and the CH_3 group attached to the carbons of the C=C group are on the same side. ✓ (Note that this type of isomerism exists because of the presence of the C=C double bond. There is a CH_3 group and a H atom attached to one carbon and a phenol group and a H atom attached to the other. When the groups with "highest priority" are on the same side it is called the *Z* isomer and if not it is an *E* isomer. Remember, *Z* for *zame!*).

134.

a. i.

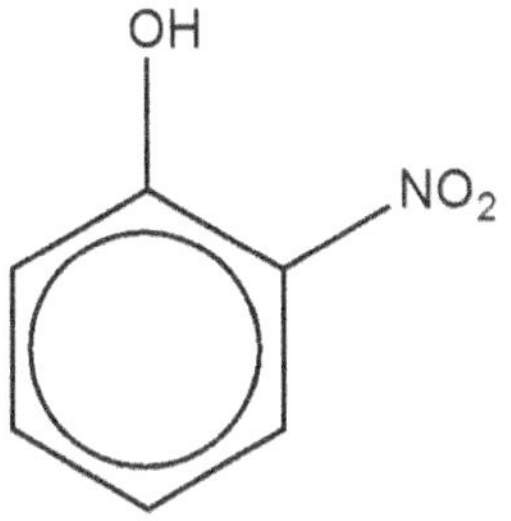

[2-nitrophenol structure ✓] [4-nitrophenol structure ✓]

ii. OH is an electron donating group ✓ and therefore has a 2- and 4- directing effect ✓. (This is why the NO_2 group has been substituted at the 2- and 4- positions).

b. Electrophilic substitution ✓. NO_2^+, the electrophile, substitutes one of the hydrogen atoms in the ring ✓.

135.

a. i. [benzene-1,3-dicarbaldehyde structure ✓]

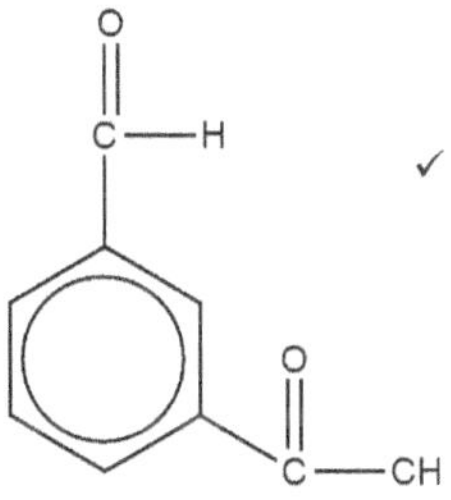

ii. [3-nitrobenzoic acid structure ✓]

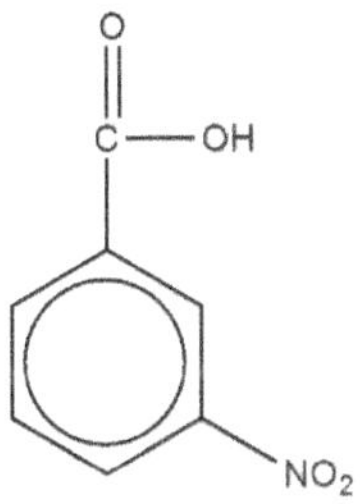

iii. [4-nitrotoluene and 2-nitrotoluene structures ✓✓]

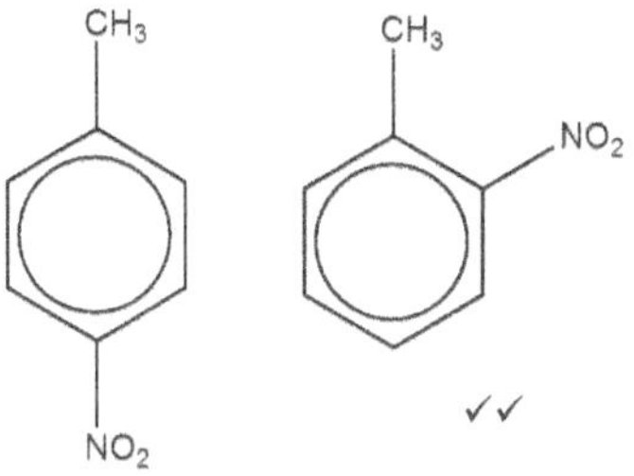

b. i. Electron-donating groups have 2- and 4- directing effects ✓. Electron-withdrawing groups have a 3-directing effect ✓.

ii. Organic synthesis. ✓

c. CN ✓ (CN⁻ can attack carbonyl compounds (aldehydes and ketones) to form hydroxynitriles.)

136.

a. Skeletal ✓

b. $C_{19}H_{24}N_2$ ✓

c. 11 ✓

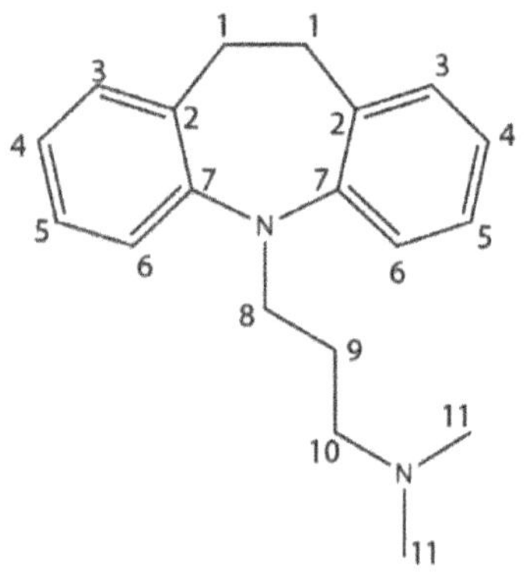

[Structure of imipramine with numbered carbons: two benzene rings fused to a central 7-membered ring containing N, with a propyl chain ending in N(CH₃)₂. Carbons labelled 1,1,2,2,3,3,4,4,5,5,6,6,7,7,8,9,10,11,11] ✓

d. The lone pair of electrons on the nitrogen atom ✓ (s) accepts protons ✓.

137.

a. i.

Major product ✓
Water formed ✓
Balancing ✓

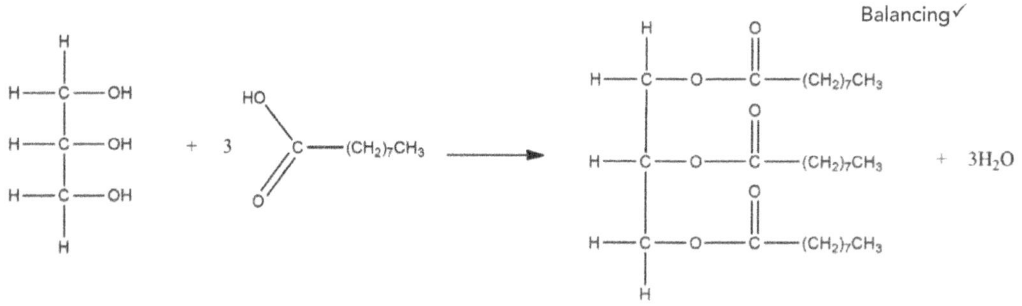

[Reaction: glycerol + 3 CH₃(CH₂)₇COOH → triester (tri-nonanoyl glycerol) + 3H₂O]

ii. Esterification ✓

iii. $CH_3(CH_2)_7COCl$ ✓

b. i.

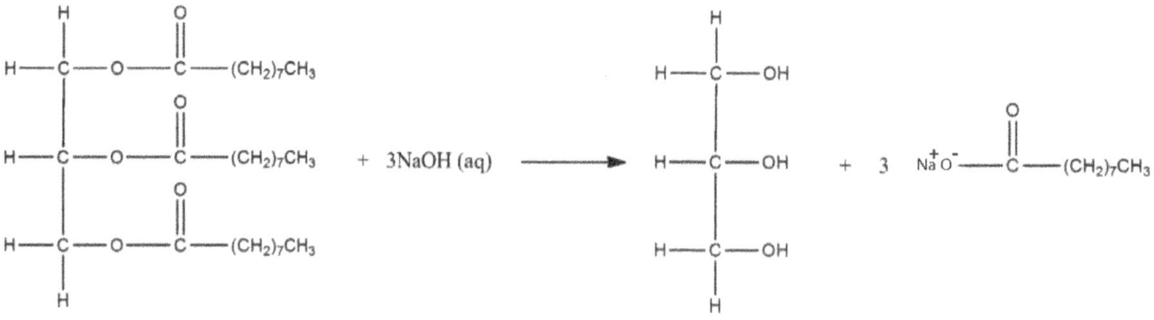

[Reaction: triester + 3NaOH(aq) → glycerol + 3 Na⁺⁻O-CO-(CH₂)₇CH₃]

Sodium salt ✓
Glycerol ✓
Balancing ✓

ii. Hydrolysis ✓

138.

a.

Two glycine molecules ✓
Cu²⁺ ✓
Dative bonds shown with arrow ✓
- Charge on O of COO ✓
Lone pairs of electrons ✓

b. i.

Reactants ✓
Products ✓

ii.

✓ ✓

139.

a.

✓ ✓

b.

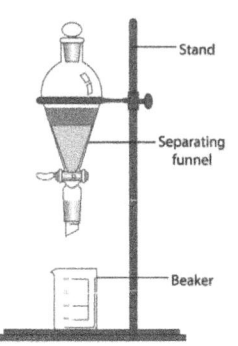

✓✓✓
Lose marks for missing:
Separating funnel
Beaker
Stand
Labels

Compound M is organic. The other product from the reaction is water (aqueous)✓. The two are immiscible✓.

c. Moles = mass (g) / molar mass = 2.56 / 156 ✓ = 0.0164 ✓ mol

ANSWERS

Organic Chemistry and Analysis

140.

a. 4-hydroxylphenyl butanoate ✓ $C_{10}H_{12}O_3$ ✓

b. i.

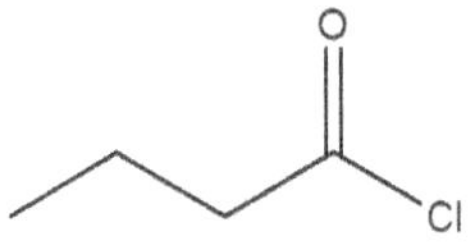

Butanoyl chloride 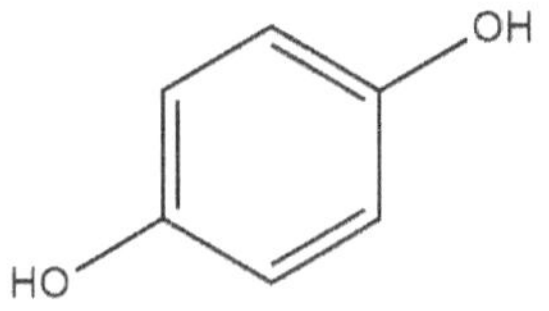 ✓✓   4-hydroxyphenol ✓✓

Note that phenols are **not** readily esterified by carboxylic acids and hence acyl chlorides are used.

ii. Add acidified dichromate ✓ (acidified potassium or sodium dichromate). When added to 4-hydroxyphenol, there will be a colour change from orange to green ✓.

iii. Moles of compound W = 4.20 / 180 = 0.02333 moles ✓.

Reactants ✓
Products ✓

Ratio is 1:1:1:1 in the equation. So 0.02333 moles of each reagent is required ✓.

Butanoyl chloride: 0.02333 x 106.5 = 2.49 g ✓

4-hydroxyphenol: 0.02333 x 110 = 2.57 g ✓

141.

a. Phenol ✓ and dilute nitric acid ✓.

b. OH is an electron-donating group ✓ and has a 2 and 4-directing effect ✓. 4-nitrophenol ✓ may also be produced.

c. i.

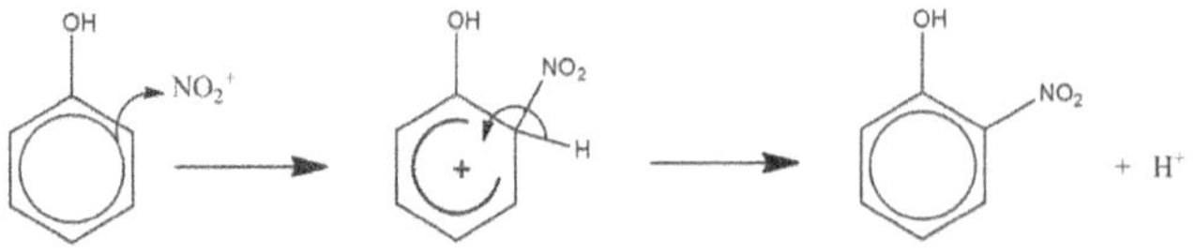

Curly arrows ✓
Intermediate ✓
H^+ formed ✓

ii. Electrophilic substitution ✓

142.

a. React butan-2-one ✓ with a mixture of NaCN or KCN and HCl ✓.

$CH_3COCH_2CH_3 + HCN \rightarrow CH_3C(OH)(CN)CH_2CH_3$ (reactants ✓ products ✓)

Organic Chemistry and Analysis — ANSWERS

b. i.

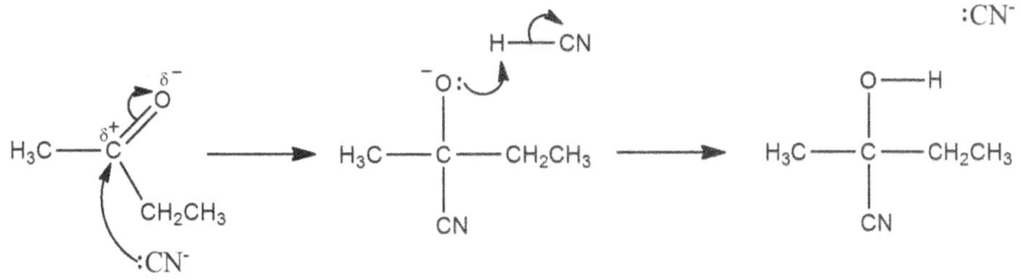

Curly arrows ✓
Dipoles ✓
Intermediate ✓
CN^- formed ✓

ii. Nucleophilic addition ✓.

c. i.

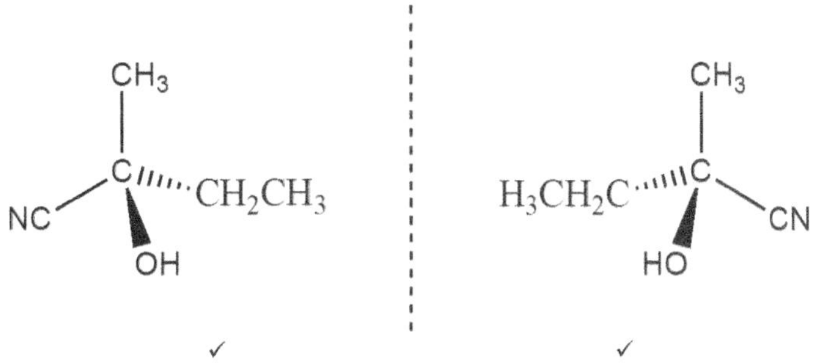

✓   ✓

ii. Optical isomerism ✓

143.

a. ✓

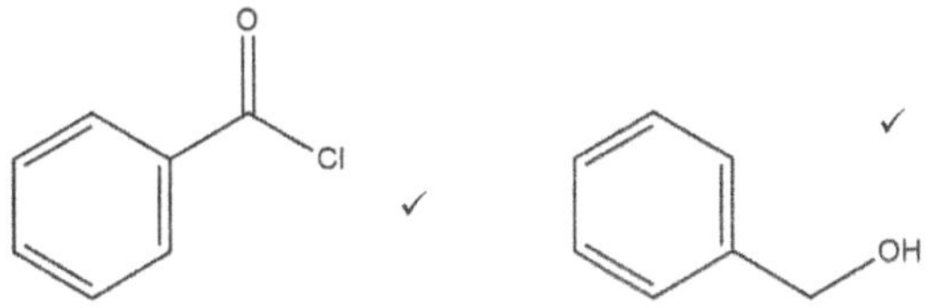

b. ✓ ✓

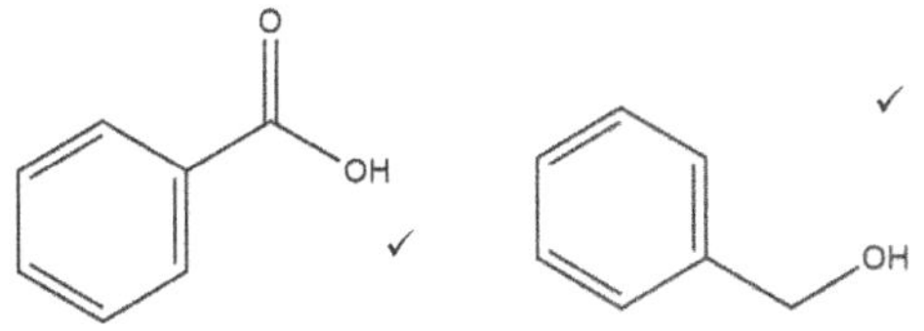

c. **Step 1:**

React benzyl alcohol ($C_6H_5CH_2OH$) with acidified potassium dichromate ✓.

+ 2[O] →

reactants ✓
products ✓

ANSWERS
Organic Chemistry and Analysis

Step 2:

React benzoic acid (C$_6$H$_5$COOH) with thionyl chloride (SOCl$_2$)✓.

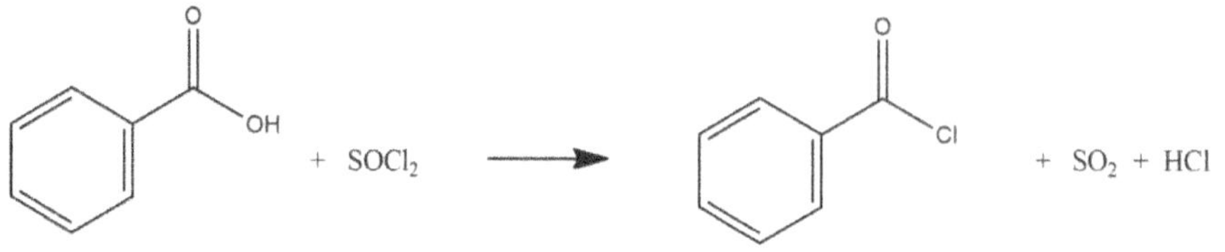

C$_6$H$_5$COOH + SOCl$_2$ ⟶ C$_6$H$_5$COCl + SO$_2$ + HCl

reactants✓
products✓

d. i. 10✓

ii.

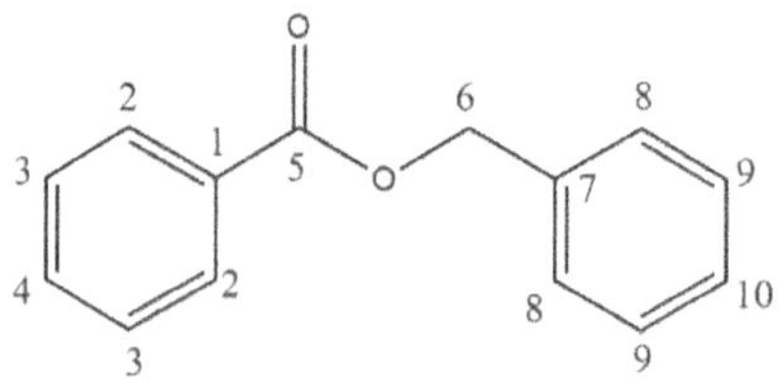

All 10 correct✓✓
5 correct✓

144.

a.

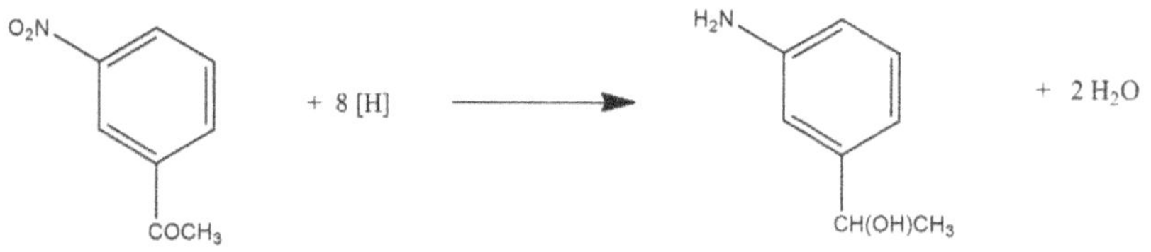

O$_2$N–C$_6$H$_4$–COCH$_3$ + 8 [H] ⟶ H$_2$N–C$_6$H$_4$–CH(OH)CH$_3$ + 2 H$_2$O

reactants✓
products✓
balancing✓

b. i.

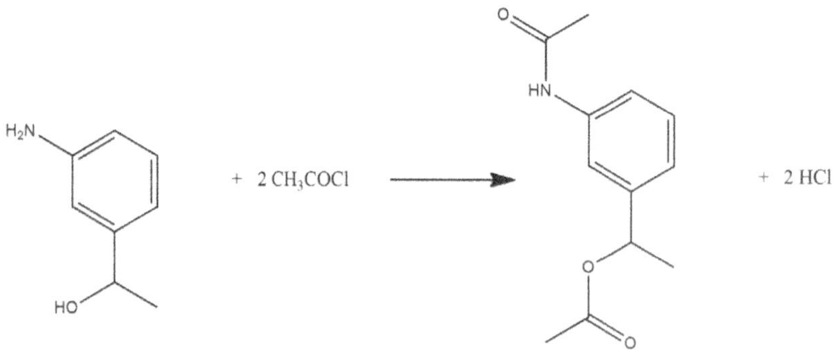

H$_2$N–C$_6$H$_4$–CH(OH)CH$_3$ + 2 CH$_3$COCl ⟶ (N-acetyl, O-acetyl product) + 2 HCl

reactants✓
products✓
balancing✓

ii. Moles = 0.20 / 137 = 0.00146✓   As the ratio of the major reactant to major product is 1:1, there will be 0.00146✓ moles of product G.   Theoretical yield in grams = moles × molar mass = 0.00146 × 221 = 0.323✓ g

As the percentage yield is, 53.2%, the actual yield = 0.323 × 0.532 = 0.17 g (answer✓ 2 s.f.✓)

The answer should be given to two significant figures as the mass (0.20g) given in the question is given to 2 s.f. So you cannot be any more accurate than 2 s.f..

145.

a.

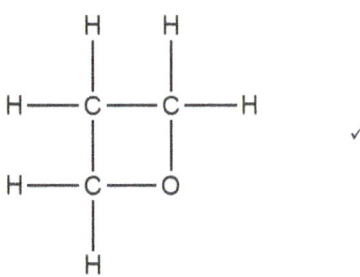

b. Molecular ion peak is at m/z = 58 as the compound has a molecular mass of 58 g mol⁻¹. ✓

$C_3H_6O \rightarrow C_3H_6O^+ + e^-$ ✓ m/z of position ion formed = 58 (molecular ion peak)

$C_3H_6O^+ \rightarrow C_2H_4^+ + CH_2O•$ ✓ m/z of position ion formed = 28

$C_3H_6O^+ \rightarrow CH_2O^+ + C_2H_4•$ ✓ m/z of position ion formed = 30

c. $(CH_2)_2$-O protons will produce a triplet as the adjacent carbon has two protons ✓

CH_2-CH_2-CH_2 protons will produce a multiplet(pentet) as the adjacent carbon has four protons ✓

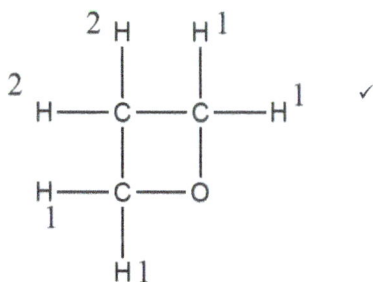

d. 2 ✓

146.

a. Peak at 1.2 min = Hexane ✓ Peak at 2.1 min = Octane ✓. Peak at 6.8 min = Decane ✓.

b. Hexane is present in the mixture and calibration data for hexane is also provided in the table. Now plot a calibration curve using the data provided in Table 2.

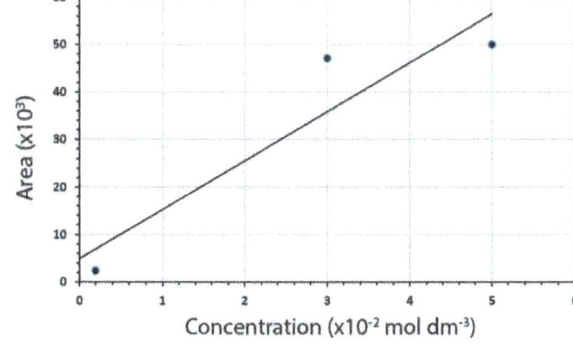

Axes labelled correctly ✓
Points plotted correctly ✓
Line of best fit ✓

The calibration curve can be used to work out the concentration of hexane in the mixture.

Hexane = 0.016 mol dm^{-3} ✓

147.

a. **Step 1**

React ethanoic acid with thionyl chloride✓.

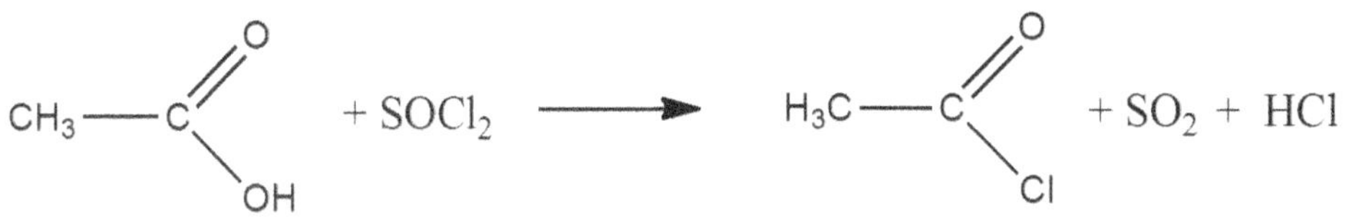

$$CH_3-C(=O)OH + SOCl_2 \longrightarrow H_3C-C(=O)Cl + SO_2 + HCl$$

reactants✓
products✓

Step 2 (Note that this is Friedel-Crafts acylation)

React ethanoyl chloride with naphthalene in the presence of a halogen carrier such as $AlCl_3$✓. The reaction mixture has to be heated at 60 °C for about 30 minutes✓.

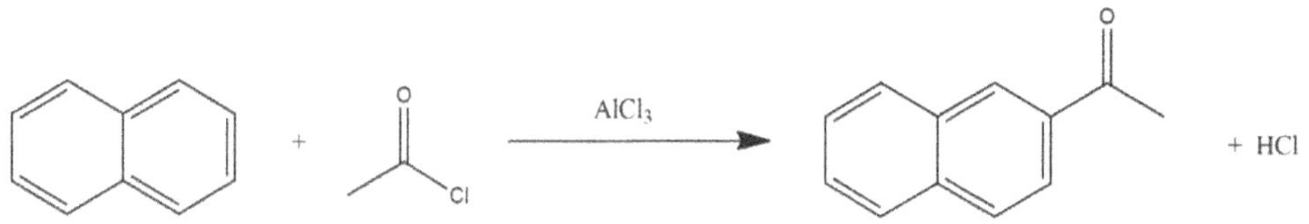

naphthalene + ethanoyl chloride →(AlCl$_3$) 1-(2-naphthyl)ethanone + HCl

reactants✓
products✓

b. Moles of ethanoic acid = 2.3 / 60 = 0.03833 moles✓. Molar ratio of ethanoic acid to 1-(2-naphthyl)ethanone between the two equations is 1:1.

The theoretical yield of 1-(2-naphthyl)ethanone in moles = 0.03833✓. This is equivalent to 0.03833 × 170 ✓ = 6.52g✓.

Let the mass in grams, required to get a percentage yield of 67% be Y:

(Y/6.52) × 100 = 67.   Therefore Y = 0.67 × 6.52 = 4.37g✓.

148.

a. Molecular ion peak is at m/z = 72 as the compound has a molecular mass of 72 g mol^{-1}.✓

$C_3H_4O_2 \rightarrow C_3H_4O_2^+ + e^-$✓   m/z of position ion formed = 72✓

$C_3H_4O_2^+ \rightarrow C_2H_3^+ + CHO_2$·✓   m/z of position ion formed = 27✓

$C_3H_4O_2^+ \rightarrow C_3H_3O^+ + OH$·✓   m/z of position ion formed = 55✓

b. i.

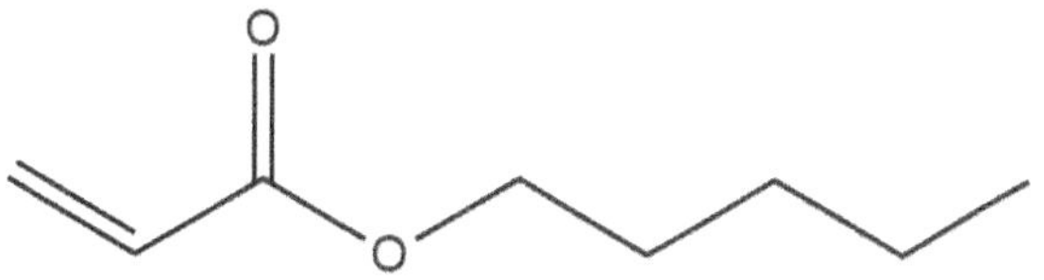

[structure: pentyl acrylate] ✓

ii.

[structure: sodium acrylate] ✓   HO—[pentyl chain] ✓

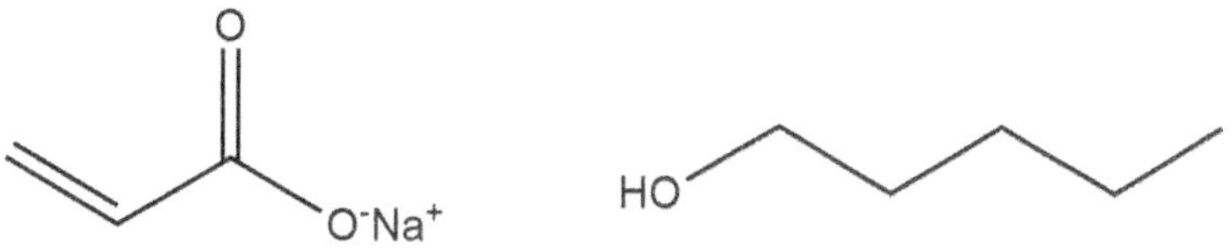

149.

a. $2CH_3CH_2COOH + Na_2CO_3 \rightarrow 2CH_3CH_2COO^-Na^+ + CO_2 + H_2O$ (✓✓ lose a mark for each mistake)

b. $2CH_3COOH + Mg \rightarrow (CH_3COO^-)_2Mg^{2+} + H_2$ (✓✓ lose a mark for each mistake)

c. $MgO + 2CH_3CH_2CH_2COOH \rightarrow (CH_3CH_2CH_2COO^-)_2Mg^{2+} + H_2O$ (✓✓ lose a mark for each mistake)

d. $Ca(OH)_2 + 2CH_3COOH \rightarrow (CH_3COO^-)_2Ca^{2+} + 2H_2O$ (✓✓ lose a mark for each mistake)

150.

a.

Reaction 1

[propanoic acid] + $SOCl_2$ → [propanoyl chloride] + SO_2 + HCl   reactants ✓ products ✓

Reaction 4

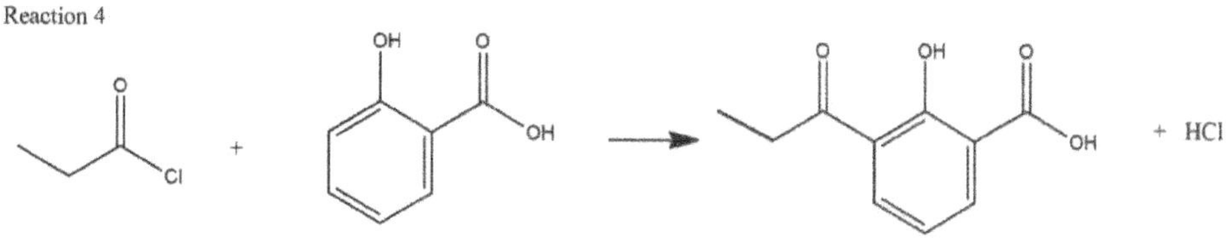

[propanoyl chloride] + [salicylic acid] → [acylated salicylic acid] + HCl   reactants ✓ products ✓

Reaction 4 requires a halogen carrier such as $AlCl_3$ ✓. The reaction mixture has to be heated at 60 °C for about 30 minutes ✓.

b.

Reaction 2 [N-propyl propanamide]   Reaction 5 [pentyl propanoate]   ✓✓✓✓

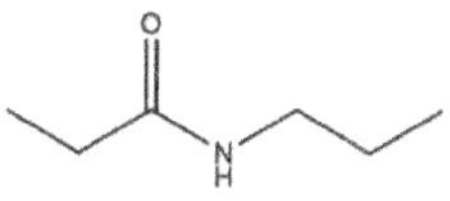

Reaction 3 [propanamide]   Reaction 6 [propanoic acid]

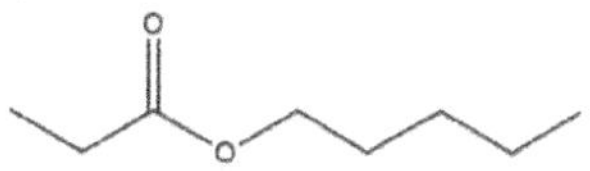

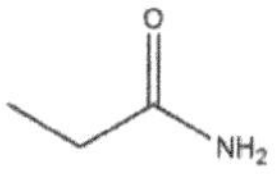

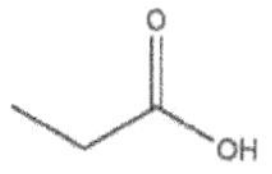

ANSWERS   Organic Chemistry and Analysis

c. Nucleophilic addition ✓

151.

a.

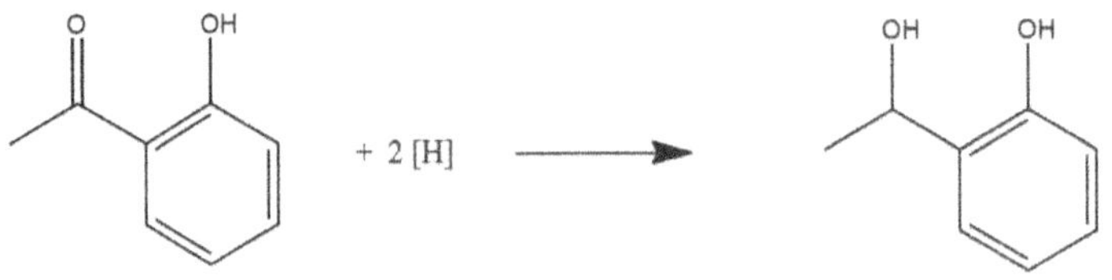

reactants ✓
products ✓

b.

Curly arrows ✓
Dipoles ✓
Intermediate ✓
OH⁻ formed ✓

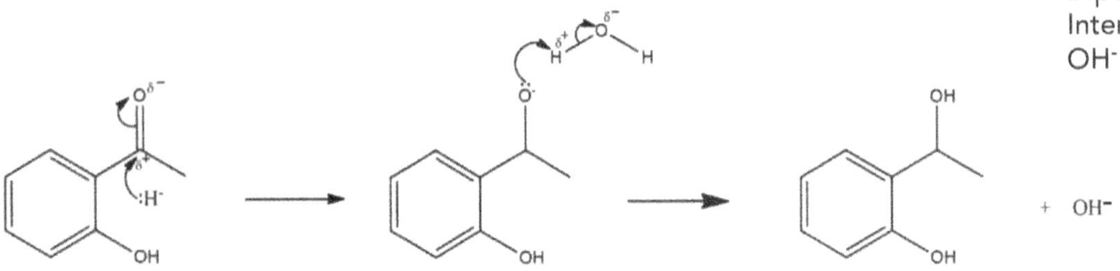

Nucleophilic addition ✓

152.

a. (Alpha) amino acid ✓

b. i. ✓   ii. ✓   iii. ✓

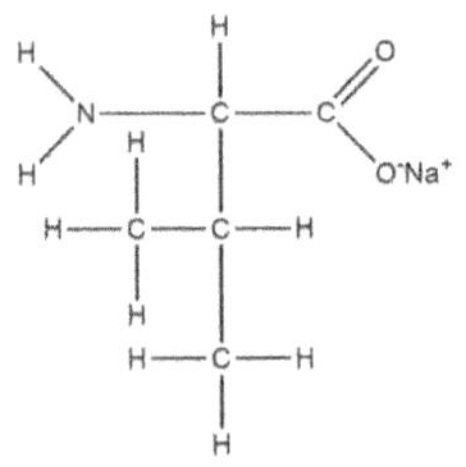

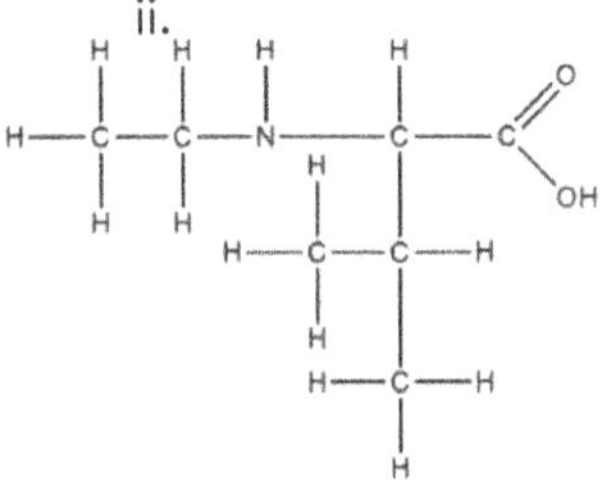

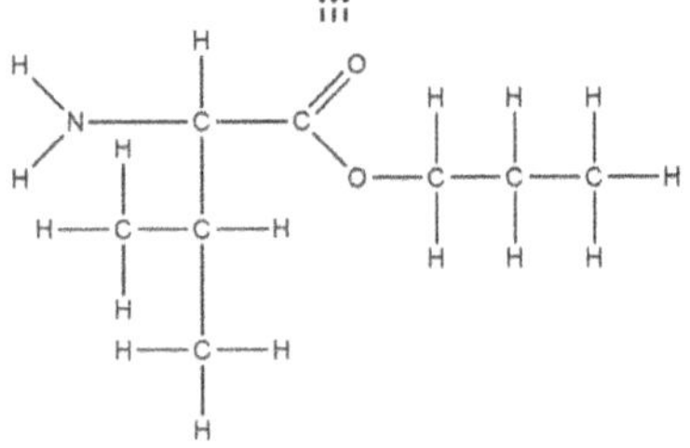

c. i. ✓

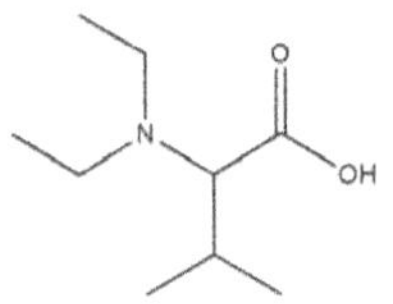

ii. 2 ✓ HBr ✓

d.

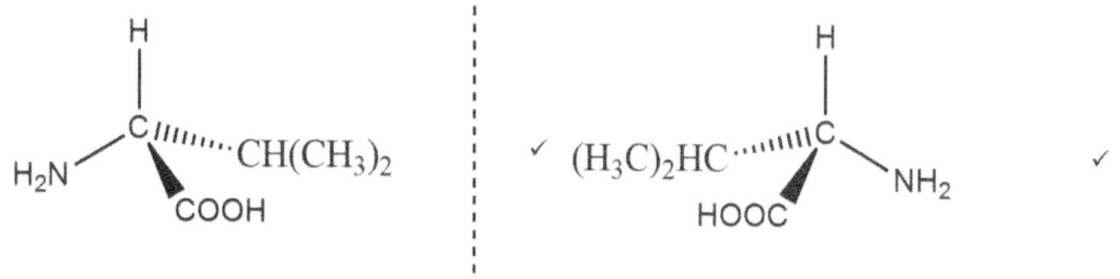

[Structures of two enantiomers of valine: H₂N-C(H)(COOH)-CH(CH₃)₂ and its mirror image (H₃C)₂HC-C(H)(NH₂)-COOH] ✓

153.

a. Reduce using a mixture of tin and concentrated hydrochloric acid✓, heated under reflux✓. The excess acid is neutralised after.

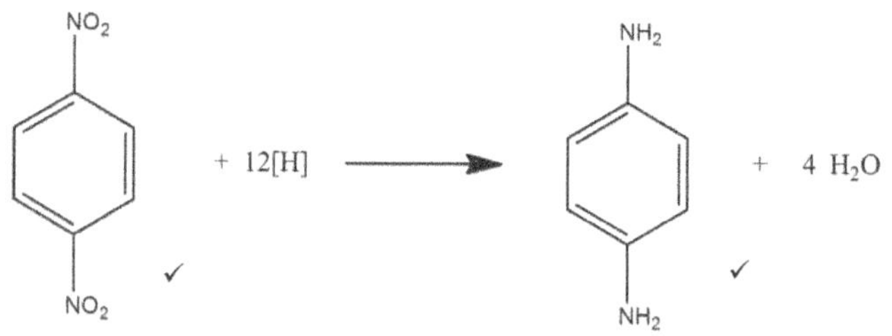

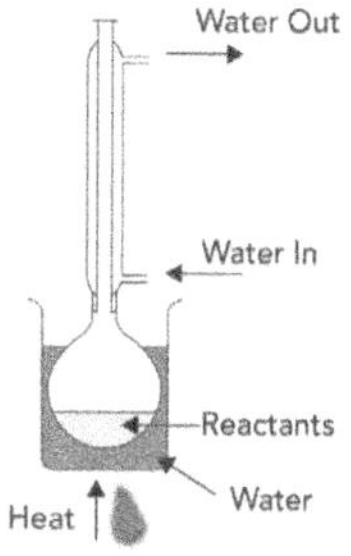

[Equation: 1,4-dinitrobenzene + 12[H] → benzene-1,4-diamine + 4 H₂O] ✓ balancing✓

Reflux condensor ✓
Heat source - water bath not essential ✓
Round bottom flask ✓
Water in / out labelled ✓

b. Benzene-1,4-diamine✓. Lone pair of electrons on the nitrogen atoms in benzene-1,4-diamine✓ is delocalised into the ring✓ and is therefore less available for protonation✓.

154.

a. $C_{18}H_{20}N_2O_4$ ✓

b. Using Avogadro's constant, $1.0 \times 10^{18} / 6.02 \times 10^{23}$ ✓ $= 1.66 \times 10^{-6}$ ✓ moles. Mass $= 1.66 \times 10^{-6} \times 328$ ✓ $= 5.4 \times 10^{-4}$ g ✓.

c. Amino acids. ✓

155.

a. Condensation / Polyamide ✓

b.

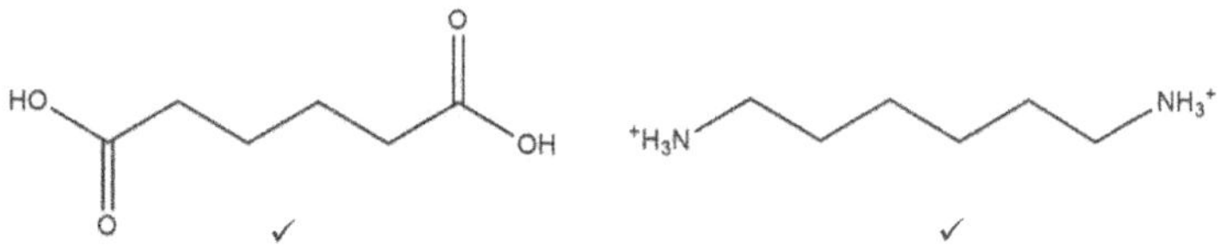

[Structures: hexanedioic acid (HOOC-(CH₂)₄-COOH) ✓ and protonated 1,6-diaminohexane (⁺H₃N-(CH₂)₆-NH₃⁺) ✓]

156.

a. Polyester ✓

ANSWERS Organic Chemistry and Analysis

b.

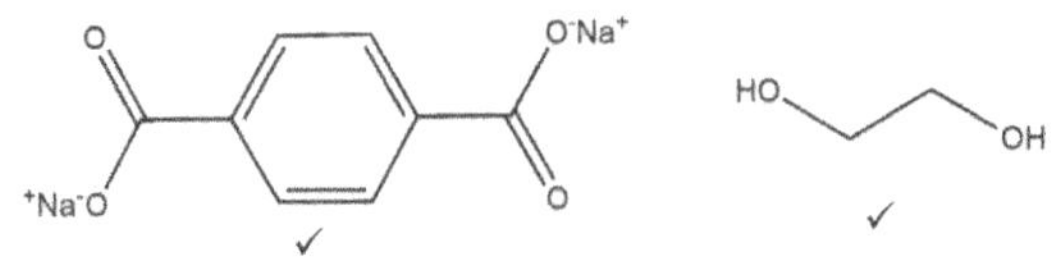

157.

a.

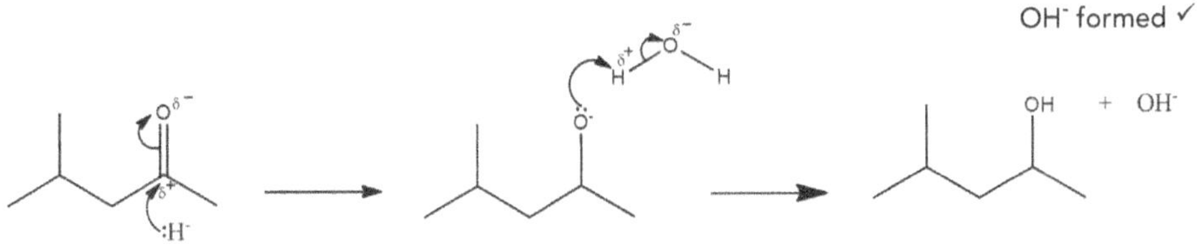

b. Nucleophilic addtion ✓

158.

a. Heat 1-chloropentane under reflux ✓ with a solution of sodium or potassium cyanide ✓ in ethanol ✓.

b.

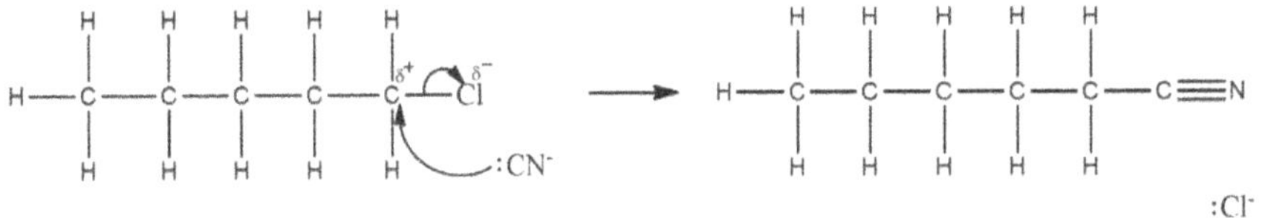

Curly arrows ✓
Dipoles ✓
Cl⁻ formed ✓

159.

a.

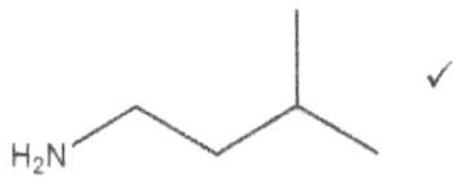

b. React 3-methylbutanenitrile/$CH_3CH(CH_3)CH_2CN$ with hydrogen gas ✓ in the presence of a Ni catalyst ✓.

$CH_3CH(CH_3)CH_2CN + 2H_2 \rightarrow CH_3CH(CH_3)CH_2CH_2NH_2$ ✓

160. React hexanoic acid with thionyl chloride ✓.

$CH_3CH_2CH_2CH_2CH_2COOH + SOCl_2 \rightarrow CH_3CH_2CH_2CH_2CH_2COOCl + SO_2 + HCl$ ✓✓

161.

a.

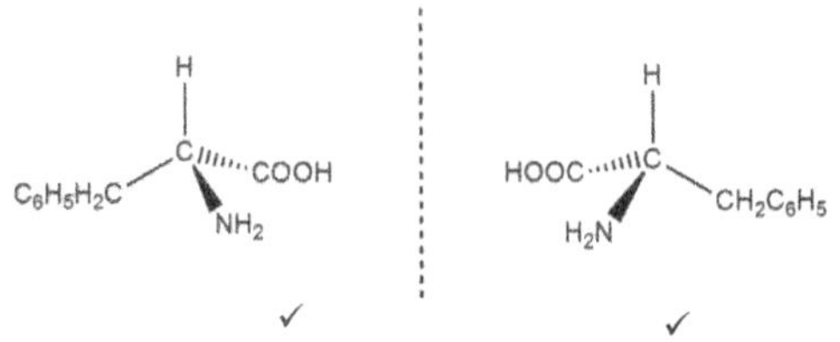

b. Optical isomerism ✓

162.

a. Heat under reflux✓ with a mixture of tin and concentrated HCl✓, followed by neutralisation of the excess acid.

$$\text{O}_2\text{N-naphthalene-NO}_2 + 12\,[\text{H}] \longrightarrow \text{H}_2\text{N-naphthalene-NH}_2 + 4\,\text{H}_2\text{O} \checkmark$$

b. Filter solution✓. Recrystallise✓

c. Three bonding pairs and one lone pair **AND** therefore shape: pyramidal ✓ with a bond angle of 107°✓.

163. Extract the product mixture. Shake and leave to settle in a separating funnel✓. Separate the layers by tapping off✓. Add some anhydrous $MgSO_4$ or $CaCl_2$ to the organic layer in a dry conical flask✓. Distil the oganic layer and collect the fraction distilling at 98 °C✓.

Reflux condensor ✓
Heat source - water bath not essential ✓
Round bottom flask ✓
Water in / out labelled ✓

6 mark questions like this will be marked based on the level of response. That is: Level 3 (5-6 marks), level 2 (3-4 marks) and Level 1(1-2 marks). If your diagram lacks labels for example, your answer will fall into Level 1.

164.

a.

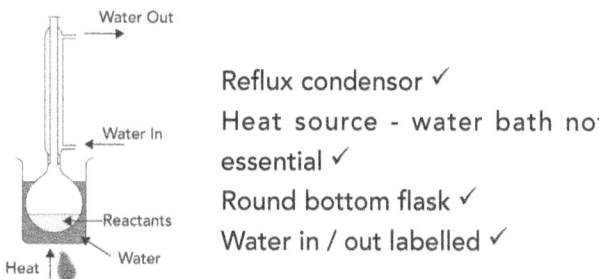

Reactants ✓
Products ✓

b. Moles = mass(g) / molar mass = 2.00 / 138✓ = 1.45 x 10⁻² mol✓.

6 ANSWERS Organic Chemistry and Analysis

c.

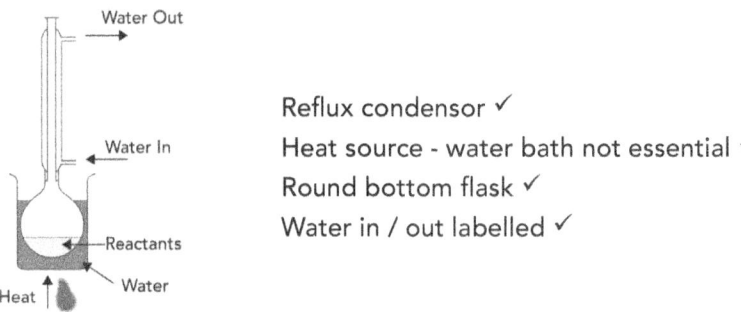

Reflux condensor ✓
Heat source - water bath not essential ✓
Round bottom flask ✓
Water in / out labelled ✓

d. Catalyst ✓

e.

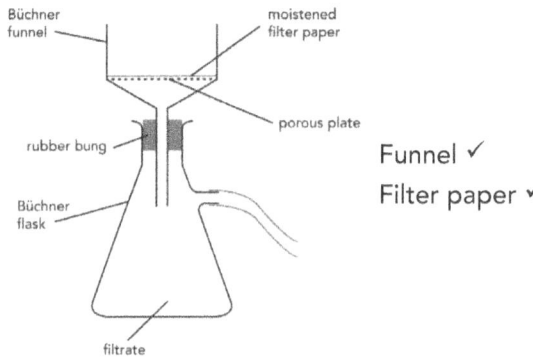

Funnel ✓
Filter paper ✓

f. Put some aspirin into the sealed end of a capillary tube ✓. Make sure it is densely packed / compacted in the capillary tube ✓. Now place it in a melting point apparatus ✓. Heat it very slowly between 115-135 °C ✓. Record the temperature at which aspirin melts ✓. Pure aspirin melts at 136 °C.

g. Moles of aspirin = 1.20 / 180 = 6.67×10^{-3} mol ✓. Moles of salicylic acid = 1.45×10^{-2} mol.

Theoretical yield of aspirin = 1.45×10^{-2} mol (Molar ratio of salicylic acid to aspirin from the balanced equation is 1:1).

Actual yield obtained = 6.67×10^{-3} mol

Percentage yield = $(6.67 \times 10^{-3} / 1.45 \times 10^{-2}) \times 100$ ✓ = 46.0% ✓

Answer scores full marks. Allow error carried forward - **ECF**. (In these type of questions, if you made an error in a previous part of the question and you used your calculation from that part in your current working, you will still get the marks for your new calculation).

165.

a.

[Structural equation: H₃C–C₆H₄–CH₂–C≡N + 2 H₂ →(Ni) H₃C–C₆H₄–CH₂–CH₂–NH₂]

[Structural equation: H₃C–C₆H₄–CH₂–CHCl– + NH₃ ✓ → H₃C–C₆H₄–CH₂–CH₂–NH₂ + HCl]

b.

166. **Step 1:** React 2-methylbutanenitrile with hot aqueous hydrochloric acid (acid hydrolysis). **Step 2:** React the carboxylic acid formed with $SOCl_2$ to form the acyl chloride.

167.

a. $CH_3CH(CH_3)_2CH_2OH + [O] \rightarrow CH_3CH(CH_3)_2COH + H_2O$ ✓   Distil ✓ the alcohol with acidified potassium dichromate ✓. Distil the product formed at 74 °C ✓.

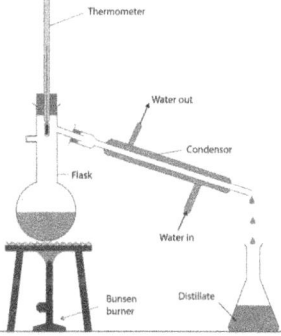

Condenser ✓
Heat source ✓
Round bottom flask ✓
Water in / out labelled ✓
Thermometer ✓

b. Redistil the solution ✓

168. **Step 1:** add HCN OR H_2SO_4 / KCN ✓ $CH_3CH_2CHO + HCN \rightarrow CH_3CH_2CH(OH)CN$ ✓

   Step 2: React with H_2 / Ni ✓ $CH_3CH_2CH(OH)CN + 2H_2 \rightarrow CH_3CH_2CH(OH)CH_2NH_2$ ✓

169.

   a. Electrophilic substitution ✓

   b. Nucleophilic substitution ✓

   c. Nucleophilic addition ✓

   d. Electrophilic substitution ✓

ANSWERS
Organic Chemistry and Analysis

170.

a. amino acid ✓

b. i.

(structure of glutamate monosodium salt: HOOC-CH2-CH2-CH(NH2)-COO⁻Na⁺) ✓

ii. Filter solution ✓ and recrystallise ✓

171. **Step 1:** React with hot aqueous HCl (or other acid) ✓   **Step 2:** React with ethanol ✓

Step 1:

CH3-CH(CH3)-CH2-CH2-C≡N + 2H2O + HCl ⟶ CH3-CH(CH3)-CH2-CH2-COOH + NH4Cl ✓

Step 2:

CH3-CH(CH3)-CH2-CH2-COOH + CH3-CH2-OH ⟶ CH3-CH(CH3)-CH2-CH2-COO-CH2-CH3 + H2O ✓

172.

a. 4-methylpentanoyl chloride ✓

b. **Step 1:** React with acidified potassium dichromate. Heat under reflux ✓

Step 2: React with thionyl chloride ✓

Step 1:

(CH3)2CH-CH2-CH2-CH2-OH + 2[O] ⟶ (CH3)2CH-CH2-CH2-COOH + H2O ✓

Step 2:

(CH3)2CH-CH2-CH2-COOH + SOCl2 ⟶ (CH3)2CH-CH2-CH2-COCl + SO2 + HCl ✓

c. The molar ratios among the equations of 4-methylpentan-1-ol to 4-methylpentanoic acid to 4-methylpentanoyl chloride is 1:1:1.

Moles of **compound K** needed = 5 / 134.5 = 0.037175✓. Provided there was 100% yield, 0.037175 moles of 4-methylpentanoic acid need to react. However, as the yield is only 73%, 0.037175 x (100 / 73) = 0.050924 moles of 4-methylpentanoic acid would be needed✓.

As the percentage yield of the first step is 53%, 0.050924 x (100 / 53) = 0.096084 moles of 4-methylpentan-1-ol is required✓. This given in grams = 0.096084 x 102 = 9.8 g (answer✓ 2 s.f.✓).

173.

a. Moles of hexanoic acid = 0.2 x 0.05 = 0.01 mol✓

The equations tell us that the ratio of hexanoic acid : carbon dioxide is 2:1.

Therefore, 0.01 mol of hexanoic acid would form 0.005 mol of carbon dioxide✓.

Volume (dm^3) = mol x 24 = 0.005 x 24 = 0.12 dm^3 ✓ = 120 cm^3 ✓.

b. hexan-1-ol✓

c. Hexanal gets oxidised to hexanoic acid✓. Silver ions are reduced to silver.✓

Answers to MCQ

1. **D** There is a bond angle of 120° around each carbon. It is clearly not 270°.

   (A) An aldehyde group can be oxidised using an oxidising agent such as acidified dichromate ions.

2. **B** The areas of the peaks in proton-NMR give information about the proportions of protons in each environment. Carbon-13 NMR spectrum of phenol will show four peaks.

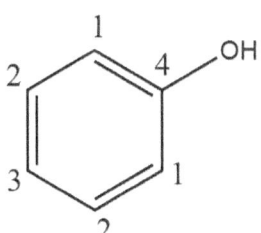

3. **B** This is an ester formed from the reaction between ethanoic acid and propanol.

4. **D** Figure 2 has a broad peak around 2500-3300 cm^{-1} which is a result of the O-H group of carboxylic acids.

5. **C** (D: When bromine water reacts with phenol, a white precipitate of 2,4,6-tribromophenol is formed)

6. **D** A primary alcohol is one where the carbon atom to which the OH group is attached is bonded only to one carbon atom. Likewise, in a tertiary alcohol, the carbon atom to which the OH group is attached is bonded to three other carbon atoms.

ANSWERS

7. **A** This compound is a ketone (the carbon atom of carbonyl group is on the 2nd carbon) with a methyl group on the 3rd carbon. The position of the carbonyl carbon has to be included in the name for a ketone unlike in aldehydes. In aldehydyes, the carbonyl carbon is always at the end.

8. **A** (3) Standard conditions here refer to 298k, 100 kPa pressure and 1 mol dm^{-3} concentration. Benzene reacts in the presence of a strong electrophile in electrophilic substitution reactions and this does not require standard conditions.

9. **B** This compound has 8 carbon atoms in total, 6 of which are in the benzene ring. There are 4 hydrogen atoms in the benzene ring.

10. **A** The carbon atom to which the CH_3 group is attached is a chiral centre as it has 4 different atoms or groups of atoms attached. A compound with a chiral centre will have optical isomers.

11. **C** A yellow or orange precipitate will only form when Brady's reagent is added to an aldehyde or ketone.

    Although carboxylic acids have the C=O group, there will be no precipitate formed when Brady's reagent is added to a carboxylic acid.

12. **A** When Brady's reagent is added to an aldehyde or ketone, it will result in an orange precipitate. Ammoniacal silver nitrate can be used to distinguish between these two, which when added to an aldehyde will result in a "silver mirror". Pentanal is the only aldehyde among the four options.

13. **B** Hydrolysis of an ester using sodium hydroxide will form the constituent alcohol and the sodium salt of the carboxylic acid. Butyl pentanoate is made from the alcohol butanol and carboxylic acid pentanoic acid. Sodium salt of pentanoic acid is sodium pentanoate.

14. **C** (3) *E/Z* isomerism and optical isomerism are both types of stereoisomerism. E/Z isomers are not optical isomers.

15. **A** When hexanal reacts with $NaBH_4$ (sodium tetrahydridoborate(iii)), hexan-1-ol will be formed. Hexan-1-ol is a primary alcohol. $NaBH_4$ can be used to reduce aldehydes and ketones to primary and secondary alcohols respectively.

16. **B**

17. **C** Adsorption is the process by which a solid holds molecules on its surface.

18. **D**

19. **D**

20. **B**

21. **A** In substitution reactions of benzene, one or more hydrogen atoms are replaced by a different atom or groups of atoms. This way the stability of the ring is preserved.

22. **A** Benzene has the molecular formula C_6H_6. In chlorobenzene, one of the hydrogen atoms has been replaced with a chlorine atom.

23. **B** Molecular weight of empirical formula = 12 + 1 = 13

    Molecular weight / 13 = 26 / 13 = 2
    2 x CH : C_2H_2

24. **D**

ANSWERS 6

Organic Chemistry and Analysis

25. **D** The peak closer to 200 ppm is due to the carbon from a C=O group. The two peaks between 120 and 140 ppm could be due to C=C carbons. The molecular mass has to be higher than 46, because four peaks means there are at least four carbon atoms.

26. **A** Different isotopes of an element have different atomic weights because they have different numbers of neutrons. (C) Stereoisomerism only applies to molecules and not individual elements like hydrogen.

27. **D** Although benzene and cyclohexane both have six carbon atoms, they are all in the same environment. The proton NMR spectra of both cyclohexane and benzene will only show one peak as their protons are in the same environment. However the chemical shifts of the peaks will be different.

28. **C**

29. **D**

30. **B**

31. **C**

32. **A** When pharmaceuticals are developed, it is important to make sure that the drug contains only one optical isomer because the other optical isomer of the same compound may cause unwanted reactions in the patient.

33. **A**

    This compound is used to reduce aldehydes and ketones to primary and secondary alcohols respectively.

34. **D**

35. **B** Hexan-3-one is a ketone. When it is reduced using $NaBH_4$, a secondary alcohol (hexan-3-ol) is formed along with a hydroxide ion.

36. **A** This is a skeletal formula. The "line ends" or "vertices" represent one carbon atom each unless there is a different atom at the end of the line. In this case there are 5 carbon atoms in the main chain and 1 sticking out from the 2nd carbon atom (the "carboxylic group" carbon atom is always numbered 1).

37. **D** The correct formula should be $CH_3CH_2CH_2CH_2NH_2$.

38. **B**

39. **D** Only aldehydes can be oxidised to carboxylic acids. In the production of aldehydes, that is why it is important to remove the aldehyde from the reaction mixture as they are formed in the oxidation of primary alcohols. This prevents further oxidation to carboxylic acids.

40. **A** Aromatic amines are amines with a ring structure such as benzene. Aliphatic amines on the other hand are chained structures. To prepare an aliphatic amine, halogenoalkanes are warmed gently with an excess of ammonia. Ethanol is used as the solvent.

41. **A**

42. **A** This is because the deuterium in D_2O exchanges with the H in OH and NH. Deuterium, unlike hydrogen does not produce a signal in proton NMR spectra.

43. **C**

ANSWERS

44. A (Nitriles are reduced to form primary amines).

45. B

46. B (Note that in acylation, almost all substitution happens in the 4th position. In 3, the same compound is drawn in a different orientation).

47. A (There are three proton environments). The molar mass of the compound is 96.0 g mol^{-1}.

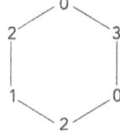

48. C (The major organic product from the reaction is 1,3-dinitrobenzene. Its carbon-13 NMR spectrum would have 4 peaks).

49. D (Refer to the data sheet. Carbon atoms in two different environments are bonded to oxygen atoms. The carbon atom in a different environment is bonded exclusively to two carbon atoms. Even though there are four carbon atoms, there are only three carbon environments because it is a symmetrical molecule).

50. B   Proton environments   Carbon environments

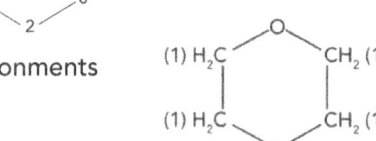 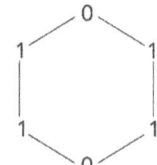

51. A (Friedel-Crafts reaction)

52. C (To answer this question, R_f value of component A needs to be calculated using chromatograph 1. If we say the R_f value of component A is X, and the distance moved by the solvent front in chromatograph 2 is Y, the distance moved by component A in chromatograph 2 would be equal to X times Y.)

53. D (OH and NH_2 are electron-donating groups. NO_2 is an electron-withdrawing group.)

54. A (NO_2 is an electron-withdrawing group and they have a "3-directing effect". That is, the "new" nitro group will be substituted on to the third carbon atom.)

55. D. The relative peak areas are equal to the **ratio of the number of protons** in each environment. That is, it is not always directly equal to the number of protons in that environment.

56. C. Friedel-Crafts reaction. Acyl chloride with aromatic compound (benzene) in the presence of a halogen carrier. The reaction mixture has to be heated.

57. A. OH has 2- and 4- directing effects.

58. B. Pentanoyl chloride, sulfur dioxide and hydrochloric acid are the products. HCl reacts with NH_3 to form NH_4Cl. Molecular mass of pentanoyl chloride is 120.5 g mol^{-1}.

Organic Chemistry and Analysis — ANSWERS

59. **A.** Carbonyl compounds react with a mixture of NaCN or KCN and HCl to form hydroxynitriles.

60. **A.** Acyl chlorides react with primary amines to form secondary amides.

61. **A** Even though there are two C=C bonds, one of each carbon in the two groups are bonded to the same atom (H). The atoms / groups have to be different for stereoisomers to exist.

62. **B** They can use acidified potassium dichromate to convert an alcohol (butan-1-ol) to a carboxylic acid (butanoic acid). So they would only need butan-1-ol and ethanol.

63. **D** Phenol reacts with sodium hydroxide to give a colourless solution but not with magnesium carbonate. Propanoic acid reacts with magnesium carbonate giving off carbon dioxide. Chloroethane reacts with aqueous silver nitrate in ethanol to form a white precipitate. The remaining unlabelled sample is propanone.

64. **B** All species are gases. Moles of a gas = volume (cm^3) / 24 000. Therefore the volumes are proportional to the number of moles. Dividing the volumes by 30 to simplify the numbers give 1:9:6. The equation can be written as: 1CH + 9O$_2$ → 6CO$_2$ + H$_2$O. As 6 C on the RHS, 6 should be present on the LHS. To balance O of LHS, 6 more need to be added to the RHS. This gives: 1C$_6$H$_{12}$ + 9O$_2$ → 6CO$_2$ + 6H$_2$O.

65. **A** Moles of (E)-4-(phenyldiazenyl)phenol required = 7.4 / 198 = 0.03737 mol. Looking at the equations, 1 mole of phenyl amine forms 1 mole of (E)-4-(phenyldiazenyl)phenol (1:1 ratio). Therefore, if the yield was 100%, we would need 0.03737 mol of phenylamine. In grams, this is = 0.03737 x 93 = 3.48g. However, as the yield is only 57%, we would need: 3.48 x 100/57 = 6.1 g of phenylamine.

66. **A** B is not correct as the product is 2-nitrophenol and not 3-nitrophenol. C shows a hydrogen bond which is not shown in skeletal formulae. D suggests a methyl group is attached to the ring.

67. **C** Note that skeletal formulae does not show H atoms unless they are bonded to an atom other than carbon. In skeletal formulae, if there is no atom indicated at the end of a bond, it is taken to be C with H attached to it: therefore CH$_3$.

68. **B** An ester is formed from the reaction. The molar mass of the ester will be: 116 + 88 - 18. You have to minus 18, because a water molecule is given off. Molar mass of water is 18. Alcohol is limiting with 0.02155 moles and carboxylic acid is in excess with 0.02386 moles. Therefore there will be 0.02155 moles of the ester which corresponds to 4.01g (theoretical yield).

69. **C** In this reaction, an alcohol is reacting with an acyl chloride to form an ester (and HCl is given off). Molar mass of the acyl chloride is found by: 178 + 36.5 - 94 = 120.5. 4.30g / 178 = 0.02416 moles of ester. Acyl chloride react with alcohol to form ester in ratio 1:1:1. Therefore if the percentage yield was to be 100%, 0.02416 moles of the other compound (acyl chloride) would be required. This corresponds to a mass of : 120.5 x 0.02416 = 2.911 g. The % yield is 67.3%. Therefore 2.911 x 100/67.3 = 4.33 g would be required. (Note 36.5 is the molar mass of HCl).

70. **A** Primary amide: Nitrogen atom is only bonded to one carbon. Secondary: bonded to two carbons.

71. **A** The molar mass of **compound S** is bigger than that of the ester. So it cannot be a carboxylic acid or an acyl chloride. It could be an acid anhydride. Any acid anhydride contains three oxygen atoms. 130- (16 x3) = 82. The structure of the acid anhydride is such that it seems to be made from two carboxylic acids joined together. Only one of these carboxylic acid parts join to form the ester. 82 / 2= 41. Now add one oxygen 41 + 16 = 57 as the carboxylic acid part contributes one oxygen to the ester. Substracting 57 from 102 will give an approximate value for the mass of the alcohol. 102 -57 = 45. The other easy way to solve this is by trial and error. Just try each option!

72. **B** Option 3 is not correct because the y-axis states "x 10^5". This means the actual abundance of one of the compounds is 188000 and not 1.88.

Data sheet — Periodic Table

The Periodic Table Of The Elements

Key
Atomic number
Symbol
Name
Relative atomic mass

1	2		3	4	5	6	7	8	9	10	11	12	13	14	15	16	17	18
1 H hydrogen 1.0																		2 He helium 4.0
3 Li lithium 6.9	4 Be beryllium 9.0												5 B boron 10.8	6 C carbon 12.0	7 N nitrogen 14.0	8 O oxygen 16.0	9 F fluorine 19.0	10 Ne neon 20.2
11 Na sodium 23.0	12 Mg magnesium 24.3												13 Al aluminium 27.0	14 Si silicon 28.1	15 P phosphorus 31.0	16 S sulfur 32.1	17 Cl chlorine 35.5	18 Ar argon 39.9
19 K potassium 39.1	20 Ca calcium 40.1		21 Sc scandium 45.0	22 Ti titanium 47.9	23 V vanadium 50.9	24 Cr chromium 52.0	25 Mn manganese 54.9	26 Fe iron 55.8	27 Co cobalt 58.9	28 Ni nickel 58.7	29 Cu copper 63.5	30 Zn zinc 65.4	31 Ga gallium 69.7	32 Ge germanium 72.6	33 As arsenic 74.9	34 Se selenium 79.0	35 Br bromine 79.9	36 Kr krypton 83.8
37 Rb rubidium 85.5	38 Sr strontium 87.6		39 Y yttrium 88.9	40 Zr zirconium 91.2	41 Nb niobium 92.9	42 Mo molybdenum 95.9	43 Tc technetium	44 Ru ruthenium 101.1	45 Rh rhodium 102.9	46 Pd palladium 106.4	47 Ag silver 107.9	48 Cd cadmium 112.4	49 In indium 114.8	50 Sn tin 118.7	51 Sb antimony 121.8	52 Te tellurium 127.6	53 I iodine 126.9	54 Xe xenon 131.3
55 Cs caesium 132.9	56 Ba barium 137.3	57–71 lanthanoids	72 Hf hafnium 178.5	73 Ta tantalum 180.9	74 W tungsten 183.8	75 Re rhenium 186.2	76 Os osmium 190.2	77 Ir iridium 192.2	78 Pt platinum 195.1	79 Au gold 197.0	80 Hg mercury 200.6	81 Tl thallium 204.4	82 Pb lead 207.2	83 Bi bismuth 209.0	84 Po polonium	85 At astatine	86 Rn radon
87 Fr francium	88 Ra radium	89–103 actinoids	104 Rf rutherfordium	105 Db dubnium	106 Sg seaborgium	107 Bh bohrium	108 Hs hassium	109 Mt meitnerium	110 Ds darmstadtium	111 Rg roentgenium	112 Cn copernicium			114 Fl flerovium		116 Lv livermorium		

Lanthanoids and Actinoids

| 57 La lanthanum 138.9 | 58 Ce cerium 140.1 | 59 Pr praseodymium 140.9 | 60 Nd neodymium 144.2 | 61 Pm promethium 144.9 | 62 Sm samarium 150.4 | 63 Eu europium 152.0 | 64 Gd gadolinium 157.2 | 65 Tb terbium 158.9 | 66 Dy dysprosium 162.5 | 67 Ho holmium 164.9 | 68 Er erbium 167.3 | 69 Tm thulium 168.9 | 70 Yb ytterbium 173.0 | 71 Lu lutetium 175.0 |
| 89 Ac actinium | 90 Th thorium 232.0 | 91 Pa protactinium | 92 U uranium 238.1 | 93 Np neptunium | 94 Pu plutonium | 95 Am americium | 96 Cm curium | 97 Bk berkelium | 98 Cf californium | 99 Es einsteinium | 100 Fm fermium | 101 Md mendelevium | 102 No nobelium | 103 Lr lawrencium |

Data sheet

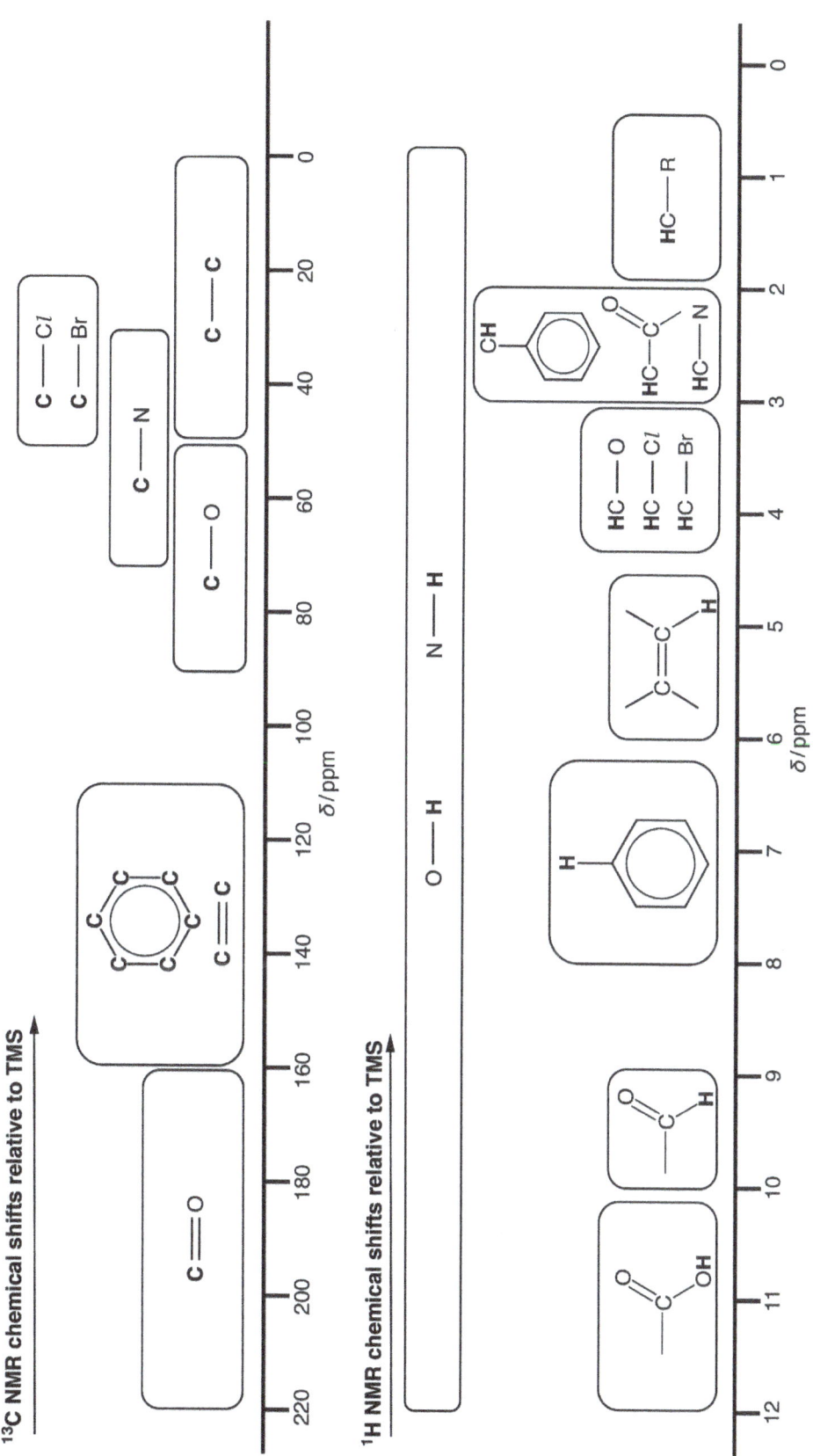

Data sheet

General Information

Avogadro constant, $N_A = 6.02 \times 10^{23}\,\text{mol}^{-1}$

1 tonne = $10^6\,\text{g}$

Arrhenius equation: $k = Ae^{-E_a/RT}$ or $\ln k = -E_a/RT + \ln A$

Molar gas volume = $24.0\,\text{dm}^3\,\text{mol}^{-1}$ at room temperature and pressure, RTP

Specific heat capacity of water, $c = 4.18\,\text{J}\,\text{g}^{-1}\,\text{K}^{-1}$

Ionic product of water, $K_w = 1.00 \times 10^{-14}\,\text{mol}^2\,\text{dm}^{-6}$ at 298 K

Gas constant, $R = 8.314\,\text{J}\,\text{mol}^{-1}\,\text{K}^{-1}$

Infrared absorptions in organic molecules

Bond	Location	Wavenumber/cm^{-1}
C–C	Alkanes, alkyl chains	750–1100
C–X	Haloalkanes (X = Cl, Br, I)	500–800
C–F	Fluoroalkanes	1000–1350
C–O	Alcohols, esters, carboxylic acids	1000–1300
C=C	Alkenes	1620–1680
C=O	Aldehydes, ketones, carboxylic acids, esters, amides, acyl chlorides and acid anhydrides	1630–1820
aromatic C=C	Arenes	Several peaks in range 1450–1650 (variable)
C≡N	Nitriles	2220–2260
C–H	Alkyl groups, alkenes, arenes	2850–3100
O–H	Carboxylic acids	2500–3300 (broad)
N–H	Amines, amides	3300–3500
O–H	Alcohols, phenols	3200–3600

Data sheet

Index

Symbols

[H+] 54
α- amino acid 129
α-amino acid 153
Δ G· *See also* free energy
Δ S· *See also* entropy
ΔS· *See also* entropy

Numbers

1,2,4-tribromobenzene 312
1st order
reactant 18
2,4-dinitrophenylhydrazine 302
2,4-dinitrophenylhydrazone 151, 302
2nd order reactant 18
2p 296
3-directing effect 119

Roman

A

absorption peak 209
acid 22, 51, 93, 95, 235
acid anhydride 126
acid-base titration 142
acid-base titrations 58
Acid dissociation constant 41, 53, 292
Acids 24, 25, 152, 211
activation energy 17
Acyl chloride 125
Acyl Chloride 139
Addition polymer 130, 304, 308
Addition polymerisation 130, 154, 309
Addition reactions 149, 207, 297, 299
adsorption 101, 207, 311, 351
alcohol 122, 151
alcohols 120
aldehyde 122, 149, 150, 153, 202, 210, 212, 301, 302, 350, 351
aldehydes 124
aliphatic 116, 128, 212, 352
alkali 21, 22, 24, 93
alkenes 117, 123, 296
Alkylation 117
aluminium 26
aluminium oxide 26
amine 128, 129, 151, 211, 302, 303, 309, 352
Amine 127
amino acid 154
Amino acid 129
ammonia 128
ammoniacal silver nitrate 205, 350
ammonium 128
amphoteric 129
aromatic 116, 128, 151, 212, 302, 352
Arrhenius equation 17, 18

B

base 21, 23, 24, 25, 95, 233, 237, 303, 307
benzene 116, 117, 146, 149, 203, 204, 207, 208, 296, 299, 300, 350, 351
bidentate 39, 64, 100, 293
bond angle 202, 350
Bonding 116
Born-Haber cycle 28, 56, 102, 294
Brady's reagent 122, 205, 301, 350
bromobenzene 148
buffer 52, 94, 237, 239, 240, 292

C

calcium hydroxide 21
Calorimeter 143
carbon- 13 NMR 137, 153, 161, 162, 163, 164, 166, 170, 206
Carbon-13 NMR Spectroscopy 134
carbon environment 312, 314
carbon monoxide 38
carbonyl 122, 150, 206, 211, 302, 350
Carbonyl 138
carbonyl compounds 121
Carbonyl compounds 120
carboxyl group 308
carboxylic 122, 127, 149, 202, 300, 301, 303, 309, 352
carboxylic acid 139
Carboxylic acid 127
carboxylic acids 124, 205, 350
carboyxlic acid 138
catalyst 20, 97, 154, 253
cells 31
chemical shift 137, 162, 312
Chemical tests 122
chiral 155, 160, 206, 308, 350
chromatography 132, 147, 160, 206, 311
chromium 35, 97
cis 300
cis-platin 39, 66, 98, 255, 257
cis-trans 39, 253, 255
Cobalt 35
complex ion 39, 65
Complex ions 40
concentration 26, 231, 234, 240
concentration-time graph 15, 92
condensation 129, 130, 154, 306
conjugate
acid-base pair 25
conjugate acid-base pair 54, 93
coordinate bond 39, 64, 253, 255, 293
coordination number 39, 63, 246, 253

copper 35, 97
covalent 27, 56, 244
Cr 293
Cu 293
curly arrow 149
cyclic 117
cyclohexa-1,3,5-triene 296, 299
cyclohexene 149, 299

D

D2O 167, 209, 212, 315, 316, 352
dative covalent bond 253
d-block 35
Decimal Places 42
delocalisation 117, 207, 208, 296
delocalised 116, 117, 146
Deuterium 137
dipeptide 129, 156
displayed formula 202
dissociate 25
dissociation 21, 22, 239
Distillation 141
double bonds 116
doublet 317
d sub-shell 292

E

electrochemical cell 31, 32, 62, 96. *See also* Fuel Cells
Electrochemical cell 32
electrode potential 32
electrode potentials 32
electron configuration 61
electron density 119, 150
Electron-donating 118
electronegative 122
electronegativity 149
electrons 34
electrophile 117, 148, 298
electrophilic addition 206, 208
electrophilic substitution 119, 148, 298, 306
empirical formula 208, 242, 351
en 39, 65, 100
enantiomers 255
endothermic 20, 30, 232, 235, 292
ENDOTHERMIC 27
end point 56
enthalpy 248
Enthalpy change 259
enthalpy change of atomisation 27
enthalpy change of formation 59
enthalpy change of hydration 55, 243
enthalpy change of hydrogenation 296
enthalpy change of neutralisation 56
enthalpy change of solution 55
enthalpy changes 27, 28

Enthalpy changes 27
entropy 29, 57, 95, 103, 245, 248, 249, 250, 292
equilibrium 20, 22, 31, 32, 34, 49, 232, 240, 244
equilibrium constant 19, 20, 41, 49, 292
equivalence point 23, 56, 244, 247
ester 124, 127, 313, 350
esterification 303
Esterification 124
esters 206
Esters 126
ethane-1,2-diamine 100, 255. *See also* en
ethanol 352
exothermic 30, 232, 235, 243, 248
Exothermic 20
EXOTHERMIC 27
E/Z isomers 206, 306, 350

F

Filtration 141
first electron affinity 26, 27, 28, 234, 294
first ionisation energy 27, 57, 234
first order 47, 92, 104, 230, 291
free energy 30, 59, 248
Friedel-Crafts acylation 139
fuel cell 34
fuel cell vehicle 101, 253. *See also* FCV
functional group 122, 132
Functional groups 122

G

gas chromatography 132, 164, 168, 209, 210, 312
Gas chromatography 131
Gas Chromatography 131
GC-MS 312
Gibbs free energy. *See also* free energy
glycine 156
Gradient 18

H

haemoglobin 100, 156, 256
half-cell 31, 61, 62, 250, 252, 293
half-life 18, 92, 230
Half-life 15
haloalkane 137
haloalkanes 123
halogen 299
Halogenation 117
halogen carrier 117, 119, 298
heptet 317
Hess's Law 58, 247
Heterogeneous 20
hexadentate 39, 100, 293
hexagonal 117
Homogeneous 20
hydrocarbon 117, 296

hydrochloric acid 21, 105
hydrogen 31
hydrogen cyanide 137
Hydrogen ion 23
hydrolysis 127, 151, 157, 158, 309, 350
Hydrolysis 126, 127
hydroxyl 299, 309
hydroxynitriles 138

I

indicator 23
infrared spectroscopy 167, 206, 209
intermediate 50
iodine 259
ioinc product of water 41, 54, 94, 237
ionic charges 56
ionic equation 62, 63, 93
ionic radii 56
ionisation 26
Iron 35
IR spectra 203
isoelectric point 304
isomers 163, 166
isotopes 209

J

Joules 29

K

k. *See* rate constant
Ka 93, 94. *See also* Acid dissociation constant
Kc. *See* equilibrium constant
kekule structure 146, 148, 208
Kelvin 30
ketone 122, 150, 153, 167, 210, 212, 301, 302, 350, 351
Ketones 121
kevlar 309
KMnO4 99
Kstab. *See also* stability constant
Kw. *See* Ionic product of water

L

lattice enthalpy 27, 56, 95, 103, 243, 244
ligand 39, 98, 253, 258, 293
ligand substitution 38, 41, 65, 67, 98, 230, 258
Ligand substitution 37
lone pair 299

M

magnetic resonance imaging 316
mass spectrometry 132, 136, 161, 206, 315
Mass spectroscopy 136
Melting point 143
metal carbonate 142
methanoic acid 127

methanol 101
MnO4- 255, 257
mobile phase 160, 311
molecular formula 209, 210
molecular ion 167, 315
molecular weight 167, 315, 351
Mole fractions 19
moles 234
Moles 41
monodentate 39
monomer 130, 309
m/z value 315

N

n+1 rule 167, 316
NaBH4 121, 206, 210, 211, 302, 351, 352. *See also* entropy
NaOH 105
negative terminal 31, 32
neutralisation 23, 24, 206
nitration 117, 146, 148
nitric acid 21
nitriles 138
nitroarenes 128
nitrobenzene 128, 148, 158, 207, 212
nitronium 148
NMR 135, 137, 208, 312, 314, 315, 317
non-superimposable 307
nucleophile 121, 128, 151
nucleophilic addition 121, 122, 138, 206, 301, 302
nucleophilic substitution 128, 137, 206, 304
nylon-6,6 309

O

octahedral 39, 64, 229, 246
optical isomer 39, 65, 154, 155, 205, 206, 210, 253, 255
orbital 116, 296
order 14, 48, 92
oxidation 68, 248, 249, 259
Oxidation 33, 120
oxidation number 254
Oxidation number 32
oxidation state 31, 99, 250
Oxidising 40
oxidising agent 40, 61, 66, 122, 150, 250, 257

P

partial pressure 19
Partial Pressures 19
peptide bond 129, 156
pH 22, 25, 52, 93, 94, 237, 238, 247, 307
pharmaceutical 351
phenol 116, 117, 119, 125, 149, 203, 300, 306
Phenol 118
phenols 124
phenylamine 309
pH meter 143

photodegradable 130, 158
pKa 53, 240
planar hexagonal 296
plane-polarised light 206, 210
polyamides 130, 156, 158, 308
polyesters 130, 157, 308, 309
poly(lactic acid) 308
polymer 130, 156, 308
Polymers 130
polypeptide 305
positive terminal 31, 32
potassium dichromate 301
potassium hydroxide 21
precipitate 64, 97, 119, 230, 293, 301, 302, 350
Precipitation reactions 36
pressure 26, 247
primary alcohol 121, 122, 204, 350, 351
primary alcohols 124
primary amine 128
proteins 129, 156, 305
proton 23, 93, 314, 316
Proton 23
Proton environment 134
proton NMR 137, 167, 170, 352
Proton NMR Spectroscopy 132
Purification 141

R

rate 232
Rate 232
rate-concentration 18
rate-concentration graph 14, 47, 104
rate constant 16, 18, 41, 229, 235, 292
rate-determining step 49, 233
rate equation 16, 48
rate of reaction 46, 229
reaction mechanism 50, 146, 232
reaction rate 14, 18, 291
recrystallisation 141
redox 59
reducing agent 40, 66, 122, 128, 250
reduction 248
Reduction 33, 121
reflux 141
relative solubility 207
resonance energy 146
retention time 132, 161, 168, 209, 210, 314, 316
Rf value 131, 132, 161, 207, 297, 298, 312

S

salt bridge 31, 96
saturated 299, 303
secondary alcohol 121, 122, 204, 352
secondary alcohols 124
secondary amine 128

second electron affinity 26, 28
second order 47, 104, 230
Separating funnel 141
shapes 40
sigma 296
silver mirror 205
silver nitrate 123, 300
singlet 317
sodium hydroxide 21
sodium tetrahydridoborate (III) 210
solvent front 132, 311
spectral database 314
spin-spin coupling 202
square planar 39, 256
standard cell potential 34
standard conditions 27, 58, 204, 243, 350
Standard conditions 25
standard electrode potential 250
standard enthalpy change of formation 27, 234
standard enthalpy change of hydration 103
Standard state 25
starch 259
State symbols 21
stationary phase 206, 311
stereoisomers 39, 63, 155, 209, 255, 305
Strong acids 21
strong base 25
strong bases 21
sub-shell 96
substitution 117, 147, 207, 298
sulfuric acid 21

T

tangent 18
temperature 26, 49, 232, 247, 296
tertiary alcohol 204, 350
Terylene 304
tetrahedral 39
tetramethylsilane 137, 162, 212, 312
thin-layer chromatography 132, 160, 311
Thin-Layer Chromatography 131
thiosulfate 100, 259
Titanium 35
titration 23
titration curve 104
titration curves 23
Titrations 142
Tollens' reagent 122, 150, 302, 303
transition element 35, 96, 246, 253
triglycerides 304
trigonal planar 117, 296

U

unsaturated 299, 303

V

Vanadium 35
vanadium(V) oxide 253
volume 234

W

water of crystallisation 142

weak acid 22, 25
withdrawing groups 118

Z

zero order 47, 92, 104, 230, 291